GENERATING RANDOM NETWORKS AND GRAPHS

Generating Random Networks and Graphs

A.C.C. Coolen

Department of Mathematics, King's College London, UK

A. Annibale

Department of Mathematics, King's College London, UK

E.S. Roberts

Department of Mathematics, King's College London, UK

OXFORD

UNIVERSITY PRESS

OXFORD
UNIVERSITY PRESS

Great Clarendon Street, Oxford, OX2 6DP,
United Kingdom

Oxford University Press is a department of the University of Oxford.
It furthers the University's objective of excellence in research, scholarship,
and education by publishing worldwide. Oxford is a registered trade mark of
Oxford University Press in the UK and in certain other countries

First Edition published in 2017

Impression: 1

Published in the United States of America by Oxford University Press
198 Madison Avenue, New York, NY 10016, United States of America

British Library Cataloguing in Publication Data

Data available

Library of Congress Control Number: 2016946811

ISBN 978–0–19–870989–3

Printed and bound by
CPI Group (UK) Ltd, Croydon, CR0 4YY

Preface

Networks have become popular tools for characterizing and visualizing complex systems with many interacting variables. In the study of such systems, it is often helpful to be able to generate *random* networks with statistical features that mimic those of the true network at hand. These random networks can serve as 'null hypotheses' in hypothesis testing, or as proxies in simulation studies. How to generate random networks with controlled topological characteristics accurately and efficiently has become an important practical challenge, and is in itself an interesting theoretical question.

Random networks result, by definition, from stochastic processes. The natural language for graph generation protocols is therefore that of network or graph ensembles. These ensembles are constrained by the statistical features that one chooses to impose. The core of the network generation problem is to construct stochastic processes or algorithms that, upon equilibration, sample the space of *all networks* that have the desired features, but are *otherwise unbiased*. Moreover, these algorithms should be able to generate large networks within practical time scales.

We set out to write a concise but self-contained text on the generation of random networks with controlled topological properties, including the mathematical derivations of relevant algorithms and their numerical implementations. Such details appear to be under-represented in the presently available books. They tend to be found in discipline-specific journals that are not easily accessible, and research articles devoted to one algorithm at a time. We have tried to bring together the most important results that are scattered over the physics, combinatorial mathematics and computer science literature, and explain with uniform notation the rationale, the potential and limitations, and application domains of each approach. We seek to explain the theoretical issues and the practical algorithms resulting from analysis of the problem in such a way that the reader is able to focus on either aspect if he or she so wishes. To achieve this, we have separated (where possible) practical recipes from theoretical arguments, to aid selective navigation driven by personal and domain-specific preferences and needs. More extensive mathematical arguments are delegated to appendices, and we have supplemented the main text with examples and exercises with solutions.

This book is aimed at established or junior researchers and their graduate or advanced undergraduate students in computer science, quantitative biology, social science, ecology, bioinformatics, applied mathematics and theoretical physics, whose work involves networks or graphs. We hope that it may serve as a useful reference, as a textbook for postgraduate lecture courses and as an inspiration for further quantitative research.

London, November 2016   Ton Coolen, Alessia Annibale and Kate Roberts

Acknowledgements

It is our great pleasure to thank the many colleagues and students with whom over the years we have enjoyed discussing the modelling and generation of graphs and networks. In particular, we would like to mention (in alphabetical order) Elena Agliari, Ginestra Bianconi, Zdisław Burda, Sun Sook Chung, Owen Courtney, Andrea De Martino, Luis Fernandes, Thomas Fink, Franca Fraternali, Clara Grácio, Alexander Hartmann, Jens Kleinjung, Reimer Kühn, Alexander Mozeika, Alessandro Pandini, Nuria Planell-Morell, Conrad Perez-Vicente and Peter Sollich.

Contents

Part I

The basics

1
Introduction

Networks and their popularity. Our ability to collect and store vast amounts of quantitative data in many areas of the natural and the man-made world has increased dramatically. These include data of a socio-economic nature, such as social links between individuals or professional collaboration networks, consumer preferences and commercial interactions, trade dependencies among corporations and credit or insurance obligations between financial institutions. They include usage and traffic data on computer and telephone communication systems, satellite networks, the Internet, electricity grids, rail, road or air travel connections and distribution networks. We also collect and process large amounts of geological and meteorological data, data on sea levels, air and water pollution, volcanic and seismic records, and sizes of polar and glacier ice sheets. Last but not least, in recent years we have seen an explosion of biomedical information, such as experimental data on biochemical processes and structures at cellular, sub-cellular and even molecular levels, the topologies of complex composite organs such as the brain or the immune system, genomic and epigenetic data (gene expression levels, DNA mutations), epidemiological data and vast numbers of patient records with clinical information.

However, one tends to collect data for a reason. This reason is in many cases the desire to understand the dynamical behaviour of the system under study to such an extent that one can predict its future evolution, and its response to perturbations or interventions. We may want this for academic reasons, for financial gain, to improve and optimize a system's efficiency, to design effective regulatory controls or (in the case of medicine) to understand and cure diseases. For small and simple systems, the translation of observed data into qualitative and quantitative understanding of function is usually not difficult. We do not have to be engineers to understand the workings of a bicycle once we have seen it in action, and after we have made an inventory of the parts and their links. In contrast, if we collect data on complex systems with millions or more of nonlinearly interacting variables, just having a list of parts and their connections and/or many observations of their collective patterns of behaviour is no longer enough to understand how these systems work. This is even more true if we do not know the detailed forces via which the components interact.

A first and often useful step in modelling and visualizing structural data on complex many-variable systems is to represent them as networks or graphs[1]. The idea is to represent each observed component as a node in the network, and each observed interaction between two components as a link. Dependent upon the application domain,

[1] In this book we will use the words 'network' and 'graph' interchangeably.

Generating Random Networks and Graphs. First Edition. A.C.C. Coolen, A. Annibale, and E.S. Roberts. © A.C.C. Coolen, A. Annibale, and E.S. Roberts 2017. Published in 2017 by Oxford University Press. DOI: 10.1093/acprof:oso/9780198709893.001.0001

the nodes of such networks may then represent anything ranging from genes, molecules or proteins (in biology), via processors or servers (in computer science), to people, airports, power plants or corporations (in social science or economics). The links could refer to biochemical reactions, wired or wireless communication channels, friendships, financial contracts, etc. The result is an appealing and intuitive representation of the system, but the price paid is a significant loss of information and ambition. Limiting ourselves to the network representation means giving up a lot of information; we only record *which* parts interact, and disregard *how and when* they interact. The rationale is that much can perhaps already be learned from the topologies of the networks alone.

For example, DNA contains the assembly instructions for complicated macromolecules called proteins. Proteins are the universal workhorses of the cell. They serve as building material, molecular factories, information readers, translators, transport vehicles, sensors and messengers. They interact with each other by forming complexes, which are semi-stable supermolecules formed of multiple proteins that attach to each other selectively. The network approach to this system would be to record – from the results of many complicated biochemical protein binding experiments – only which protein species are present (to be represented by network nodes) and which of these species can interact (to be represented by *nondirected* links between the nodes of interacting species, i.e. links without arrows); see Figure 1.1 (based on data from [149]). Typical databases contain some 70,000–100,000 reported interactions – still only about half of the interactions believed to exist in the human proteome – and the problem for the modeller is how to visualize and utilize all this information. Similarly, in the analysis of financial markets, we might represent each financial institution or large market player (banks, hedge funds, insurers, pension funds, regulators, large commodity traders, central banks, market makers, etc.) as a network node, and connect those nodes that are known to have mutual contractual or regulatory obligations, or that have a trading relationship. In the analysis of ecosystems, we could represent each plant or animal species by a node, and connect those nodes that have either a direct dependency in the food chain, or a symbiotic relationship or that interact indirectly by creating environmental conditions that affect each other's survival likelihood. In all those cases, one would then proceed to analyze topological features of the created networks, and try to find whether these correlate with important functional characteristics of the complex systems they are meant to represent. In these last two examples, we would use *directed* networks (i.e. with links that carry arrows), since the mutual influences need not be reciprocal. The relevant characteristics to be correlated with network topology will vary from system to system, and will also be dependent upon the interests of the modeller. In financial markets and ecosystems, the main interests would probably be in understanding the origin and nature of anomalous fluctuations, the topological factors that influence overall stability and large crashes or wipeouts, and in determining whether and how the dynamics of these systems can be predicted and controlled. In large distribution networks of, e.g. food, electricity, fuel, people or information, one would be interested in questions relating to path lengths (since these determine transport costs), vulnerability against spontaneous of maliciously induced breakdowns, and how to design the networks to minimize this vulnerability.

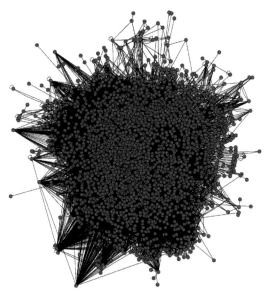

ALDH1A1	00001	NP_000680.2	ALDH1A1	00001	NP_000680.2	in vivo;yeast 2hybrid	12081471,16189514
ITGA7	02761	NP_002197.1	CHRNA1	00007	NP_001034612.1	in vivo	10910772
PPP1R9A	16000	NP_060120.2	ACTG1	00017	NP_001605.1	in vitro;in vivo	9362513,12052877
SRGN	01513	NP_002718.2	CD44	00115	NP_000601.3	in vivo	9334256
GRB7	03311	NP_005301.2	ERBB2	01281	NP_004439.2	in vitro;in vivo	9079677
PAK1	03995	AAC24716.1	ERBB2	01281	NP_004439.2	in vivo	9774445
DLG4	04199	NP_001356.1	ERBB2	01281	NP_004439.2	in vivo;yeast 2hybrid	10839362
PIK3R2	04404	NP_005018.1	ERBB2	01281	NP_004439.2	in vivo	1334406
PTPN18	05961	NP_055184.2	ERBB2	01281	NP_004439.2	in vitro	14660651
ERBB2IP	06090	AAK69431.1	ERBB2	01281	NP_004439.2	in vitro;in vivo	11278603,10878805
SMURF2	06901	NP_073576.1	ARHGAP5	04060	NP_001025226.1	yeast 2hybrid	15231748
NF2	06980	NP_000259.1	ERBB2	01281	NP_004439.2	in vitro;in vivo	12118253
CD82	09004	NP_002222.1	ERBB2	01281	NP_004439.2	in vivo	14576349
ERRFI1	09218	NP_061821.1	ERBB2	01281	NP_004439.2	in vitro;yeast 2hybrid	11003669
MMP7	01525	NP_002414.1	CD44	00115	NP_000601.3	in vitro;in vivo	11825873
TOB1	09273	NP_005740.1	ERBB2	01281	NP_004439.2	in vitro;in vivo	8632892
MUC4	11829	NP_060876.3	ERBB2	01281	NP_004439.2	in vivo	12434309,11687512,11598901
PICK1	16176	NP_036539.1	ERBB2	01281	NP_004439.2	in vitro;in vivo;yeast 2hybrid	11278603
SMURF2	06901	NP_073576.1	TXNIP	05964	NP_006463.2	yeast 2hybrid	15231748
DDX20	05859	NP_009135.3	ETV3	01282	NP_005231.1	in vitro	12007404
TLE1	02557	NP_005068.2	FOXG1B	01283	NP_005240.3	in vivo	11238932
TLE3	02558	NP_005069.1	FOXG1B	01283	NP_005240.3	in vivo	11238932
HDAC1	03143	NP_004955.2	FOXG1B	01283	NP_005240.3	in vitro;in vivo	11238932
SMAD1	03356	NP_005891.1	FOXG1B	01283	NP_005240.3	in vivo	11387330
JARID1B	09251	NP_006609.3	FOXG1B	01283	NP_005240.3	in vivo;yeast 2hybrid	12657635
CD4	01740	NP_000607.1	CD44	00115	NP_000601.3	in vitro	7539755
SMURF2	06901	NP_073576.1	DAB2	03139	NP_001334.2	yeast 2hybrid	15231748
CBL	01320	NP_005179.2	VAV1	01284	NP_005419.2	in vitro;in vivo	9200440,11133830
PLCG1	01398	NP_002651.2	VAV1	01284	NP_005419.2	in vitro;in vivo	9891995,16467851
PRLR	01457	NP_000940.1	VAV1	01284	NP_005419.2	in vivo	7768923
TYK2	01490	NP_003322.2	VAV1	01284	NP_005419.2	in vivo	10673353
SMURF2	06901	NP_073576.1	DGCR2	08999	NP_005128.1	yeast 2hybrid	15231748
MAPK1	01496	NP_002736.3	VAV1	01284	NP_005419.2	in vivo	9013873,8900182
VAV1	01284	NP_005419.2	RHOG	01537	NP_001656.2	in vitro	12376551
HRAS	01813	NP_005334.1	VAV1	01284	NP_005419.2	in vivo	8554611
BTK	02248	NP_000052.1	VAV1	01284	NP_005419.2	in vivo	9201297
TGFBR1	01822	NP_004603.1	CD44	00115	NP_000601.3	in vitro;in vivo	12145287
SYK	02514	NP_003168.2	VAV1	01284	NP_005419.2	in vitro;in vivo;yeast 2hybrid	8986718
SHB	02635	NP_003019.2	VAV1	01284	NP_005419.2	in vitro;in vivo	12084069
PRKCQ	02710	NP_006248.1	VAV1	01284	NP_005419.2	in vivo	10725744
TEC	02786	NP_003206.2	VAV1	01284	NP_005419.2	in vitro;in vivo	7651724
IL6ST	02824	NP_002175.2	VAV1	01284	NP_005419.2	in vivo	9013873

Fig. 1.1 Bottom: Extract of raw data on human protein–protein interactions as reported in biological databases, being lists of pairs of protein species that were observed to interact, together with their formal codes and information on the experiments where their interaction was observed. Each line reports on one pair interaction. Top: The network representation of the data. Each protein is represented by a node, and each pair interaction, i.e. each line in the table, is represented by a link that connects the two nodes concerned.

Network literature. There are many excellent books on the analysis and applications of networks and graphs. Some touch briefly upon certain aspects of graph generation, usually by mentioning the generation of Erdös–Rényi graphs, and selected existing algorithms that would produce specific classes of graphs (often without derivations). Here we limit ourselves to mentioning a few popular books, and apologize to the authors of those that were left out. For example, [26] is a new edition of an influential classic, which covers mainly the formal mathematical theory of random graphs. The volume [63] is a popular text, somewhat less mathematical, but with many real-world examples of networks. A large collection of chapters written by different experts on networks and their applications is found in [30]. The monograph [65] deals mainly with the mathematical analysis and application of random graphs processes on graphs. The most comprehensive text on networks presently available is probably [131]. Other recommended books are [5], [23] and [40].

The role of random networks. With larger and larger datasets, it becomes crucial to have computationally practical and analytically well-understood algorithms for generating networks. This is especially so because the ability to test a hypothesis against a properly specified control case (a *null model*) is at the heart of the scientific method. Hence, knowledge on how to generate samples from unbiased and well-controlled random graph ensembles is vital for the maturing field of network science. The null model approach is appealing in its conceptual simplicity. It effectively provides synthetic 'data', which can be analyzed in the same way as the real dataset. One can then learn which observed properties are particular to the real dataset (i.e. 'special' and informative), and which are common within the wider ensemble of which the observed network and the null model are members. Applications of network null models are consequently wide-ranging and central to network science. For instance, [164] applies null models to identify over-represented *motifs* in the transcriptional regulation network of *E. coli*; [180] discusses adapting the Watts–Strogatz method to generating random networks to model power grids; [136] explores motifs found within an inter-firm network; [132] uses network null models to study social networks; [115] compares topological properties of interaction and transcription regulatory networks in yeast with randomized null model networks and postulates that links between highly connected proteins are suppressed in protein interaction networks; finally [81] discusses the challenges of specifying a suitable matrix null model in the field of ecology.

Non-trivial aspects of graph generation. The key characteristics of network generation algorithms are their flexibility (e.g. the extent to which the topological features can be controlled), their uniformity (i.e. the extent to which they generate graphs with the correct probabilities), and their computational efficiency. These three are usually difficult to optimize simultaneously. The most commonly used methods for generating random networks with a given degree sequence (i.e. the full specification of the number of interaction partners of each node) are variations on the so-called stub-joining and edge-swap algorithms. The former is based on 'growing' links, starting from a set of disconnected nodes. The edge-swap method starts with a network with the desired degree sequence, and then randomizes it by repeatedly swapping pairs of edges. In both algorithms, the main problem lies in accurately controlling the probabilities with which graphs are generated, while keeping computational demands within feasible limits. Another class of models is that of exponential random graph ensembles. Here the graph probabilities are from the start written in the so-called Boltzmann form, with a specially constructed energy function assigned to each possible graph realization. These are particularly popular in the social sciences, and can be generated by the Metropolis–Hastings algorithm, where edges are added or removed at random. For such networks the probabilities are controlled trivially, although at high computational cost, but building in 'hard' structural constraints is a difficult problem. Finally, many models have been proposed which are specifically designed to mimic real-world networks of a certain type, of which Barabási's preferential attachment network growth model is a well-known example. Systems biology uses graph randomization protocols to identify the elementary bricks (motifs, i.e. small subgraphs that may be over-represented) characterizing cellular networks [164, 93]. Many other examples can be found in economics, ecology and the social sciences; see, e.g. [165, 166]. In all cases, it is crucial that the synthetic networks generated as null models are representative of the underlying ensembles. Any inadvertent bias in the network generation process may invalidate the hypothesis test. Generating such bias-free networks is known to be a non-trivial problem, that has produced much inspiring work and some hard open questions [17, 128, 131]. Using biased algorithms for producing null models is fundamentally unsound, and unacceptable when there are rigorous unbiased alternatives. When more complicated observables than the degree sequence are involved (like degree correlations, clustering or loop statistics), the situation rapidly becomes more difficult, both mathematically and numerically.

Overview of this book. An algorithm is a dynamical process. An algorithm that generates a *random graph* is by construction a stochastic process. Sometimes one defines random graphs by a desired probability distribution, and one needs to find a corresponding process that leads to this distribution, which can then serve as an algorithm. Alternatively, dependent upon one's problem domain, it could be more natural to define random graphs directly in terms of a process; one then needs to understand the probability distribution to which this process evolves in order to predict or have control over the end result. Although it may sometimes seem when reading the literature that these two approaches to random graphs (and their user groups) are quite distinct, it should be clear that they simply represent different entry routes into the same arena. At the core of the problem always lies the need to establish the mathematical connection between stochastic processes for graphs (usually taking the form of Markov chains) and the stationary probability distributions to which these processes evolve.

This unified view of the graph generation problem allows us to bring most approaches to graph generation under the same conceptual and notational umbrella, and is reflected in the structure of this book. In Part 1, we start with notation definitions, and introduce basic concepts in quantifying network topologies and in describing tailored families of random graphs probabilistically. This material will be largely familiar to those who have worked on network theory before. In Part 2, we discuss random graphs that are specified by a desired probability distribution, with soft or hard structural constraints, in the language of so-called graph ensembles. In Part 3, we discuss the construction Markov chain algorithms that upon equilibration will generate graphs from specified graph ensembles. In Part 4, we discuss those cases where one starts from the opposite side, i.e. from a stochastic process specification (i.e. an algorithm), and where one seeks to understand and predict the stationary probabilities to which this process will evolve. Part 5 is devoted to further topics and applications of random graphs. Most appendices deal with technical issues which, although important, interrupt the flow of the book. The final appendix provides a collection of algorithms derived and discussed in the main text.

2
Definitions and concepts

2.1 Definitions of graphs and their local characteristics

Overview: In this section, we will define and explain the following concepts and associated notation: nodes, links, degree, adjacency matrix and clustering coefficient.

Notation, nodes and links. We denote by N the number of nodes (or vertices) in a graph, and we label these N nodes with lower-case Roman indices $i, j, k \in \{1, \ldots, N\}$. For each node pair (i, j) we next define $A_{ij} = 1$ if a link (or edge) $j \to i$ is present in the graph, and $A_{ij} = 0$ if it is not. The set of N^2 variables $A_{ij} \in \{0, 1\}$ thus specifies the graph in full, and is abbreviated as $\mathbf{A}$. One can also interpret the binary numbers $\{A_{ij}\}$ as the entries of an $N \times N$ matrix, which is called the graph's adjacency matrix. Nondirected graphs are those in which $A_{ij} = A_{ji}$ for all (i, j), so the links in such graphs carry no arrows. The number of links L in a directed graph, where node pairs exist with $A_{ij} \neq A_{ji}$, equals $L = \sum_{i,j \leq N} A_{ij}$. In a nondirected graph, we do not count $A_{ij} = 1$ and $A_{ji} = 1$ separately, so here the number of links is $L = \frac{1}{2} \sum_{i \neq j \leq N} A_{ij} + \sum_{i \leq N} A_{ii}$. In this book, we will usually limit ourselves to graphs without self-interacting nodes, so $A_{ii} = 0$ for all i. These are called *simple* graphs.

Microscopic structure characterization in terms of local features. To characterize graph topologies more intuitively, we next inspect simple quantities that reveal aspects of the structure of the graph in the vicinity of individual nodes. The first of these quantities are the in- and out-degrees $k_i^{\text{out}}(\mathbf{A}) \in \mathbb{N}$ and $k_i^{\text{out}}(\mathbf{A}) \in \mathbb{N}$ of each node i in graph $\mathbf{A}$. They count, respectively, the number of arrows flowing into and out of a node i

$$k_i^{\text{in}}(\mathbf{A}) = \sum_{j=1}^{N} A_{ij}, \qquad k_i^{\text{out}}(\mathbf{A}) = \sum_{j=1}^{N} A_{ji} \qquad (2.1)$$

We will denote the pair of degrees for a node i as $\vec{k}_i(\mathbf{A}) = (k_i^{\text{in}}(\mathbf{A}), k_i^{\text{out}}(\mathbf{A})) \in \mathbb{N}^2$. We see that the total number of links in a directed graph can now be written as either $L(\mathbf{A}) = \sum_{i=1}^{N} k_i^{\text{in}}(\mathbf{A})$ or $L(\mathbf{A}) = \sum_{i=1}^{N} k_i^{\text{out}}(\mathbf{A})$. The average in-degree and the average out-degree in a graph are always identical, which reflects the simple fact that all arrows flowing out of a node will inevitably flow into another node. In nondirected graphs, where $A_{ij} = A_{ji}$ for all (i, j), we find that $k_i^{\text{in}}(\mathbf{A}) = k_i^{\text{out}}(\mathbf{A})$ for all i. Here we can drop the superscripts and simply refer to the degree $k_i(\mathbf{A}) = \sum_j A_{ij} \in \mathbb{N}$ of each node i. In nondirected graphs without self-interactions, the number of links is then seen to be $L(\mathbf{A}) = \frac{1}{2} N \bar{k}(\mathbf{A})$, where $\bar{k}(\mathbf{A}) = N^{-1} \sum_{i \leq N} k_i(\mathbf{A})$ is the average degree in the graph.

Generating Random Networks and Graphs. First Edition. A.C.C. Coolen, A. Annibale, and E.S. Roberts. © A.C.C. Coolen, A. Annibale, and E.S. Roberts 2017. Published in 2017 by Oxford University Press. DOI: 10.1093/acprof:oso/9780198709893.001.0001

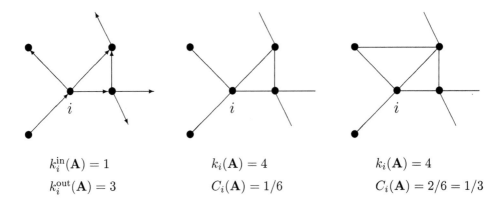

$$k_i^{\text{in}}(\mathbf{A}) = 1$$
$$k_i^{\text{out}}(\mathbf{A}) = 3$$

$$k_i(\mathbf{A}) = 4$$
$$C_i(\mathbf{A}) = 1/6$$

$$k_i(\mathbf{A}) = 4$$
$$C_i(\mathbf{A}) = 2/6 = 1/3$$

Fig. 2.1 Left: Illustration for directed graphs of the meaning of in- and out-degrees $k_i^{\text{in}}(\mathbf{A})$ and $k_i^{\text{out}}(\mathbf{A})$, i.e. the number of arrows flowing into and out of the given node. Middle and right: Illustration for nondirected graphs of degrees $k_i(\mathbf{A})$ and clustering coefficients $C_i(\mathbf{A})$. The latter gives the number of distinct neighbour pairs of i that are themselves connected, divided by the total number of such pairs. Due to the absence of link directionality, there is no distinction in nondirected graphs between in- and out-degrees.

There are several ways to characterize a graph's local structure beyond the level of counting links from a node i to its immediate neighbours. For instance, for nondirected graphs without self-links, one defines the clustering coefficient $C_i(\mathbf{A}) \in [0, 1]$, which gives the fraction of nodes linked to i that are themselves mutually connected

$$C_i(\mathbf{A}) = \frac{\text{number of \textit{connected} node pairs among neighbours of } i}{\text{number of node pairs among neighbours of } i}$$
$$= \frac{\sum_{j,k=1}^{N}(1-\delta_{jk})A_{ij}A_{jk}A_{ik}}{\sum_{j,k=1}^{N}(1-\delta_{jk})A_{ij}A_{ik}} = \frac{2T_i(\mathbf{A})}{k_i(\mathbf{A})[k_i(\mathbf{A})-1]} \qquad (2.2)$$

Here $T_i(\mathbf{A}) = \frac{1}{2}\sum_{j,k=1}^{N} A_{ij}A_{jk}A_{ki} \in \mathbb{N}$ counts the number of distinct triangles, i.e. the number of nondirected loops of length three in the graph in which node i participates [163]. The factor $\frac{1}{2}$ in $T_i(\mathbf{A})$ corrects for overcounting: any nondirected triangle starting from node i can be drawn with two possible orientations. We also used Kronecker's shorthand δ_{ab}, defined as

$$\delta_{ab} = \begin{cases} 1 & \text{if } a = b \\ 0 & \text{if } a \neq b \end{cases} \qquad \delta_{ab} = \int_{-\pi}^{\pi} \frac{d\omega}{2\pi}\, e^{i\omega(a-b)} \qquad (2.3)$$

The mathematically equivalent integral formula for δ_{ab} on the right will turn out to be a helpful tool for simplifying various calculations. Clearly, the clustering coefficient can only be defined for nodes i with degree $k_i(\mathbf{A}) \geq 2$. In Fig. 2.1, we illustrate the meaning of degrees and clustering coefficients with some simple examples.

The above quantities can be generalized in obvious ways. For instance, the generalized degrees count the number of distinct paths of a given length $\ell \geq 1$ that either flow out of, or into a node i:

$$k_i^{(\ell)\text{in}}(\mathbf{A}) = \sum_{j_1=1}^{N} \cdots \sum_{j_\ell=1}^{N} A_{ij_1} A_{j_1j_2} \dots A_{j_{\ell-1}j_\ell} \tag{2.4}$$

$$k_i^{(\ell)\text{out}}(\mathbf{A}) = \sum_{j_1=1}^{N} \cdots \sum_{j_\ell=1}^{N} A_{j_\ell j_{\ell-1}} \dots A_{j_2 j_1} A_{j_1 i} \tag{2.5}$$

For $\ell = 1$, these formulae reduce to the previous expressions (2.1). We can similarly introduce quantities that generalize the triangle counters $T_i(\mathbf{A})$ by counting closed paths in the graph $\mathbf{A}$ of arbitrary lengths $\ell \geq 3$ that pass through node i.

2.2   Macroscopic characterizations of graphs

Overview: In this section, we define and explain the following concepts and associated notation: degree distribution, degree–degree correlation and adjacency matrix spectra.

Macroscopic characterization based on statistics of local features. The quantities defined so far provide microscopic structural information in the vicinity of individual nodes. We can characterize the structure of the full graph by listing for all N nodes the values of the locally defined quantities. In nondirected graphs, we would then obtain, for instance, the degree sequence $\mathbf{k}(\mathbf{A}) = (k_1(\mathbf{A}), \dots, k_N(\mathbf{A})) \in \mathbb{N}^N$. In directed graphs, where each node has two degrees $\vec{k}_i(\mathbf{A}) = (k_i^{\text{in}}(\mathbf{A}), k_i^{\text{out}}(\mathbf{A}))$, this degree sequence would take the form $\mathbf{k}(\mathbf{A}) = (\vec{k}_1(\mathbf{A}), \dots, \vec{k}_N(\mathbf{A})) \in \mathbb{N}^{2N}$. We can make similar lists of quantities such as the clustering coefficients.

For large graphs, or when comparing graphs of different sizes, it will be more helpful to define quantities that are intrinsically macroscopic in nature. The simplest of these are histograms of the observed values of the previously defined local features. If we divide, for each possible value, how often this value is observed by the total number of observations (i.e. the number of nodes), we obtain the empirical distribution of the given feature in the graph. For example, we can define the degree distribution of a nondirected graph via

$$\forall k \in \mathbb{N}: \quad p(k|\mathbf{A}) = \frac{1}{N} \sum_{i=1}^{N} \delta_{k,k_i(\mathbf{A})} \tag{2.6}$$

The quantity $p(k|\mathbf{A})$ represents the probability that a randomly drawn node in the nondirected graph $\mathbf{A}$ will have degree k. Similarly we can define the joint distribution $p(k, T|\mathbf{A})$ of degrees and triangle numbers

$$\forall k, T \in \mathbb{N}: \quad p(k, T|\mathbf{A}) = \frac{1}{N} \sum_{i=1}^{N} \delta_{k,k_i(\mathbf{A})} \delta_{T,T_i(\mathbf{A})} \tag{2.7}$$

Now $p(k, T|\mathbf{A})$ is the probability that a randomly drawn node will have degree k and participate in T triangles. For directed graphs, the equivalent definition is

$$\forall (k^{\text{in}}, k^{\text{out}}) \in \mathbb{N}^2 : \quad p(k^{\text{in}}, k^{\text{out}} | \mathbf{A}) = \frac{1}{N} \sum_{i=1}^{N} \delta_{k^{\text{in}}, k_i^{\text{in}}(\mathbf{A})} \delta_{k^{\text{out}}, k_i^{\text{out}}(\mathbf{A})} \tag{2.8}$$

Here $p(k^{\text{in}}, k^{\text{out}} | \mathbf{A})$ represents the probability that a randomly drawn node in the graph will have the pair $(k^{\text{in}}, k^{\text{out}})$ of in- and out-degrees.

Exercise 2.1 Calculate the average degree $\langle k \rangle = \sum_{k \geq 0} p(k) k$ and the degree variance $\sigma_k^2 = \langle k^2 \rangle - \langle k \rangle^2$ for the degree distribution $p(k) = e^{-\gamma k} / C_N$ for $k \in \mathbb{N}$, with $\gamma > 0$. Next, consider the following scale-free distribution: $p(k) = 0$ for $k = 0$ or $k > N$, and $p(k) = k^{-\gamma} / C_N$ for $0 < k \leq N$, again with $\gamma > 0$. Give a formula for C_N. For which γ is $p(k)$ normalizable as $N \to \infty$? Give formulae for the average degree and the degree variance. For which γ values are the average degree and the variance finite as $N \to \infty$? Give an estimate of the average and variance for $\gamma = 2.5$ and $N = 10{,}000$, using the approximation $\sum_{k=1}^{N} k^{-\lambda} \approx \int_1^N \mathrm{d}k\, k^{-\lambda}$.

Macroscopic characterization involving degree correlations. The interpretation of degree distributions in terms of drawing nodes in the graph at random suggests additional macroscopic routes for characterizing graph structures. For instance, for nondirected graphs $\mathbf{A}$ without self-interactions, we can define the probability $W(k, k' | \mathbf{A})$ that a randomly drawn link will connect nodes with specified degrees k and k' as

$$W(k, k' | \mathbf{A}) = \frac{\sum_{i,j=1}^{N} \delta_{k, k_i}(\mathbf{A}) A_{ij} \delta_{k', k_j}(\mathbf{A})}{\sum_{i,j=1}^{N} A_{ij}}$$

$$= \frac{1}{N\bar{k}(\mathbf{A})} \sum_{i,j=1}^{N} \delta_{k, k_i}(\mathbf{A}) A_{ij} \delta_{k', k_j}(\mathbf{A}) \tag{2.9}$$

Clearly $W(k, k' | \mathbf{A}) = W(k', k | \mathbf{A})$ for all $(k, k') \in \mathbb{N}^2$, and $W(k, k' | \mathbf{A}) = 0$ if $k = 0$ or $k' = 0$ (or both). From (2.9) follows also the *assortativity* $a(\mathbf{a})$ [129], which is defined as the overall Pearson correlation between the degrees of connected node pairs:

$$a(\mathbf{A}) = \frac{\sum_{k,k'>0} W(k, k' | \mathbf{A}) k k' - \left(\sum_{k>0} W(k | \mathbf{A}) k \right)^2}{\sum_{k>0} W(k | \mathbf{A}) k^2 - \left(\sum_{k>0} W(k | \mathbf{A}) k \right)^2} \in [-1, 1] \tag{2.10}$$

with the marginal distribution $W(k | \mathbf{A}) = \sum_{k'>0} W(k, k' | \mathbf{A})$. If $a(\mathbf{A}) > 0$, there is a preference in the graph for linking high-degree nodes to high-degree nodes and low-degree nodes to low-degree nodes; such graphs are called *assortative*. If $a(\mathbf{A}) < 0$, the preference is for linking high-degree nodes to low-degree ones; these graphs are called *dissortative*. Upon summing the definition (2.9) over k', we see that the marginal $W(k | \mathbf{A})$ follows directly from the degree distribution:

$$W(k | \mathbf{A}) = \sum_{k'>0} W(k, k' | \mathbf{A}) = \frac{1}{N\bar{k}(\mathbf{A})} \sum_{i=1}^{N} \delta_{k, k_i}(\mathbf{A}) k_i(\mathbf{A})$$

$$= \frac{k}{\bar{k}(\mathbf{A})} p(k|\mathbf{A}) \qquad (2.11)$$

The reason why $W(k|\mathbf{A}) \neq p(k|\mathbf{A})$ is that in $W(k|\mathbf{A})$ the degree likelihood of nodes is conditioned on these nodes coming up when picking links at random; this favours nodes with many links over those with few. In those graphs where there are no correlations between the degrees of connected nodes, one would find that the joint distribution (2.9) is simply the product of the respective marginals (2.11), $W(k, k'|\mathbf{A}) = W(k|\mathbf{A})W(k'|\mathbf{A})$ for all $k, k' > 0$. Hence, a useful quantity to characterize correlations is the ratio

$$\Pi(k, k'|\mathbf{A}) = \frac{W(k, k'|\mathbf{A})}{W(k|\mathbf{A})W(k'|\mathbf{A})} = \frac{\bar{k}^2(\mathbf{A})}{kk'} \frac{W(k, k'|\mathbf{A})}{p(k|\mathbf{A})p(k'|\mathbf{A})} \qquad (2.12)$$

This ratio is by definition equal to one for graphs without degree correlations. Any deviation from $\Pi(k, k'|\mathbf{A}) = 1$ will signal the presence of degree correlations.

One can easily convince oneself that the degree correlations captured by $\Pi(k, k'|\mathbf{A})$ can provide valuable new information that is not contained in the degree distribution $p(k|\mathbf{A})$. For instance, in Fig. 2.2 we show two networks that have nearly identical degree distributions, but are nevertheless seen to be profoundly different at the level of their degree correlations. This shows us that for large graphs there are still many distinct possible wirings that are consistent with an imposed degree sequence. The result of calculating the macroscopic characteristics $p(k|\mathbf{A})$ and $\Pi(k, k'|\mathbf{A})$ for the example protein interaction data of Fig. 1.1 is shown in Fig. 2.3.

For directed networks, the degree correlations are described by a similar function $W(\vec{k}, \vec{k}'|\mathbf{A})$, where $\vec{k} = (k^{\text{in}}, k^{\text{out}})$ and $\vec{k}' = (k'^{\text{in}}, k'^{\text{out}})$. This reflects the need in directed graphs to distinguish between in-in degree correlations, out-out degree correlations and in-out degree correlations:

$$W(\vec{k}, \vec{k}'|\mathbf{A}) = \frac{\sum_{i,j=1}^{N} \delta_{\vec{k}, \vec{k}_i(\mathbf{A})} A_{ij} \delta_{\vec{k}', \vec{k}_j(\mathbf{A})}}{\sum_{i,j=1}^{N} A_{ij}}$$

$$= \frac{1}{N\bar{k}(\mathbf{A})} \sum_{i,j=1}^{N} \delta_{\vec{k}, \vec{k}_i(\mathbf{A})} A_{ij} \delta_{\vec{k}', \vec{k}_j(\mathbf{A})} \qquad (2.13)$$

In contrast to the kernel $W(k, k'|\mathbf{A})$ for nondirected graphs, which was symmetric under interchanging k and k', here we will generally have $W(\vec{k}, \vec{k}'|\mathbf{A}) \neq W(\vec{k}', \vec{k}|\mathbf{A})$.

Modularity. Sometimes the prominent structure of a network is *modularity*, i.e. the nodes can be divided into groups such that the links preferentially connect node pairs within the same group. Finding the optimal assignment of a string $(x_1, \ldots, x_N)$ of module labels to the nodes in such a network is a common problem in network applications. To quantify the extent to which a simple nondirected graph is modular, one compares the number $L_{\text{intra}}(\mathbf{A}) = \frac{1}{2} \sum_{ij} A_{ij} \delta_{x_i, x_j}$ of intra-modular links in $\mathbf{A}$ (the factor $\frac{1}{2}$ reflects its nondirected nature) to what would have been found if the wiring had been random. In a random N-node graph $\mathbf{A}'$ with the same degree sequence $(k_1(\mathbf{A}), \ldots, k_N(\mathbf{A}))$ as $\mathbf{A}$, we would find $\langle L_{\text{intra}}(\mathbf{A}') \rangle = \frac{1}{2} \sum_{i,j=1}^{N} \langle A'_{ij} \rangle \delta_{x_i, x_j}$. Now

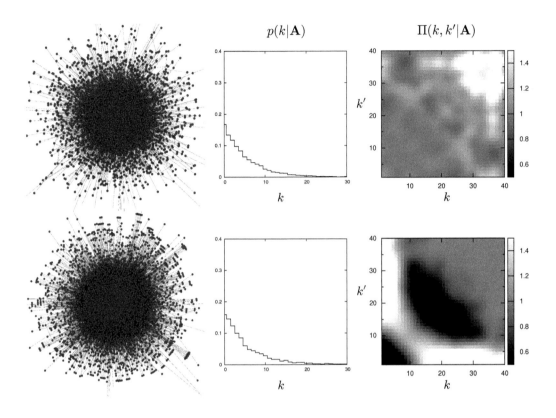

Fig. 2.2 Illustration of the limitations of using only degree statistics to characterize graphs. The two $N = 5000$ graphs $\mathbf{A}$ shown here (top row versus bottom row) look similar and have nearly indistinguishable degree distributions $p(k|\mathbf{A})$ (shown as histograms). However, they differ profoundly at the level of degree correlations, which is visible only after calculating the functions $\Pi(k, k'|\mathbf{A})$, which are shown as greyscale heat maps on the right.

$$\langle A'_{ij} \rangle = \sum_{A'_{ij} \in \{0,1\}} A'_{ij} \, p(A'_{ij}|\mathbf{k}(\mathbf{A})) \ = p(1|\mathbf{k}(\mathbf{A})) \qquad (2.14)$$

Here $p(A'_{ij}|\mathbf{k})$ is the probability that in a random graph with degree sequence $\mathbf{k} = (k_1, \ldots, k_N)$ one finds nodes i and j connected. This must be proportional to k_i and k_j, so we estimate that $\langle A'_{ij} \rangle \approx k_i(\mathbf{A}) k_j(\mathbf{A}) / C(\mathbf{A})$. The value of $C(\mathbf{A})$ is determined by summing both sides over i and j, giving $\sum_{i,j \leq N} \langle A'_{ij} \rangle = [\sum_{i \leq N} k_i(\mathbf{A})]^2 / C(\mathbf{A})$. Hence $C(\mathbf{A}) = N\bar{k}(\mathbf{A})$, and we find our estimate $\langle A'_{ij} \rangle = k_i(\mathbf{A}) k_j(\mathbf{A}) / N\bar{k}(\mathbf{A})$. Therefore

$$\langle L_{\text{intra}}(\mathbf{A}') \rangle \approx \frac{1}{2} \sum_{i,j=1}^{N} \frac{k_i(\mathbf{A}) k_j(\mathbf{A})}{N\bar{k}(\mathbf{A})} \delta_{x_i, x_j} \qquad (2.15)$$

One now defines the measure of modularity $Q(\mathbf{A})$, apart from an overall scaling factor,

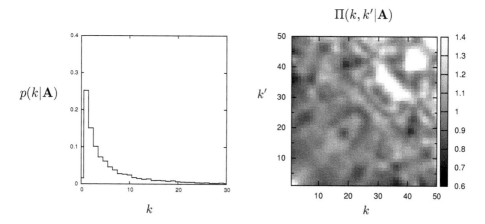

Fig. 2.3 The degree distribution $p(k|\mathbf{A})$ (2.6) (left), and the normalized degree correlation kernel $\Pi(k, k'|\mathbf{A})$ (2.12) (right, shown as a greyscale heatmap) for the protein interaction network of Fig. 1.1. Here $N \approx 9000$ and $\bar{k}(\mathbf{A}) \approx 7.5$. Significant deviations from $\Pi(k, k'|\mathbf{A}) \approx 1$ mark non-trivial structural properties beyond those captured by the degree distribution.

as the difference between the number of intra-modular links in $\mathbf{A}$ and the number we expect to find by accident in a random graph $\mathbf{A}'$ with the same degrees

$$Q(\mathbf{A}) = \frac{1}{N\bar{k}(\mathbf{A})} \sum_{i,j=1}^{N} \left(A_{ij} - \frac{k_i(\mathbf{A})k_j(\mathbf{A})}{N\bar{k}(\mathbf{A})} \right) \delta_{x_i, x_j} \tag{2.16}$$

This definition is applicable only if the graph has a non-zero number of links, so $\bar{k}(\mathbf{A}) > 0$. The chosen scaling factor ensures that $Q(\mathbf{A})$ has the convenient size-independent bounds $-1 \leq Q(\mathbf{A}) \leq 1$. Non-modular graphs give $Q(\mathbf{A}) \approx 0$.

Exercise 2.2 Prove that the modularity (2.16) obeys the general bounds $-1 \leq Q(\mathbf{A}) \leq 1$. Give an example of an N-node simple nondirected graph for which $Q(\mathbf{A}) = 1$.

Other structure characterizations. The properties discussed above have been selected because they are important network metrics, with a long history in the field of random graphs. They measure the statistics of local topological properties, where local is either interpreted from a node-centred point of view, as with $p(k|\mathbf{A})$ and $p(\vec{k}|\mathbf{A})$, or from a link-centred point of view, as with $W(k, k'|\mathbf{A})$ and $W(\vec{k}, \vec{k}'|\mathbf{A})$. We will see that it is indeed generally much easier to incorporate topological characteristics as constraints into a random graph ensembles if these relate to (distributions of) local properties.

Alternative approaches to quantifying graph topologies involve measures based on counting typical and minimal path lengths, and the eigenvalue spectrum $\varrho(\lambda|\mathbf{A})$ of the adjacency matrix. For nondirected graphs the adjacency matrix $\mathbf{A}$ is symmetric, and

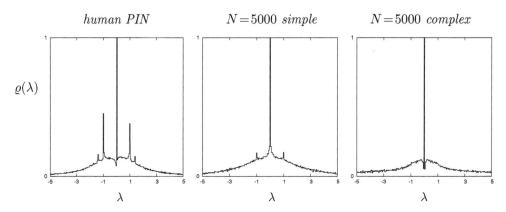

Fig. 2.4 Adjacency matrix eigenvalue distributions $\varrho(\lambda|\mathbf{A})$ of the three graphs $\mathbf{A}$ shown and quantified in previous figures. Left: Eigenvalue distribution for the human protein interaction graph (see Figs. 1.1 and 2.3). Middle and right: Eigenvalue distributions for the two graphs in Fig. 2.2. The middle histogram refers to the graph with only weak degree correlations (top line in Fig. 2.2), and the right histogram refers to the graph with strong degree correlations (bottom line in Fig. 2.2). The eigenvalue spectra of the last two graphs are seen to be significantly different, in spite of their nearly identical degree distributions.

hence it will have N real-valued eigenvalues $\{\lambda_1(\mathbf{A}), \dots, \lambda_N(\mathbf{A})\}$ (which need not be distinct). One defines the distribution of these eigenvalues as

$$\varrho(\lambda|\mathbf{A}) = \frac{1}{N} \sum_{i=1}^{N} \delta[\lambda - \lambda_i(\mathbf{A})] \tag{2.17}$$

(see Appendix A for the definition and properties of the δ-distribution). It is easy to prove that the smallest and largest eigenvalues $\lambda_{\min}(\mathbf{A})$ and $\lambda_{\max}(\mathbf{A})$ of simple nondirected graphs $\mathbf{A}$ obey various inequalities related to the degrees, such as $\lambda_{\min}(\mathbf{A}) \leq \bar{k}(\mathbf{A})$ or $\bar{k}(\mathbf{A}) \leq \lambda_{\max}(\mathbf{A}) \leq \max_{i=1\dots N} k_i(\mathbf{A})$. The eigenvalue spectrum of a nondirected graph contains information on the numbers of closed paths of all possible lengths in this graph, since for any $m \in \mathbb{N}$ we may write, using some simple linear algebra,

$$\int_{-\infty}^{\infty} d\lambda \, \lambda^m \varrho(\lambda|\mathbf{A}) = \frac{1}{N} \sum_{i=1}^{N} \lambda_i^m(\mathbf{A}) = \frac{1}{N} \sum_{i=1}^{N} (\mathbf{A}^m)_{ii}$$

$$= \frac{1}{N} \sum_{i_1=1}^{N} \sum_{i_2=1}^{N} \cdots \sum_{i_m=1}^{N} A_{i_1 i_2} A_{i_2 i_3} \cdots A_{i_{m-1} i_m} A_{i_m i_1} \tag{2.18}$$

Apart from the factor N, the last line gives us the total number of closed walks of length m in the graph $\mathbf{A}$ (including self-intersecting and backtracking ones). Some examples of adjacency matrix eigenvalue spectra for nondirected graphs are shown in Fig. 2.4. For further examples, and more details on the mathematical relation between the adjacency matrix spectra of nondirected random graphs and characteristics such as their degree distributions and degree correlations, see, e.g. [158].

2.3 Solutions of exercises

Exercise 2.1

One calculates C_N (which follows from the normalization requirement $\sum_{k\geq 0} p(k) = 1$), $\langle k \rangle$ and $\langle k^2 \rangle$ using the geometric series $\sum_{\ell \geq 0} x^\ell = (1-x)^{-1}$:

$$C_N = \sum_{k\geq 0} e^{-\gamma k} = \sum_{k\geq 0} (e^{-\gamma})^k = (1-e^{-\gamma})^{-1} \tag{2.19}$$

$$\langle k \rangle = -\frac{1}{C_N}\frac{d}{d\gamma}\sum_{k\geq 0}(e^{-\gamma})^k = -\frac{1}{C_N}\frac{d}{d\gamma}C_N = -\frac{d}{d\gamma}\log C_N = \frac{e^{-\gamma}}{1-e^{-\gamma}} \tag{2.20}$$

$$\langle k^2 \rangle = \frac{1}{C_N}\frac{d^2}{d\gamma^2}\sum_{k\geq 0}(e^{-\gamma})^k = \frac{1}{C_N}\frac{d^2}{d\gamma^2}C_N = -\frac{1}{C_N}\frac{d}{d\gamma}\frac{e^{-\gamma}}{(1-e^{-\gamma})^2}$$

$$= \frac{e^{-\gamma}(1-e^{-\gamma})^2 + 2e^{-2\gamma}(1-e^{-\gamma})}{(1-e^{-\gamma})^3} = \frac{e^{-\gamma}(1+e^{-\gamma})}{(1-e^{-\gamma})^2} \tag{2.21}$$

Hence, the degree variance is

$$\sigma_k^2 = \frac{e^{-\gamma}(1+e^{-\gamma})}{(1-e^{-\gamma})^2} - \frac{e^{-2\gamma}}{(1-e^{-\gamma})^2} = \frac{e^{-\gamma}}{(1-e^{-\gamma})^2} \tag{2.22}$$

Next we turn to the scale-free degree distribution. Again, C_N follows from probability normalization, giving $C_N = \sum_{k=1}^{N} k^{-\gamma}$. In the limit $N \to \infty$ the series $\sum_k k^{-\gamma}$ converges if and only if $\gamma > 1$. The formulae for $\langle k \rangle$ and the variance σ_k^2 become

$$\langle k \rangle = C_N^{-1}\sum_{k=1}^{N} k^{1-\gamma}, \qquad \sigma_k^2 = C_N^{-1}\sum_{k=1}^{N} k^{2-\gamma} - \left(C_N\sum_{k=1}^{N} k^{1-\gamma}\right)^2 \tag{2.23}$$

The series for $\langle k \rangle$ converges for $N \to \infty$ if $\gamma > 2$. The series for $\langle k^2 \rangle$ (which is the first term of σ_k^2) converges if $\gamma > 3$. Hence the degree variance is finite if and only if $\gamma > 3$. Now we choose $\gamma = 2.5$ and $N = 10^4$, and approximate the summations as indicated:

$$\sum_{k=1}^{N} k^{-\lambda} \approx \int_1^N dk\, k^{-\lambda} = \frac{1}{1-\lambda}\left[k^{1-\lambda}\right]_1^N = \frac{1}{1-\lambda}\left(N^{1-\lambda}-1\right)$$

With this we can calculate C_N (by choosing $\lambda = \gamma$), $\langle k \rangle$ (by choosing $\lambda = \gamma - 1$), and $\langle k^2 \rangle$ (by choosing $\lambda = \gamma - 2$):

$$\langle k \rangle \approx \frac{\frac{1}{2-\gamma}\left(N^{2-\gamma}-1\right)}{\frac{1}{1-\gamma}\left(N^{1-\gamma}-1\right)} = 3\frac{N^{-\frac{1}{2}}-1}{N^{-\frac{3}{2}}-1} = 3\frac{1-0.01}{1-0.000001} \approx 2.97$$

$$\langle k^2 \rangle = -3 \, \frac{N^{\frac{1}{2}} - 1}{N^{-\frac{3}{2}} - 1} = 3 \, \frac{99}{1 - 0.000001} \approx 297.00$$

Hence the degree variance is approximately $\sigma_k^2 \approx 297 - 9 = 288$. The degree distribution thus has average $\langle k \rangle \approx 3$ and width $\sigma_k \approx 17$. Although the width of power-law distributed degree distributions with $2 < \gamma < 3$ diverges as $N \to \infty$, we see that for realistic network sizes this divergence is a red herring. $\square$

Exercise 2.2

We start with the lower bound for (2.16). Since both $A_{ij} \in \{0, 1\}$ and $\delta_{x_i, x_j} \in \{0, 1\}$, we may write

$$Q(\mathbf{A}) \geq -\frac{1}{N\bar{k}(\mathbf{A})} \sum_{i,j=1}^{N} \frac{k_i(\mathbf{A}) k_j(\mathbf{A})}{N\bar{k}(\mathbf{A})} \delta_{x_i, x_j}$$

$$\geq -\frac{1}{N\bar{k}(\mathbf{A})} \sum_{i,j=1}^{N} \frac{k_i(\mathbf{A}) k_j(\mathbf{A})}{N\bar{k}(\mathbf{A})} = -\frac{1}{N\bar{k}(\mathbf{A})} \frac{N^2 \bar{k}^2(\mathbf{A})}{N\bar{k}(\mathbf{A})} = -1 \qquad (2.24)$$

The upper bound for (2.16) is proved in an even simpler way:

$$Q(\mathbf{A}) \leq \frac{1}{N\bar{k}(\mathbf{A})} \sum_{i,j=1}^{N} A_{ij} \delta_{x_i, x_j} \leq \frac{1}{N\bar{k}(\mathbf{A})} \sum_{i,j=1}^{N} A_{ij} = \frac{N\bar{k}(\mathbf{A})}{N\bar{k}(\mathbf{A})} = 1 \qquad (2.25)$$

The upper bound is seen to be satisfied, for instance, by the trivial choice $A_{ij} = 1$ for all $(i \neq j)$ and $x_i = x_j$ for all (i, j). $\square$

3
Random graph ensembles

Definition and use of graph ensembles. A graph is defined in full by giving the number N of its nodes and its adjacency matrix $\mathbf{A}$; hence, in order to specify a *random N-node graph*, we must give the set $G \subseteq \{0,1\}^{N^2}$ of allowed graphs (the configuration space), together with a probability distribution $p(\mathbf{A})$ over this set[1]. This combination $\{G, p\}$ of a graph set G with associated probabilities is called a *random graph ensemble*.

Apart from their use in formalizing what we mean by random graphs, thinking in terms of random graph ensembles is very fruitful in the modelling of large complex systems. In such systems, it is usually hard to measure and incorporate all microscopic parameters (e.g. topology, forces) in full detail. Luckily, macroscopic characteristics can often be written as sums over a huge number of microscopic variables, and here statistical laws tend to take over: if the system is sufficiently large[2], the values of macroscopic characteristics can often be calculated with high accuracy already from the *distribution* of the microscopic parameters, since all microscopic states with non-negligible probabilities give nearly identical values. This property is called *self-averaging*, and is very common in statistical physics. For example, one of the key results of kinetic theory is that the macroscopic properties of a gas, such as its pressure, do not depend on the detailed motion of each individual particle, but only on the *distribution* of their velocities, which, in equilibrium, is given by Maxwell's formula. This formula only depends on a few measurable control parameters, such as the temperature of the gas.

Similarly, in complex networks a detailed knowledge of all the microscopic entries $\{A_{ij}\}$ of the adjacency matrix is often not accessible, and not even useful for the purpose of efficiently extracting information about the macroscopic system behaviour. Hence, one normally assumes a suitable probability distribution $p(\mathbf{A})$ for the microscopic variables $\mathbf{A} = \{A_{ij}\}$, and one calculates average values of macroscopic quantities over this distribution. These averages over $p(\mathbf{A})$ are expected to give predictions that are correct even for individual graphs generated from $p(\mathbf{A})$, if N is sufficiently large. Like the Maxwell distribution, one expects moreover that suitable distributions $p(\mathbf{A})$ can be written in terms of only a few parameters – details should not matter!

The crucial question will now be how to choose suitable probabilities $p(\mathbf{A})$, i.e. how to 'tailor' random graph ensembles to mimic real-world graphs. In this section, we define families of networks which satisfy specified topological conditions. This includes a theoretical treatment of how to tailor an ensemble to meet hard or soft constraints.

---

[1] Equivalently, we could always take $G = \{0,1\}^{N^2}$ and assign $p(\mathbf{A}) = 0$ to all disallowed graphs $\mathbf{A}$. Which route to choose is a matter of taste.

[2] In practice this can even mean just a few hundreds of interacting variables.

Generating Random Networks and Graphs. First Edition. A.C.C. Coolen, A. Annibale, and E.S. Roberts. © A.C.C. Coolen, A. Annibale, and E.S. Roberts 2017. Published in 2017 by Oxford University Press. DOI: 10.1093/acprof:oso/9780198709893.001.0001

3.1 The Erdös–Rényi graph ensemble

Overview: In this section, we define the simplest random graph ensemble, the Erdös–Rényi model (ER). We derive some of its properties to gain intuition and as a warm-up.

Definition. The classical and simplest random graph ensemble is the one proposed by Erdös and Rényi [152]. Here G is the set of all simple nondirected N-node graphs, and the $\frac{1}{2}N(N-1)$ individual links A_{ij} between the pairs of nodes (i,j) are for all $\mathbf{A} \in G$ drawn independently, according to $\text{Prob}(A_{ij}=1) = p$ and $\text{Prob}(A_{ij}=0) = 1 - p$:

$$p(\mathbf{A}) = \prod_{i<j} \left[p\,\delta_{A_{ij},1} + (1 - p)\delta_{A_{ij},0} \right] \tag{3.1}$$

Since graphs generated according to this prescription will have *on average*, a mean degree $\langle k \rangle = p(N - 1)$, we can also write the graph probabilities (3.1) equivalently as

$$p(\mathbf{A}|\langle k \rangle) = \prod_{i<j} \left[\frac{\langle k \rangle}{N-1}\delta_{A_{ij},1} + \left(1 - \frac{\langle k \rangle}{N-1}\right)\delta_{A_{ij},0} \right] \tag{3.2}$$

Note that $\langle k \rangle$ is the average *over the ensemble* of the mean connectivity $\bar{k}(\mathbf{A})$ of the individual graphs, i.e. $\langle k \rangle = \langle \bar{k}(\mathbf{A}) \rangle$, where the average $\langle \ldots \rangle$ is taken over (3.1).

Exercise 3.1 Show that the mean degree $\bar{k}(\mathbf{A})$ of individual N-node graphs **a** drawn from the ER ensemble (3.1) is *on average* equal to $p(N-1)$.

We see furthermore that, as a consequence of $A_{ij} \in \{0,1\}$, the probabilities (3.1) can be written in the alternative form:

$$p(\mathbf{A}) = \prod_{i<j} \left[p^{A_{ij}}(1-p)^{1-A_{ij}} \right] = p^{\sum_{i<j} A_{ij}}(1-p)^{\frac{1}{2}N(N-1)-\sum_{i<j} A_{ij}} \tag{3.3}$$

From this, it follows that the dependence of $p(\mathbf{A})$ on $\mathbf{A}$ can be expressed fully in terms of the number $L(\mathbf{A}) = \sum_{i<j} A_{ij}$ of links in the graph $\mathbf{A}$, via the binomial formula

$$p(\mathbf{A}) = p^{L(\mathbf{A})}(1-p)^{N(N-1)/2-L(\mathbf{A})} \tag{3.4}$$

Hence, we may conclude that the Erdös–Rényi ensemble assigns equal probabilities to all graphs with the same number of links.

Exercise 3.2 Use the self-averaging assumption to show that for $N \to \infty$ the degree distribution $p(k|\mathbf{A}) = N^{-1}\sum_i \delta_{k_i(\mathbf{A}),k}$ of an N-node graph drawn from the Erdös–Rényi ensemble (3.2), in which $\langle k \rangle$ remains fixed and finite[3], will become the Poissonian distribution

$$p(k) = e^{-\langle k \rangle}\langle k \rangle^k/k! \tag{3.5}$$

---

[3]This specific scaling regime of parameters, where $\langle k \rangle$ is kept finite and $N \to \infty$, is called the *sparse* or *finite connectivity* regime.

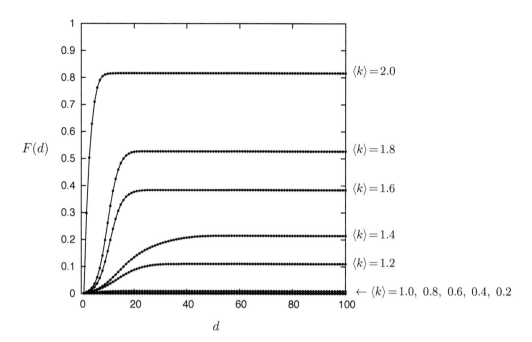

Fig. 3.1 The fraction $F(d)$ of node pairs (i,j) in randomly generated ER graphs that are found to have a distance $d_{ij}(\mathbf{A}) \leq d$, shown versus d. Here $N = 1000$ and $\langle k \rangle \in [0,2]$. One observes clearly that there is a sharp transition at $\langle k \rangle \approx 1$. For $\langle k \rangle < 1$, the fraction of node pairs at a finite distance from each other is vanishingly small. For $\langle k \rangle > 1$, this fraction suddenly becomes finite, which marks the emergence of the giant component in the graph.

Clustering and degree correlations. The average clustering coefficient $C_i = \langle C_i(\mathbf{A}) \rangle$ for graphs from the ER ensemble (3.1) is found to be (see Appendix B):

$$C_i = p\left[1 - (1-p)^{N-1} - (N-1)p(1-p)^{N-2}\right] \tag{3.6}$$

If we keep the parameter $p > 0$ finite and fixed, while sending $N \to \infty$, we obtain

$$\lim_{N \to \infty} C_i = p \tag{3.7}$$

If, alternatively, we set $p = \langle k \rangle/(N-1)$ as in definition (3.2), and keep $\langle k \rangle$ finite and fixed while taking $N \to \infty$, we find

$$\begin{aligned}
C_i &= \frac{\langle k \rangle}{N-1}\left[1 - (1 - \frac{\langle k \rangle}{N-1})^{N-1} - \langle k \rangle(1 - \frac{\langle k \rangle}{N-1})^{N-2}\right] \\
&= \frac{\langle k \rangle}{N}\left[1 - e^{-\langle k \rangle} - \langle k \rangle e^{-\langle k \rangle}\right] + \mathcal{O}(N^{-2})
\end{aligned} \tag{3.8}$$

So if $\langle k \rangle$ is forced to be finite, then all clustering coefficients of typical Erdös–Rényi graphs will vanish for $N \to \infty$. Similarly, one can show (see again Appendix B) that large random graphs generated from the Erdös–Rényi ensemble (3.2) with fixed $\langle k \rangle$ will on average have vanishing degree correlations,

$$\text{for all } (k, k') : \quad \lim_{N \to \infty} \langle W(k, k'|\mathbf{A}) \rangle = \langle W(k|\mathbf{A}) \rangle \langle W(k'|\mathbf{A}) \rangle \qquad (3.9)$$

with $W(k, k'|\mathbf{A})$ as defined in (2.9).

Path lengths and the giant component. A further prominent feature of the the Erdös–Rényi ensemble (3.2) with fixed $\langle k \rangle$ is that for large N it exhibits a percolation transition at $\langle k \rangle = 1$. For values $\langle k \rangle < 1$, the typical graphs generated from the ensemble will be highly fractured, and this is reflected in the observation that for the majority of node pairs (i, j) there will not be any path in the graph that leads from i to j. In contrast, as soon as $\langle k \rangle > 1$, a finite fraction of the N nodes *will* be connected, in what is called a *giant component*. This effect can be observed, for instance, by measuring the fraction $F(d)$ of node pairs (i, j) in the graph that have as their shortest mutual distance $d_{ij}(\mathbf{A}) \leq d$. Here $d_{ij}(\mathbf{A})$ is the smallest value of $\ell \in \mathbb{N}$ (if any) such that $(\mathbf{A}^\ell)_{ij} > 0$; if i and j are not connected, we put $d_{ij}(\mathbf{A}) = \infty$. See Figure 3.1, where $N = 1000$. The percolation transition at $\langle k \rangle = 1$ becomes increasingly sharp for large values of N.

Shortcut: *The Erdös–Rényi ensemble is the simplest non-trivial graph ensemble, but many of its properties are rather different from those observed in real (social, biological, commercial or technological) networks. The latter often have large clustering coefficients, correlated degrees, so-called small-world properties, and non-Poissonian scale-free degree distributions with fat tails.*

3.2 Graph ensembles with hard or soft topological constraints

Overview: *In this section, we present theoretical tools for defining random graph ensembles systematically via soft or hard topological constraints.*

Building topological features into graph ensembles. Given an observed network $\mathbf{A}^\star$, we now want to define systematically random graph ensembles $\{G, p\}$ that are tailored to generate graphs that are structurally similar to $\mathbf{A}^\star$. To achieve this, we must first define what we mean by similar. We quantify the relevant features that we want our graphs to inherit from $\mathbf{A}^\star$ in terms of specific measurements $\Omega_\mu(\mathbf{A}^\star)$, $\mu = 1 \ldots K$, and then demand either that these same measurements are reproduced by *all* graphs generated in the ensemble (i.e. via a hard constraint), or that they are reproduced *on average* (i.e. via a soft constraint) . Upon abbreviating the set of K chosen features as $\mathbf{\Omega}(\mathbf{A}) = (\Omega_1(\mathbf{A}), \ldots, \Omega_K(\mathbf{A}))$, this can be written as

hard constraint : *all* generated graphs must have the chosen features,

$$\mathbf{\Omega}(\mathbf{A}) = \mathbf{\Omega}(\mathbf{A}^\star) \quad \text{for all } \mathbf{A} \in G \text{ with } p(\mathbf{A}) > 0 \qquad (3.10)$$

soft constraint : generated graphs must have the chosen features on average,

$$\sum_{\mathbf{A} \in G} p(\mathbf{A}) \mathbf{\Omega}(\mathbf{A}) = \mathbf{\Omega}(\mathbf{A}^\star) \qquad (3.11)$$

Our chosen features $\Omega_\mu(\mathbf{A})$ will normally represent topological characteristics of the network that are important to the process that runs on it.

Both (3.10) and (3.11) give us conditions on the possible random graph ensembles, but they do not yet specify the probabilities $p(\mathbf{A})$ in full. The unique specification of $p(\mathbf{A})$ for both types of constraints will require information-theoretic arguments [53]. In tailoring our random graphs we want to construct an ensemble $\{G, p\}$ that incorporates the chosen features $\Omega(\mathbf{A})$ but is otherwise strictly unbiased. Given a choice for the set G, this will be true only for the distribution $p(\mathbf{A})$ that, subject to the chosen constraints (3.10) or (3.11), maximizes the Shannon entropy

$$S[p, G] = -\sum_{\mathbf{A} \in G} p(\mathbf{A}) \log p(\mathbf{A}) \tag{3.12}$$

The Shannon entropy defines the information content of a typical sample from the distribution $p(\mathbf{A})$. If we were to choose any distribution other than the one for which the Shannon entropy is maximal (subject to the chosen constraints), we would by definition already know something about our graphs in addition to the information that is contained in the constraints. In other words: we would introduce an *ad hoc* bias into the probability distribution $p(\mathbf{A})$, and hence into our graph ensembles. We show below how maximization of (3.12), subject to the applicable constraints, leads us to unique formulas for our tailored graph ensembles.

Microcanonical ensembles: tailored random graphs with hard constraints. If we opt for the hard constraint (3.10), we need only define the probabilities $p(\mathbf{A})$ for all those graphs $\mathbf{A}$ that obey $\Omega(\mathbf{A}) = \Omega(\mathbf{A}^\star)$. All other graphs are forbidden anyway. We choose the collection of allowed graphs to be our set G, and we take the functions $\Omega_\mu(\mathbf{A})$ to be integer-valued for simplicity (similar results can be obtained for non-integer definitions). Maximization of (3.12) subject to the constraint $\sum_{\mathbf{A} \in G} p(\mathbf{A}) = 1$, using Lagrange's method, then gives the following equations that are to be solved simultaneously for $\{p(\mathbf{A})\}$ and the Lagrange parameter λ

$$\frac{\partial}{\partial p(\mathbf{A})}\left[S[p, G] - \lambda\left(1 - \sum_{\mathbf{A}' \in G} p(\mathbf{A}')\right)\right] = 0 \tag{3.13}$$

Working out the above partial derivatives leads to $\log p(\mathbf{A}) = \lambda - 1$ for all $\mathbf{A} \in G$. The value of the Lagrange parameter λ follows from the normalization condition $\sum_{\mathbf{A} \in G} p(\mathbf{A}) = 1$, and the end result is the following maximum entropy ensemble, in which all graphs $\mathbf{A}$ that meet our hard constraints are equally probable

$$\forall \mathbf{A} \in G: \quad p(\mathbf{A}|\Omega) = \frac{1}{Z(\Omega)} \tag{3.14}$$

$Z(\Omega) \geq 1$ is the number of graphs in $\{0, 1\}^{N^2}$ that have $\Omega(\mathbf{A}) = \Omega(\mathbf{A}^\star)$. We could also choose to work with the set $G = \{0, 1\}^{N^2}$ of all N-node graphs, and forbid those graphs $\mathbf{A}$ that do not satisfy the constraints by assigning to them $p(\mathbf{A}) = 0$. This gives formulae of the form

$$\forall \mathbf{A} \in G: \quad p(\mathbf{A}|\mathbf{\Omega}) = \frac{\delta_{\mathbf{\Omega}(\mathbf{A}),\mathbf{\Omega}}}{Z(\mathbf{\Omega})} \quad Z(\mathbf{\Omega}) = \sum_{\mathbf{A} \in G} \delta_{\mathbf{\Omega}(\mathbf{A}),\mathbf{\Omega}} \quad (3.15)$$

Global constraints such as non-directionality or absence of self-interactions can in principle be incorporated via the quantities $\Omega_\mu(\mathbf{A})$, but in practice it is often more transparent to build such constraints into the definition of G. For instance, we can limit ourselves to simple nondirected graphs by choosing in definition (3.15) the set

$$G = \{\mathbf{A} \in \{0,1\}^{N^2}|\; A_{ij} = A_{ji} \text{ for all } (i,j) \text{ and } A_{ii} = 0 \text{ for all } i\} \quad (3.16)$$

The hard-constrained ensembles (3.15) are also called *microcanonical* ensembles.

Canonical ensembles: tailored random graphs with soft constraints. If we opt to tailor our ensemble via the soft topological constraints (3.11), we have to build any global constraints into the set G, and maximize the entropy (3.12) subject to $\sum_{\mathbf{A} \in G} p(\mathbf{A}) = 1$ and $\sum_{\mathbf{A} \in G} p(\mathbf{A})\Omega_\mu(\mathbf{A}) = \Omega_\mu(\mathbf{A}^\star)$ for all $\mu = 1 \ldots K$. The Lagrange maximization method will now involve $K + 1$ Lagrange parameters $\{\lambda_\mu\}$, K of which correspond to the imposed constraints, and λ_0 corresponds to the normalization requirement $\sum_{\mathbf{A} \in G} p(\mathbf{A}) = 1$. Our equations for $\{p(\mathbf{A})\}$ then become

$$\frac{\partial}{\partial p(\mathbf{A})} \left[S[p, G] - \lambda_0 \left(1 - \sum_{\mathbf{A}' \in G} p(\mathbf{A}') \right) - \sum_{\mu=1}^{K} \lambda_\mu \left(\Omega_\mu(\mathbf{A}^\star) - \sum_{\mathbf{A}' \in G} p(\mathbf{A}')\Omega_\mu(\mathbf{A}') \right) \right] = 0 \quad (3.17)$$

The solution of these latter equations takes the form

$$p(\mathbf{A}) = \frac{e^{\sum_{\mu=1}^{K} \lambda_\mu \Omega_\mu(\mathbf{A})}}{Z(\mathbf{\Omega})}, \quad Z(\mathbf{\Omega}) = \sum_{\mathbf{A} \in G} e^{\sum_{\mu=1}^{K} \lambda_\mu \Omega_\mu(\mathbf{A})} \quad (3.18)$$

in which the parameters $\{\lambda_1, \ldots, \lambda_K\}$ are found by solving the coupled equations

$$\forall \mu \in \{1, \ldots, K\}: \quad \Omega_\mu(\mathbf{A}^\star) = \frac{1}{Z(\mathbf{\Omega})} \sum_{\mathbf{A} \in G} e^{\sum_{\mu=1}^{K} \lambda_\mu \Omega_\mu(\mathbf{A})} \Omega_\mu(\mathbf{A}). \quad (3.19)$$

One can show that equations (3.19) have a unique solution, so again we arrive at a well-defined tailored random graph ensemble. Solving (3.19), whether analytically or numerically, can unfortunately be quite difficult. It is sometimes possible to proceed analytically, if one chooses the observables $\Omega_\mu(\mathbf{A})$ wisely and if N is large. The soft-constrained ensembles (3.18) are also called *exponential* or *canonical* ensembles.

In the above derivations of the hard-constrained and the soft-constrained ensembles, we should in principle have also included the inequality constraints $p(\mathbf{A}) \geq 0$ for all $\mathbf{A} \in G$ when maximizing the Shannon entropy (3.12) – after all, probabilities are not allowed to be negative. However, it turned out in both cases that even without imposing them explicitly, the inequality constraints are satisfied automatically by the maximum entropy distributions we obtained. Hence, these further constraints are obsolete.

Exercise 3.3 Show that the distribution $p(\mathbf{A})$ which maximizes the Shannon entropy (3.12) for simple nondirected graphs, subject to the average degree constraint $\sum_{\mathbf{A} \in G} p(\mathbf{A}) \bar{k}(\mathbf{A}) = \langle k \rangle$ and subject to the normalization $\sum_{\mathbf{A} \in G} p(\mathbf{A}) = 1$, is the Erdös–Rényi ensemble (3.2).

Chapters 4 and 6 will explore soft- and hard-constrained ensembles in greater detail, including the mechanism for including 'hidden' constraints and a discussion of how to combine both hard and soft constraints simultaneously in a single ensemble.

Shortcut: To find the correct probability weights $p(\mathbf{A})$ to define an ensemble of graphs that satisfy certain (hard or soft) topological constraints but are otherwise unbiased, it is necessary to maximize the Shannon entropy $S = -\sum_{\mathbf{A} \in G} p(\mathbf{A}) \log p(\mathbf{A})$. One finds upon doing so that for hard constraints every valid configuration is given the same probability weight $1/|G|$, where $|G|$ represents the number of allowed configurations. For soft-constrained ensembles, the solution is more subtle, and may need to be approached via numerical simulations or $N \to \infty$ type limiting arguments.

3.3   The link between ensembles and algorithms

Overview: In this section, we discuss the link between graph ensembles, i.e. the statistical specifications of desired topological features of our random graphs, and algorithms, i.e. the dynamical processes for the evolution of graphs.

Markovian graph generation algorithms. A graph generation algorithm is a discrete time[4] dynamical process for the adjacency matrix $\mathbf{A}$. Virtually all graph generation algorithms are stochastic. The evolution involves at each step one or more random numbers, so that one cannot write deterministic dynamical equations for the graph $\mathbf{A}$ itself, but only equations relating to the *probability* of finding a certain graph at a given time step. Most algorithms are of the so-called Markovian form, or can be transformed into that form, which means that the graph probabilities at the next iteration step depend only on those of the previous step (as opposed to depending also on states at even earlier stages of the process). Deterministic algorithms are just special cases of stochastic ones, in which the relevant state probabilities at any time can take only the values 0 or 1. Some algorithms take the form of growth processes, where nodes are being added dynamically, so not only the adjacency matrix $\mathbf{A} \in \{0, 1\}^{N^2}$ but also the graph size N itself evolves in time. It follows that the generic graph generation problem can be summarized as in Fig. 3.2.

The quantities $\mathcal{W}\big[(N', \mathbf{A}') \to (N, \mathbf{A})\big] \in [0, 1]$ in Fig. 3.2 represent the probabilities to move to a new state $(N, \mathbf{A})$ at the next iteration if the present state is $(N', \mathbf{A}')$, and they therefore contain the complete specification of an algorithm. Any applicable hard topological constraints must be built into the algorithm by forbidding all moves into states that violate these constraints. We note that probability conservation requires that $\sum_{N \geq 0} \sum_{\mathbf{A} \in \{0,1\}^{N^2}} \mathcal{W}\big[(N', \mathbf{A}') \to (N, \mathbf{A})\big] = 1$ for all $(N', \mathbf{A}')$. In addition

---

[4]Time is inevitably discrete if an algorithm runs on a conventional digital computer.

to choosing the transition probabilities, one also needs to initialize the algorithm, i.e. one must specify $p_0(N, \mathbf{A})$ for all $(N, \mathbf{A})$, subject to the condition that $p_0(N, \mathbf{A}) = 0$ unless $\mathbf{A} \in \{0, 1\}^{N^2}$. The stationary graph sampling weights $p(N, \mathbf{A})$ to which an algorithm evolves, together with the set G of graphs for which these probabilities will be non-zero, will then define our random graph ensemble in the sense of the previous section.

Note that the link between graph generation algorithms and the graph sampling weights found upon their equilibration (the stationary probabilities) in Fig. 3.2 need not be unique. For each choice of stationary probabilities there exist multiple stochastic processes that would have these probabilities as stationary states; this is often a welcome feature, since in practical applications where we just want to generate graphs with specified probabilities, any process that equilibrates to our desired graph ensemble would do. The more algorithms there are that will do our job, the easier it should be to find and implement one. In contrast, if we start by defining a graph generation algorithm, then any potential non-uniqueness of the associated stationary probabilities is problematic. Here we have to ensure that our algorithm is *ergodic*, i.e. that any state $(N, \mathbf{A})$ that does not violate applicable hard constraints can in principle be reached by our stochastic process with non-zero probability in a finite number of iteration steps. Proving ergodicity of stochastic processes can be hard, but if ergodicity is violated we run the danger that the graph probabilities after equilibration will depend on our choice of initialization (which introduces a dangerous bias into our sampling).

The above picture represents a mathematical ideal that in practice cannot be realized fully on a computer. First, no truly random numbers ever emerge from computer code, only numbers that are approximately random. No numerical algorithm will therefore be really random, and when graphs are large, one has to be careful that the number of random number calls made by our code does not approach the period of our random number generator. Secondly, true equilibration of a stochastic process generally requires $t \to \infty$, but in practice we terminate our algorithm after a specified number of steps which we expect to be sufficient. Moreover, algorithms in which the number of nodes increases have to be stopped by hand simply to end with a graph of finite size. The choice of termination time has to be substantiated somehow, either via analytical calculation of process relaxation times (which is usually hard), or via monitoring of the Hamming distance[5] between the graph $\mathbf{A}_0$ at time $t = 0$ and the current graph $\mathbf{A}_t$ (for algorithms that keep the size N fixed). Only in very simple cases will we work with graph generation algorithms that equilibrate truly within a finite number of steps; for instance, one could interpret (3.1) trivially as a process that keeps N fixed and converges within one iteration step.

Randomization versus growth. In writing this book, we have found it helpful to distinguish between the following two classes of graph generation algorithms, as they tend to be used and studied in distinct communities and give rise to distinct questions:

- *Graph randomization algorithms.*
  These are algorithms with $\mathcal{W}\big[(N', \mathbf{A}') \to (N, \mathbf{A})\big] = 0$ unless $N = N'$, so the

---

[5]This is the number of entries for which the two matrices $\mathbf{A}_0$ and $\mathbf{A}_t$ differ.

GRAPH GENERATION ALGORITHM:
evolving probabilities

$$p_{t+1}(N, \mathbf{A}) = \sum_{N' \geq 0} \sum_{\mathbf{A}' \in \{0,1\}^{N'^2}} \mathcal{W}\big[(N', \mathbf{A}') \to (N, \mathbf{A})\big] \, p_t(N', \mathbf{A}')$$

GRAPH SAMPLING WEIGHTS:
stationary probabilities

$$p(N, \mathbf{A}) = \lim_{t \to \infty} p_t(N, \mathbf{A})$$

Fig. 3.2 The central representation of graph generation as the result of a discrete time Markovian stochastic process, i.e. a Markov chain. The core of the graph generation problem is to establish the precise link between specified generation algorithms and the stationary probabilities to which these algorithms will converge. This representation captures virtually all algorithms in the literature today. It depends on the context of one's study and the preferences in one's research community whether one enters this box from the top or from the bottom. If we enter at the top, we *define our random graphs by the algorithm*, and then need to know the associated stationary probabilities in order to anticipate the features of the graphs that will be generated upon termination of our algorithm. If we enter from the bottom, we *specify our random graphs by the desired sampling weights* (i.e. by the desired random graph ensemble), and our problem is to construct an algorithm that will have these weights as its stationary probabilities upon convergence.

number of nodes in the graph is never changed. One usually starts from some seed graph $\mathbf{A}^* \in \{0,1\}^{N^2}$, and hard topological constraints are built in by choosing a seed graph in which these constraints are satisfied, in combination with a set of allowed process moves that leave invariant the values of the chosen topological features to be constrained. Since N is fixed, we can now simplify the algorithm's Markov chain equation, using the implicit convention that always $\mathbf{A}, \mathbf{A}' \in \{0,1\}^{N^2}$, to

$$p_{t+1}(\mathbf{A}) = \sum_{\mathbf{A}'} \mathcal{W}[\mathbf{A}' \to \mathbf{A}] \, p_t(\mathbf{A}'), \qquad p_0(\mathbf{A}) = \delta_{\mathbf{A}, \mathbf{A}^*} \qquad (3.20)$$

- *Graph growth algorithms.*
  These are algorithms with $\mathcal{W}\big[(N', \mathbf{A}') \to (N, \mathbf{A})\big] = 0$ unless $N > N'$. Here, at each iteration step one increases the number of nodes in the graph, and one generates new links to connect the new nodes to the existing ones. The initialization would typically be a set of disconnected nodes.

3.4 Solutions of exercises

Exercise 3.1

We have:

$$\langle \bar{k}(\mathbf{A}) \rangle = \sum_{\mathbf{A} \in G} p(\mathbf{A}) \bar{k}(\mathbf{A}) = \frac{1}{N} \sum_{i} \sum_{\mathbf{A} \in G} p(\mathbf{A}) k_i(\mathbf{A})$$

$$= \frac{1}{N} \sum_{i \neq j} \sum_{\mathbf{A} \in G} p(\mathbf{A}) A_{ij} \qquad (3.21)$$

Upon defining the single link probabilities

$$p(A_{k\ell}) = p\delta_{A_{k\ell},1} + (1-p)\delta_{A_{k\ell},0} \qquad (3.22)$$

we may write

$$\langle \bar{k}(\mathbf{A}) \rangle = \frac{1}{N} \sum_{i \neq j} \sum_{\mathbf{A} \in G} \left[\prod_{k < \ell} p(A_{k\ell}) \right] A_{ij}$$

$$= \frac{1}{N} \sum_{i \neq j} \left[\sum_{A_{ij} \in \{0,1\}} p(A_{ij}) A_{ij} \right] \times \prod_{(k,\ell) \neq (i,j)} \left[\sum_{A_{k\ell} \in \{0,1\}} p(A_{k\ell}|p) \right] = \frac{1}{N} \sum_{i \neq j} p$$

$$= p(N-1) \qquad (3.23)$$

Note that in the penultimate line we used factorization over links of the sum $\sum_{\mathbf{A} \in G} = \prod_{k < \ell} \sum_{A_{k\ell} \in \{0,1\}}$, we isolated the contribution from A_{ij}, and we used the two identities $\sum_{A_{k\ell} \in \{0,1\}} p(A_{k\ell}) = 1$ and $\sum_{A_{k\ell} \in \{0,1\}} p(A_{k\ell}) A_{k\ell} = p$. $\qquad \square$

Exercise 3.2

The assumed self-averaging property of macroscopic observables such as $p(k|\mathbf{A})$ for large graphs allows us to write

$$\lim_{N \to \infty} p(k|\mathbf{A}) = \lim_{N \to \infty} \sum_{\mathbf{A} \in G} p(k|\mathbf{A}) p(\mathbf{A}|\langle k \rangle) \qquad (3.24)$$

Using the integral representation of the Kronecker delta symbol (2.3), we obtain

$$\lim_{N \to \infty} p(k|\mathbf{A}) = \lim_{N \to \infty} \frac{1}{N} \sum_{i} \int_{-\pi}^{\pi} \frac{d\omega}{2\pi} e^{i\omega k} \sum_{\mathbf{a} \in G} p(\mathbf{A}|\langle k \rangle) e^{-i\omega \sum_{\ell} A_{i\ell}} \qquad (3.25)$$

At this point we can use the symmetry of the adjacency matrix and the identity $\sum_{\ell} A_{i\ell} = \sum_{k\ell} A_{k\ell} \delta_{ki} = \sum_{k < \ell} A_{k\ell}(\delta_{ki} + \delta_{\ell i})$ to write

$$\lim_{N \to \infty} p(k|\mathbf{A}) = \lim_{N \to \infty} \frac{1}{N} \sum_{i} \int_{-\pi}^{\pi} \frac{d\omega}{2\pi} e^{i\omega k} \sum_{\mathbf{A} \in G} p(\mathbf{A}|\langle k \rangle) e^{-\frac{1}{2} i\omega \sum_{k < \ell} A_{k\ell}(\delta_{ki} + \delta_{\ell i})} \qquad (3.26)$$

The next step is to insert the probabilities (3.2):

$$\lim_{N\to\infty} p(k|\mathbf{A})$$

$$= \lim_{N\to\infty} \frac{1}{N} \sum_i \int_{-\pi}^{\pi} \frac{d\omega}{2\pi} e^{i\omega k} \sum_{\mathbf{A}\in G} \prod_{k<\ell} \left[\frac{\langle k\rangle}{N-1}\delta_{A_{k\ell},1} + \left(1 - \frac{\langle k\rangle}{N-1}\right)\delta_{A_{k\ell},0}\right] e^{-i\omega A_{k\ell}(\delta_{ki}+\delta_{\ell i})}$$

$$= \lim_{N\to\infty} \frac{1}{N} \sum_i \int_{-\pi}^{\pi} \frac{d\omega}{2\pi} e^{i\omega k} \prod_{k<\ell} \sum_{A_{k\ell}\in\{0,1\}} \left[\frac{\langle k\rangle}{N-1} e^{-i\omega(\delta_{ki}+\delta_{\ell i})}\delta_{A_{k\ell},1} + \left(1 - \frac{\langle k\rangle}{N-1}\right)\delta_{A_{k\ell},0}\right]$$

$$= \lim_{N\to\infty} \frac{1}{N} \sum_i \int_{-\pi}^{\pi} \frac{d\omega}{2\pi} e^{i\omega k} \prod_{k<\ell} \left[1 + \frac{\langle k\rangle}{N-1}\left(e^{-i\omega(\delta_{ki}+\delta_{\ell i})} - 1\right)\right] \tag{3.27}$$

where the last two steps were achieved by multiplying through the exponential, and eliminating terms that must certainly be equal to 1 by the action of the Kronecker delta function. The next stage uses the expansion $1 + x = \exp(x - \frac{1}{2}x^2 + \mathcal{O}(x^3))$ for $x \simeq 0$

$$\lim_{N\to\infty} p(k|\mathbf{A})$$

$$= \lim_{N\to\infty} \frac{1}{N} \sum_i \int_{-\pi}^{\pi} \frac{d\omega}{2\pi} e^{i\omega k} \prod_{k<\ell} e^{\frac{\langle k\rangle}{N-1}(e^{-i\omega(\delta_{ki}+\delta_{\ell i})}-1) - \frac{\langle k\rangle^2}{2N^2}(e^{-i\omega(\delta_{ki}+\delta_{\ell i})}-1)+\mathcal{O}(N^{-3})}$$

$$= \lim_{N\to\infty} \frac{1}{N} \sum_i \int_{-\pi}^{\pi} \frac{d\omega}{2\pi} e^{i\omega k} \prod_{k<\ell} e^{(\delta_{ki}+\delta_{\ell i})\left[\frac{\langle k\rangle}{N-1}(e^{-i\omega}-1) - \frac{\langle k\rangle^2}{2N^2}(e^{-i\omega}-1)\right]+\mathcal{O}(N^{-3})}$$

$$= \lim_{N\to\infty} \frac{1}{N} \sum_i \int_{-\pi}^{\pi} \frac{d\omega}{2\pi} e^{i\omega k + \frac{1}{2}\sum_{\ell\neq k}(\delta_{ki}+\delta_{\ell i})\left[\frac{\langle k\rangle}{N-1}(e^{-i\omega}-1) - \frac{\langle k\rangle^2}{2N^2}(e^{-i\omega}-1)\right]+\mathcal{O}(N^{-3})}$$

$$= \lim_{N\to\infty} \frac{1}{N} \sum_i \int_{-\pi}^{\pi} \frac{d\omega}{2\pi} e^{i\omega k + \langle k\rangle(e^{-i\omega}-1)+\mathcal{O}(N^{-1})} \tag{3.28}$$

After writing the factor $\exp[\langle k\rangle e^{-i\omega}]$ as a Taylor series, we then obtain a form in which we can recognize and exploit once more the integral representation of the Kronecker delta symbol:

$$\lim_{N\to\infty} p(k|\mathbf{A}) = e^{-\langle k\rangle} \int_{-\pi}^{\pi} \frac{d\omega}{2\pi} e^{i\omega k} \sum_{r\geq 0} \frac{\langle k\rangle^r}{r!} e^{-ir\omega} = e^{-\langle k\rangle} \sum_{r\geq 0} \frac{\langle k\rangle^r}{r!} \delta_{k,r}$$

$$= e^{-\langle k\rangle} \langle k\rangle^k / k! \tag{3.29}$$

which gives the Poissonian degree distribution (3.5), as claimed. □

Exercise 3.3

We apply the general results (3.18, 3.19) for soft-constrained maximum entropy ensembles. Upon selecting $\bar{k}(\mathbf{A}) = N^{-1}\sum_i k_i(\mathbf{A})$ for our observable $\Omega(\mathbf{A})$, we find (3.18) reducing to

$$p(\mathbf{A}) = \frac{e^{\lambda N^{-1}\sum_{i,j=1}^N A_{ij}}}{Z(\langle k \rangle)}, \qquad Z(\langle k \rangle) = \sum_{\mathbf{A}\in G} e^{\lambda N^{-1}\sum_{i,j=1}^N A_{ij}} \qquad (3.30)$$

We can rewrite this probability distribution, using the symmetry of $\mathbf{A}$:

$$p(\mathbf{A}) = \frac{e^{2\lambda N^{-1}\sum_{i<j} A_{ij}}}{Z(\langle k \rangle)} = \frac{1}{Z(\langle k \rangle)}\prod_{i<j}\left[e^{2\lambda/N}\delta_{A_{ij},1} + \delta_{A_{ij},0}\right] \qquad (3.31)$$

The normalization sum $Z(\langle k \rangle)$ becomes

$$Z(\langle k \rangle) = \sum_{\mathbf{A}\in G}\prod_{i<j}\left[e^{2\lambda/N}\delta_{A_{ij},1} + \delta_{A_{ij},0}\right]$$
$$= \prod_{i<j}\sum_{A_{ij}\in\{0,1\}}\left[e^{2\lambda/N}\delta_{A_{ij},1} + \delta_{A_{ij},0}\right] = \prod_{i<j}\left[1 + e^{2\lambda/N}\right] \qquad (3.32)$$

In combination we can now write

$$p(\mathbf{A}) = \prod_{i<j}\left[\frac{e^{2\lambda/N}}{1+e^{2\lambda/N}}\delta_{A_{ij},1} + \frac{1}{1+e^{2\lambda/N}}\delta_{A_{ij},0}\right] \qquad (3.33)$$

The equation (3.19) from which to solve λ simplifies to

$$\langle k \rangle = \frac{1}{N}\sum_{\ell r}\sum_{\mathbf{A}\in G} A_{\ell r}\prod_{i<j}\left[\frac{e^{2\lambda/N}}{1+e^{2\lambda/N}}\delta_{A_{ij},1} + \frac{1}{1+e^{2\lambda/N}}\delta_{A_{ij},0}\right]$$
$$= \frac{2}{N}\sum_{\ell<r}\sum_{A_{\ell r}\in\{0,1\}} A_{\ell r}\left[\frac{e^{2\lambda/N}}{1+e^{2\lambda/N}}\delta_{A_{\ell r},1} + \frac{1}{1+e^{2\lambda/N}}\delta_{A_{\ell r},0}\right]$$
$$= \frac{2}{N}\sum_{\ell<r}\frac{e^{2\lambda/N}}{1+e^{2\lambda/N}} = (N-1)\frac{e^{2\lambda/N}}{1+e^{2\lambda/N}} \qquad (3.34)$$

This allows us to eliminate λ and show that (3.33) is indeed equivalent to

$$p(\mathbf{A}) = \prod_{i<j}\left[\frac{\langle k \rangle}{N-1}\delta_{A_{ij},1} + \left(1 - \frac{\langle k \rangle}{N-1}\right)\delta_{A_{ij},0}\right] \qquad (3.35)$$

□

Part II

Random graph ensembles

4

Soft constraints: exponential random graph models

Introduction. Exponential random graphs are those sampled from ensembles of the canonical form (3.18), which was derived by combining maximum entropy arguments with soft topological constraints, and involves graph probabilities $p(\mathbf{A})$ that depend *exponentially* on the quantities used to formulate the constraints.

Exponential random graph models (ERGMs) provide conceptually simple recipes for generating random graphs and for estimating the values of observables of random graphs, given only the values of some other observables. However, they are not without difficulties. While they work well for linear constraints, such as an imposed degree sequence, they can exhibit degenerate behaviour in more complicated cases. In this chapter, we illustrate such behaviour, focusing on the phenomenology of so-called phase transitions. These are abrupt quantitative changes in the macroscopic characteristics of the sampled graphs, which become increasingly prominent as the graph size N increases, and which make it very difficult in practice to control the values of relevant ensemble averages in the required manner.

We work out several examples of ERGMs, hopefully in enough detail to not only provide an understanding of the models and their limitations, but also to give the reader the opportunity to extend these results to develop their own bespoke ERGMs. We also show that the ERGM approach can be used to define network ensembles with community structure, the so-called stochastic block models (SBMs).

4.1 Definitions and basic properties of ERGMs

Overview: We define ERGMs as soft-constrained ensembles with parametrized probabilities $p(\mathbf{A})$, in which we exploit the exponential dependence on the parameters.

ERGMs as exponentially parametrized ensembles. In the previous chapter, we introduced the concept of network or graph ensembles. Each ensemble is defined by a set G of all allowed microscopic graph realizations $\mathbf{A}$, together with a probability distribution $p(\mathbf{A})$ over the set. Exponential random graph models are those where the probabilities are of the general canonical form (3.18). They were first introduced in [88], and further developed in [76] and in several later studies such as [170, 181, 159, 13, 165]. They are nowadays widely used, especially in the social sciences. Computational tools to analyze and simulate networks based on ERG models are widely available on the Internet.

Generating Random Networks and Graphs. First Edition. A.C.C. Coolen, A. Annibale, and E.S. Roberts. © A.C.C. Coolen, A. Annibale, and E.S. Roberts 2017. Published in 2017 by Oxford University Press. DOI: 10.1093/acprof:oso/9780198709893.001.0001

In ERGMs we often regard the probabilities $p(\mathbf{A})$ as parametrized directly by $\boldsymbol{\lambda} = \{\lambda_1, \dots, \lambda_K\}$. We no longer view these latter parameters as unknown complicated functions of the values $\boldsymbol{\Omega}$ of the soft-constrained topological features $\boldsymbol{\Omega}(\mathbf{A})$ that have to be solved and eliminated via (3.11), and we simply write

$$p(\mathbf{A}) = \frac{e^{\sum_{\mu=1}^{K} \lambda_\mu \Omega_\mu(\mathbf{A})}}{Z(\boldsymbol{\lambda})}, \qquad Z(\boldsymbol{\lambda}) = \sum_{\mathbf{A} \in G} e^{\sum_{\mu=1}^{K} \lambda_\mu \Omega_\mu(\mathbf{A})} \qquad (4.1)$$

The graph set G will normally only code for basic topological features, such as non-directedness. We have seen earlier that ensembles of the form (4.1) are maximally random, in the sense of maximizing the Shannon entropy $S = -\sum_{\mathbf{A} \in G} p(\mathbf{A}) \log p(\mathbf{A})$ given the imposed values of the ensemble averages of $\{\Omega_1(\mathbf{A}), \dots, \Omega_K(\mathbf{A})\}$. In this chapter, we focus on those properties of ERGMs that derive from their exponential form alone, and various subtleties which one will encounter when sampling from such ensembles.

Averages, fluctuations and susceptibilities. Once the parameter values $\boldsymbol{\lambda}$ of an ERGM are known, one can use the corresponding probability distribution $p(\mathbf{A})$ to estimate the value of any observable $f(\mathbf{A})$ by calculating the ensemble average

$$\langle f(\mathbf{A}) \rangle = \frac{\sum_{\mathbf{A} \in G} f(\mathbf{A}) e^{\sum_{\mu=1}^{K} \lambda_\mu \Omega_\mu(\mathbf{A})}}{Z(\boldsymbol{\lambda})} \qquad (4.2)$$

In particular, the expectation values for the key observables $\Omega_\mu(\mathbf{A})$ can all be written as partial derivatives of the quantity $F(\boldsymbol{\lambda}) = \log Z(\boldsymbol{\lambda})$, which apparently acts as a generating function[1]:

$$\langle \Omega_\mu(\mathbf{A}) \rangle = \frac{1}{Z(\boldsymbol{\lambda})} \sum_{\mathbf{A} \in G} \Omega_\mu(\mathbf{A}) e^{\sum_\nu \lambda_\nu \Omega_\nu(\mathbf{A})}$$

$$= \frac{1}{Z(\boldsymbol{\lambda})} \frac{\partial}{\partial \lambda_\mu} Z(\boldsymbol{\lambda}) = \frac{\partial}{\partial \lambda_\mu} F(\boldsymbol{\lambda}) \qquad (4.3)$$

The same turns out to be true for the fluctuations in the key observables, since

$$\langle \Omega_\mu^2(\mathbf{A}) \rangle - \langle \Omega_\mu(\mathbf{A}) \rangle^2 = \frac{1}{Z(\boldsymbol{\lambda})} \sum_{\mathbf{A} \in G} \Omega_\mu^2(\mathbf{A}) e^{\sum_\nu \lambda_\nu \Omega_\nu(\mathbf{A})} - \left(\frac{1}{Z(\boldsymbol{\lambda})} \frac{\partial}{\partial \lambda_\mu} Z(\boldsymbol{\lambda}) \right)^2$$

$$= \frac{1}{Z(\boldsymbol{\lambda})} \frac{\partial^2 Z(\boldsymbol{\lambda})}{\partial \lambda_\mu^2} - \frac{1}{Z^2(\boldsymbol{\lambda})} \left(\frac{\partial Z(\boldsymbol{\lambda})}{\partial \lambda_\mu} \right)^2$$

$$= \frac{\partial}{\partial \lambda_\mu} \left(\frac{1}{Z(\boldsymbol{\lambda})} \frac{\partial Z(\boldsymbol{\lambda})}{\partial \lambda_\mu} \right) = \frac{\partial^2 F(\boldsymbol{\lambda})}{\partial \lambda_\mu^2} \qquad (4.4)$$

---

[1] In the terminology of statistical physics the quantity $F(\boldsymbol{\lambda})$ would, apart from an overall multiplicative constant, be called the *free energy*. Similarly, one would call the normalizing constant $Z(\boldsymbol{\lambda})$ the *partition function* and the exponent of the probabilities, apart from an overall factor, the *Hamiltonian* of the system: $H(\mathbf{A}) = -\sum_{\mu=1}^{K} \Omega_\mu(\mathbf{A})$.

Identity (4.3) tells us that, in general, we will be able to calculate the ensemble parameters $\boldsymbol{\lambda}$ from the equations for the constraints if we can calculate the (logarithm of) the normalizing constant $Z(\boldsymbol{\lambda})$. From the above relations, it also follows that

$$\langle \Omega_\mu^2(\mathbf{A}) \rangle - \langle \Omega_\mu(\mathbf{A}) \rangle^2 = \frac{\partial}{\partial \lambda_\mu} \langle \Omega_\mu(\mathbf{A}) \rangle \tag{4.5}$$

The latter quantity is called a *susceptibility*; it measures the sensitivity of $\langle \Omega_\mu(\mathbf{A}) \rangle$ to changes in its associated control variable λ_μ.

All identities derived so far rely solely on the exponential dependence of the probabilities on the control parameters. Several popular models of random graphs involve parametrized probabilities $p(\mathbf{A})$ that can be written in the exponential form for a suitable choice of the key observables $\Omega_\mu(\mathbf{A})$, some of which will be discussed below.

Finding parameters in terms of imposed ensemble averages. The more sophisticated the observables $\Omega_\mu(\mathbf{A})$, the harder it will be, it appears, to solve the required values of the ensemble's Lagrange parameters from their equations (3.11). In fact, this is not the case; a simple argument reveals that solving (3.11) numerically will always be easy. Let us write the required values for the K ensemble averages $\langle \Omega_\mu(\mathbf{A}) \rangle$ simply as $\Omega_\mu^\star$. We then define the function

$$\mathcal{L}(\boldsymbol{\lambda}) = \log Z(\boldsymbol{\lambda}) - \sum_{\mu=1}^{K} \lambda_\mu \Omega_\mu^\star \tag{4.6}$$

Using (4.3), we find that this function has the following properties:

$$\frac{\partial \mathcal{L}(\boldsymbol{\lambda})}{\partial \lambda_\mu} = \langle \Omega_\mu(\mathbf{A}) \rangle - \Omega_\mu^\star \tag{4.7}$$

$$\frac{\partial^2 \mathcal{L}(\boldsymbol{\lambda})}{\partial \lambda_\mu \partial \lambda_\nu} = \sum_{\mathbf{A} \in G} \Omega_\mu(\mathbf{A}) \Omega_\nu(\mathbf{A}) \frac{e^{\sum_\rho \lambda_\rho \Omega_\rho(\mathbf{A})}}{Z(\boldsymbol{\lambda})} - \left(\sum_{\mathbf{A} \in G} \Omega_\mu(\mathbf{A}) e^{\sum_\rho \lambda_\rho \Omega_\rho(\mathbf{A})} \right) \frac{\partial}{\partial \lambda_\nu} \frac{1}{Z(\boldsymbol{\lambda})}$$
$$= \langle \Omega_\mu(\mathbf{A}) \Omega_\nu(\mathbf{A}) \rangle - \langle \Omega_\mu(\mathbf{A}) \rangle \langle \Omega_\nu(\mathbf{A}) \rangle \tag{4.8}$$

Expression (4.7) tells us that the stationary points of $\mathcal{L}(\boldsymbol{\lambda})$ are precisely those $\boldsymbol{\lambda} \in \mathbb{R}^K$ that solve our equations for the Lagrange parameters. Expression (4.8) tells us that the function $\mathcal{L}(\boldsymbol{\lambda})$ is convex[2], hence the requested solution(s) of the K coupled equations $\langle \Omega_\mu(\mathbf{A}) \rangle = \Omega_\mu^\star$ (if they exist) can always be found by *minimizing* the function $\mathcal{L}(\boldsymbol{\lambda})$. Moreover, if the curvature matrix $M_{\mu\nu}(\boldsymbol{\lambda}) = \partial^2 \mathcal{L}(\boldsymbol{\lambda})/\partial \lambda_\mu \partial \lambda_\nu$ is positive definite (i.e. it has no zero eigenvalues, which will be true as soon as the K chosen functions $\Omega_\mu(\mathbf{A})$ that define our graph ensemble are not linearly dependent), there will be at most one solution[3], and it will be the unique minimum of $\mathcal{L}(\boldsymbol{\lambda})$. Finding the minima of convex functions is much easier than solving coupled nonlinear equations, and there are many simple and fast algorithms available [150] that would do the job.

---

[2]This follows from the fact that the curvature matrix $M_{\mu\nu}(\boldsymbol{\lambda}) = \partial^2 \mathcal{L}(\boldsymbol{\lambda})/\partial \lambda_\mu \partial \lambda_\nu$ has the form of a covariance matrix, and must therefore be non-negative definite. To see this explicitly, note that for any $\mathbf{x} \in \mathbb{R}^K$ one will find $\sum_{\mu\nu} x_\mu M_{\mu\nu}(\boldsymbol{\lambda}) x_\nu = \langle [\sum_\mu x_\mu \Omega_\mu(\mathbf{A})]^2 \rangle - \langle \sum_\mu x_\mu \Omega_\mu(\mathbf{A}) \rangle^2 \geq 0$.

[3]Note that it may be possible that no solution of (3.11) exists, which would happen, e.g. if we were to impose extreme average values for our observables that are impossible to achieve.

4.2 ERGMs that can be solved exactly

Overview: *We introduce three examples of ERGMs in which the observables $\Omega_\mu(\mathbf{A})$ depend linearly on the adjacency matrix. In these examples, one can easily work out the values of the parameters $\{\lambda_\mu\}$ analytically.*

Example 1: Nondirected ERGM with controlled average number of links. We first demonstrate that the Erdös–Rényi ensemble (3.1) can be regarded as the simplest non-trivial ERGM. If we are given only the total number of links in a simple nondirected network, so that $\Omega(\mathbf{A}) = \sum_{i<j} A_{ij}$ and $G = \{\mathbf{A} \in \{0,1\}^{N^2} \mid A_{ij} = A_{ji} \ \forall (i,j), \ A_{ii} = 0 \ \forall i\}$, the probabilities (4.1) will become

$$p(\mathbf{A}) = \frac{e^{\lambda \sum_{i<j} A_{ij}}}{Z(\lambda)}, \qquad Z(\lambda) = \sum_{\mathbf{A} \in G} e^{\lambda \sum_{i<j} A_{ij}} \tag{4.9}$$

Since the probability $p(\mathbf{A})$ factorizes over the links, we can easily work out the generating function $F(\lambda) = \log Z(\lambda)$:

$$F(\lambda) = \log \sum_{\mathbf{A} \in G} e^{\lambda \sum_{i<j} A_{ij}} = \log \prod_{i<j} \left(\sum_{A_{ij}} e^{\lambda A_{ij}} \right)$$

$$= \log \prod_{i<j} (1 + e^\lambda) = \frac{1}{2} N(N-1) \log(1 + e^\lambda) \tag{4.10}$$

Working out (4.3) then gives us the average number of links

$$L = \Big\langle \sum_{i<j} A_{ij} \Big\rangle = \frac{\partial F(\lambda)}{\partial \lambda} = \frac{1}{2} N(N-1) \frac{e^\lambda}{1 + e^\lambda}. \tag{4.11}$$

Since the links in (4.9) are distributed independently and identically, it is clear that (4.9) is the Erdös–Rényi ensemble (3.1). To determine the individual link probabilities p of (3.1), we can rewrite the graph probabilities $p(\mathbf{A})$ in (4.9) as

$$p(\mathbf{A}) = \prod_{i<j} \left(\frac{e^\lambda}{1 + e^\lambda} \delta_{A_{ij},1} + \frac{1}{1 + e^\lambda} \delta_{A_{ij},0} \right) \tag{4.12}$$

Thus, $p = e^\lambda / (1 + e^\lambda)$, and (4.11) correctly reduces to $\langle \sum_{i<j} A_{ij} \rangle = \frac{1}{2} N(N-1)p$. We conclude that the Erdös–Rényi ensemble is indeed an ERGM. Here, we have found that the ensemble parameter λ can be calculated explicitly in terms of the relevant observable, since (4.11) can be rewritten as

$$\lambda = \log \left(\frac{L}{N(N-1)/2 - L} \right) \tag{4.13}$$

Example 2: Nondirected ERGM with controlled individual degrees. If we are given the expectation values in simple nondirected graphs of all N degrees, i.e. we prescribe

$k_i = \langle \sum_j A_{ij} \rangle$ for all i, then our ERGM will have N parameters $\boldsymbol{\lambda} = \{\lambda_1, \ldots, \lambda_N\}$, and the graph probabilities in (4.1) specialize to

$$p(\mathbf{A}) = \frac{e^{\sum_{i=1}^{N} \lambda_i \sum_j A_{ij}}}{Z(\boldsymbol{\lambda})} = \frac{e^{\sum_{i<j} A_{ij}(\lambda_i + \lambda_j)}}{Z(\boldsymbol{\lambda})} \tag{4.14}$$

The generating function $F(\boldsymbol{\lambda}) = \log Z(\boldsymbol{\lambda})$ now becomes

$$F(\boldsymbol{\lambda}) = \log \sum_{\mathbf{A} \in G} e^{\sum_{i<j}(\lambda_i + \lambda_j)A_{ij}}$$

$$= \log \prod_{i<j}(1 + e^{\lambda_i + \lambda_j}) = \sum_{i<j} \log(1 + e^{\lambda_i + \lambda_j}) \tag{4.15}$$

Working out (4.3) leads us to the following N equations:

$$k_i = \frac{\partial F(\boldsymbol{\lambda})}{\partial \lambda_i} = \sum_j \frac{e^{\lambda_i + \lambda_j}}{1 + e^{\lambda_i + \lambda_j}} = \sum_j \frac{1}{1 + e^{-\lambda_i - \lambda_j}} \tag{4.16}$$

Although the links in (4.14) are again distributed independently, they are no longer distributed identically. Expression (4.14) can be rewritten as

$$p(\mathbf{A}) = \frac{1}{Z(\boldsymbol{\lambda})} \prod_{i<j} \left(e^{\lambda_i + \lambda_j} \delta_{A_{ij},1} + \delta_{A_{ij},0} \right)$$

$$= \prod_{i<j} \left(\frac{e^{\lambda_i + \lambda_j}}{1 + e^{\lambda_i + \lambda_j}} \delta_{A_{ij},1} + \frac{1}{1 + e^{\lambda_i + \lambda_j}} \delta_{A_{ij},0} \right) \tag{4.17}$$

So the probability to have a link between nodes i and j is $p_{ij} = \langle A_{ij} \rangle = (1 + e^{-\lambda_i - \lambda_j})^{-1}$, which is clearly consistent with (4.16).

Solving our N parameters $\{\lambda_1, \ldots, \lambda_N\}$ from the coupled nonlinear equations (4.16) is generally not possible analytically. However, in the sparse regime it can be done. If all degrees k_i remain finite and $N \to \infty$, we know from (4.16) that $e^{\lambda_i + \lambda_j} = \mathcal{O}(1/N)$. Now we can deduce that

$$k_i = \sum_j \left(e^{\lambda_i + \lambda_j} + \mathcal{O}(N^{-2}) \right) = e^{\lambda_i} \sum_j e^{\lambda_j} + \mathcal{O}(N^{-1}) \tag{4.18}$$

Summing this expression over i gives $N^{-1}[\sum_j e^{\lambda_j}]^2 = \langle k \rangle + \mathcal{O}(N^{-1})$, and hence

$$\sum_j e^{\lambda_j} = \sqrt{N\langle k \rangle} + \mathcal{O}(N^{-1/2}) \tag{4.19}$$

Insertion into (4.18) then brings us to an explicit leading order formula for the λ_i:

$$e^{\lambda_i} = \frac{k_i + \mathcal{O}(N^{-1})}{\sum_j e^{\lambda_j}} = \frac{k_i + \mathcal{O}(N^{-1})}{\sqrt{N\langle k \rangle} + \mathcal{O}(N^{-1/2})} = \frac{k_i}{\sqrt{N\langle k \rangle}} + \mathcal{O}(N^{-3/2}) \tag{4.20}$$

We conclude that

$$p_{ij} = \frac{k_i k_j}{N\langle k \rangle} + \mathcal{O}(N^{-2}) \tag{4.21}$$

For large N and finite degrees, the $\mathcal{O}(N^{-2})$ corrections in this expression for the link probabilities become negligible, and we find that (4.14) reduces to the formula of [47]:

$$p(\mathbf{A}) = \prod_{i<j} \left[\frac{k_i k_j}{N\langle k \rangle} \delta_{A_{ij},1} + \left(1 - \frac{k_i k_j}{N\langle k \rangle} \right) \delta_{A_{ij},0} \right] \tag{4.22}$$

These probabilities define graph ensembles which are tailored to produce graphs with any desired soft-constrained degree sequence[4]. In particular, they provide simple recipes for generating graphs with power-law (or scale-free) degree distributions $p(k) \sim k^{-\gamma}$, with $2 < \gamma < 3$, which are often observed in nature. One simply draws each of the N target degrees k_i randomly and independently from the target degree distribution $p(k)$, inserts their values into the probabilities (4.22) and uses the so-called Monte Carlo algorithms (to be discussed in Chapter 6) to sample from (4.22).

Shortcut: *Large sparse random graphs in which the full degree sequence* $\{k_1, \dots, k_N\}$ *is imposed as soft constraints, but that are otherwise unbiased, are generated by the probabilities* $p(\mathbf{A}) = \prod_{i<j} \left[\frac{k_i k_j}{N\langle k \rangle} \delta_{A_{ij},1} + \left(1 - \frac{k_i k_j}{N\langle k \rangle} \right) \delta_{A_{ij},0} \right]$.

Graphicality of degree sequences. A potential difficulty with the previous route is that not all degree sequences that one could write down are actually graphical: some choices for $\{k_1, \dots, k_N\}$ cannot be realized in nondirected networks without introducing self-loops or multiple links. Although our degree constraints have so far been 'soft', so that we will still always produce random graphs, one expects the precision of the method to suffer when the requested degree values are impossible to realize directly. The simplest example of non-graphical degree sequences are those where $\sum_i k_i$ is odd . A sufficient condition for a degree sequence to be graphical is provided by the Erdös–Gallai theorem [67], which states that a degree sequence, with degrees ranked in decreasing order, i.e. $k_1 \geq k_2 \geq \dots \geq k_N$, is graphical if $\sum_i k_i$ is even and it satisfies the following inequality for any integer n in the range $1 \leq n \leq N-1$:

$$\sum_{i=1}^{n} k_i \leq n(n-1) + \sum_{i=n+1}^{N} \min\{n, k_i\} \tag{4.23}$$

Graphicality conditions can be violated quite easily. A recent study [58] found that degree sequences that are distributed according to a power-law distribution of the form $p(k) \sim k^{-\gamma}$, over a broad range of values bounded by $k_{\min} \leq k \leq k_{\max}$, are graphical for $k_{\max} = N - 1$ and $k_{\min} = 1$ only if $\gamma > 2$. It was argued in [12] that for $\gamma < 2$ only those degree sequences that scale with N as $k_{\max} < N^{1/\gamma}$ are graphical.

---

[4]Note that this result (4.22) also justifies *a posteriori* the earlier 'handwaving' arguments in the derivation of $\langle A_{ij} \rangle = k_i(\mathbf{A}) k_j(\mathbf{A})/N\bar{k}(\mathbf{A})$ for random graphs with specified degrees, upon which we based the construction of the modularity measure $Q(\mathbf{A})$ in (2.16).

In view of the graphically issue, it may be desirable to generate graphs with a prescribed degree distribution $p(k)$ *directly*, as opposed to working with ensembles that seek to target a particular degree sequence drawn from that distribution. An algorithm able to sample graphs directly from a known degree distribution is discussed in Chapter 7. This entails repeated rewirings of the links until the right distribution is attained. The number of links must be set to the value matching the average of the targeted degree distribution, otherwise complex behaviour will arise for $\gamma > 2$.

A completely different approach to generating networks with power-law distributed degrees, not involving any sampling from known distributions, is by means of appropriate growth algorithms. These will be discussed in Chapter 8. However, there is a definite advantage in using recipe (4.22), or the algorithms presented in subsequent chapters dealing with hard constraints, in that they both allow for a full control over the exponent γ of scale-free degree distributions. In contrast, we will see that growth processes tend to generate only degree distributions with specific values of γ.

Exercise 4.1 Define the directed version of the ERGM with controlled individual degrees $(\vec{k}_1, \ldots, \vec{k}_N)$, where $\vec{k}_i = (k_i^{\text{in}}, k_i^{\text{out}})$. Derive the equations from which to solve the ensemble parameters. Analyze these equations in the sparse regime, in leading order for large N, and show that, disregarding $\mathcal{O}(N^{-2})$ corrections, the graph probabilities will be given by

$$p(\mathbf{A}) = \prod_{ij} \left[\frac{k_i^{\text{in}} k_j^{\text{out}}}{N \langle k \rangle} \delta_{A_{ij},1} + \left(1 - \frac{k_i^{\text{in}} k_j^{\text{out}}}{N \langle k \rangle} \right) \delta_{A_{ij},0} \right] \qquad (4.24)$$

Example 3: ERGM with controlled directed link numbers. The reciprocity model was the first directed network model formulated as an ERGM . It was introduced in [88], in the context of sociometric studies. For simple directed graphs, the relevant graph set is $G = \{\mathbf{A} \in \{0, 1\}^{N(N-1)} | A_{ii} = 0 \, \forall i\}$. Below we show how reciprocation of directed edges can be controlled within the formalism of ERGMs by considering a simpler version of the reciprocity model, with homogeneous ensemble parameters, due to [140]. In this latter model version, we have two key network observables: the total number $\Omega_1(\mathbf{A})$ of directed links and the number $\Omega_2(\mathbf{A})$ of reciprocated (two-way) links:

$$\Omega_1(\mathbf{A}) = \sum_{i<j} (A_{ij} + A_{ji}), \qquad \Omega_2(\mathbf{A}) = 2 \sum_{i<j} A_{ij} A_{ji} \qquad (4.25)$$

The graph probabilities (4.1) are now

$$p(\mathbf{A}) = \frac{e^{\sum_{i<j} [\lambda_1 (A_{ij} + A_{ji}) + 2\lambda_2 A_{ij} A_{ji}]}}{Z(\lambda_1, \lambda_2)}, \qquad (4.26)$$

$$Z(\lambda_1, \lambda_2) = \sum_{\mathbf{A} \in G} e^{\sum_{i<j} [\lambda_1 (A_{ij} + A_{ji}) + 2\lambda_2 A_{ij} A_{ji}]} \qquad (4.27)$$

This gives for the generating function $F(\lambda_1, \lambda_2) = \log Z(\lambda_1, \lambda_2)$:

$$F(\lambda_1, \lambda_2) = \log \prod_{i<j} \sum_{A_{ij}, A_{ji}} e^{\lambda_1 (A_{ij} + A_{ji}) + 2\lambda_2 A_{ij} A_{ji}}$$

$$= \log \ \prod_{i<j}(1 + 2e^{\lambda_1} + e^{2(\lambda_1+\lambda_2)})$$

$$= \frac{1}{2}N(N-1)\log(1 + 2e^{\lambda_1} + e^{2(\lambda_1+\lambda_2)}) \qquad (4.28)$$

Upon taking the partial derivatives of $F(\lambda_1, \lambda_2)$ with respect to λ_1 and λ_2, we can via (4.3) express the expected number $L = \langle \sum_{i<j}(A_{ij} + A_{ji}) \rangle$ of links and $R = \langle \sum_{i<j} 2A_{ij}A_{ji} \rangle$ of reciprocal links in terms of the ensemble parameters λ_1 and λ_2:

$$L = \frac{\partial F(\lambda_1, \lambda_2)}{\partial \lambda_1} = N(N-1)\frac{e^{\lambda_1} + e^{2(\lambda_1+\lambda_2)}}{1 + 2e^{\lambda_1} + e^{2(\lambda_1+\lambda_2)}} \qquad (4.29)$$

$$R = \frac{\partial F(\lambda_1, \lambda_2)}{\partial \lambda_2} = N(N-1)\frac{e^{2(\lambda_1+\lambda_2)}}{1 + 2e^{\lambda_1} + e^{2(\lambda_1+\lambda_2)}} \qquad (4.30)$$

From these two expressions then follows the typical fraction of reciprocated links in graphs generated from the ensemble:

$$\frac{R}{L} = \frac{e^{\lambda_1+2\lambda_2}}{1 + e^{\lambda_1+2\lambda_2}} = \frac{1}{2} + \frac{1}{2}\tanh\left(\frac{1}{2}\lambda_1 + \lambda_2\right) \qquad (4.31)$$

Although the exponent of the probabilities $p(\mathbf{A})$ in the present example ERGM is not strictly linear, the model is apparently still sufficiently simple to allow us to control precisely and independently both the overall number and the reciprocity of links, by choosing appropriate values of the ensemble parameters.

4.3 ERGMs with phase transitions: the two-star model

Overview: When our controlled features $\{\Omega_\mu(\mathbf{A})\}$ become more complicated, ERGMs have a tendency to generate distinct graphs in extreme phases (e.g. very dense versus very sparse), and we become increasingly ineffective in finding parameter values that produce graphs with intermediate values of the controlled features. The two-star model is a relatively simple ERGM in which such phenomenology can be illustrated.

We illustrate some of the limitations and potential pitfalls in using ERGMs to control more complicated network observables. We investigate two relatively simple ERGMs that exhibit phase transitions, namely the two-star model and the Strauss (or random triangle) model. In the first model, one seeks to control as soft constraints the number of links and the number of two-step paths. In the second model, one seeks to control the number of links and the number of triangles. Phase transitions in graph ensembles are a serious practical problem, not just a curiosity for the entertainment of the theorist. They prevent us from generating graphs with the quantitative features that we require, and, if we are not aware of their phenomenology, they may make us waste a lot of time questioning in vain the correctness of our computer code. The material in this section is somewhat more technical than the rest of the chapter, and relies on statistical mechanics approaches and techniques, but the message can be appreciated without following in detail the technicalities. Background material on statistical mechanics techniques can be found in textbooks such as [135, 72].

Definition of the two-star model. The two-star model is an ERGM of simple nondirected graphs in which we seek to control via soft constraints the total number of links and the number of two-stars (i.e. paths of length two), so the key observables are

$$\Omega_1(\mathbf{A}) = \sum_{i<j} A_{ij} \tag{4.32}$$

$$\Omega_2(\mathbf{A}) = \frac{1}{2}\sum_{ijk}(1-\delta_{ik})A_{ij}A_{jk} = \frac{1}{2}\sum_{ijk}A_{ij}A_{jk} - \sum_{i<j}A_{ij} \tag{4.33}$$

Note that both can alternatively be written in term of the degrees $k_i(\mathbf{A}) = \sum_j A_{ij}$, as

$$\Omega_1(\mathbf{A}) = \frac{1}{2}\sum_i k_i(\mathbf{A}), \qquad \Omega_2(\mathbf{A}) = \frac{1}{2}\sum_i \left(k_i^2(\mathbf{A}) - k_i(\mathbf{A})\right) \tag{4.34}$$

Hence, for this model the graph probabilities (4.1) become

$$p(\mathbf{A}) = \frac{1}{Z(\lambda_1,\lambda_2)} e^{\frac{1}{2}(\lambda_1-\lambda_2)\sum_i k_i(\mathbf{A}) + \frac{1}{2}\lambda_2 \sum_i k_i^2(\mathbf{A})} \tag{4.35}$$

We may simplify our notation by choosing new parameters $\alpha = \frac{1}{2}(\lambda_1 + \lambda_2)$ and $\beta = \frac{1}{2}\lambda_2$, and denote by G the set of all N-node nondirected simple graphs. This allows us to write our ERGM's graph probabilities as

$$p(\mathbf{A}) = \frac{1}{Z(\alpha,\beta)} e^{\alpha \sum_i k_i(\mathbf{A}) + \beta \sum_i k_i^2(\mathbf{A})} \tag{4.36}$$

$$Z(\alpha,\beta) = \sum_{\mathbf{A}\in G} e^{\alpha \sum_i k_i(\mathbf{A}) + \beta \sum_i k_i^2(\mathbf{A})} \tag{4.37}$$

Solving the model means calculating those values for the parameters α and β for which we achieve specified average values $L = \langle \sum_{i<j} A_{ij} \rangle$ of links and $S = \langle \frac{1}{2}\sum_{ijk}(1 - \delta_{ik})A_{ij}A_{jk} \rangle$ of two-stars (provided these can be attained). As always in ERGMs, these parameter values are to be solved from equations (4.3), which here become

$$2L = \frac{\partial}{\partial \alpha} F(\alpha,\beta), \qquad 2S + 2L = \frac{\partial}{\partial \beta} F(\alpha,\beta) \tag{4.38}$$

with

$$F(\alpha,\beta) = \log \sum_{\mathbf{A}\in G} e^{\alpha \sum_i k_i(\mathbf{A}) + \beta \sum_i k_i^2(\mathbf{A})} \tag{4.39}$$

We will now solve this model in various scaling regimes for $N \to \infty$, and using different methods – some approximate and some exact.

Mean-field approximation of the solution of the two-star model. Despite its simplicity, the mean-field approach (which is inspired by the analysis of simple models of magnetism in many-particle physics) will show us clearly how the two-star model gets into trouble when we tune the parameters in an attempt to significantly increase the number of two-step paths by increasing β. The mean-field method is inspired by the observation that sums over graphs are trivial when the quantity over which the sum is taken factorizes over the $\{A_{ij}\}$. Here, the relevant sum over graphs is that in $Z(\alpha, \beta)$, from which we obtain $F(\alpha, \beta) = \log Z(\alpha, \beta)$. We note that we can write

$$Z(\alpha, \beta) = \sum_{A \in G} e^{\alpha \sum_{ij} A_{ij} + \beta \sum_{ij\ell} A_{ij} A_{i\ell}}$$

$$= \sum_{A \in G} e^{\sum_{i<j} A_{ij} \left(2\alpha + \beta \sum_{\ell} (A_{i\ell} + A_{j\ell})\right)} \tag{4.40}$$

The mean-field approximation consists of replacing the quantity inside the brackets by their ensemble averages, i.e. $\sum_{\ell} (A_{i\ell} + A_{j\ell}) \to \sum_{\ell} \langle A_{i\ell} + A_{j\ell} \rangle = k_i + k_j$, in which $k_i = \langle k_i(\mathbf{A}) \rangle$. Consistency demands that the ensemble-averaged degrees are then to be calculated within the same approximation scheme, so

$$k_i = \Big\langle \sum_{j \neq i} A_{ij} \Big\rangle = \sum_{j \neq i} \frac{\sum_{A \in G} A_{ij} e^{\sum_{\ell < m} A_{\ell m} (2\alpha + \beta k_\ell + \beta k_m)}}{\sum_{A \in G} e^{\sum_{\ell < m} A_{\ell m} (2\alpha + \beta k_\ell + \beta k_m)}}$$

$$= \sum_{j \neq i} \frac{\sum_{A_{ij}} A_{ij} e^{A_{ij}(2\alpha + \beta k_i + \beta k_j)}}{\sum_{A_{ij}} e^{A_{ij}(2\alpha + \beta k_i + \beta k_j)}} = \sum_{j \neq i} \frac{e^{2\alpha + \beta k_i + \beta k_j}}{1 + e^{2\alpha + \beta k_i + \beta k_j}} \tag{4.41}$$

The simplest solution of (4.41) is the one where all edges in the model are equivalent, so $k_i = \langle k_i(\mathbf{A}) \rangle$ is independent of i, and we may replace the above expression by

$$\langle k \rangle = \frac{(N-1)e^{2\alpha + 2\beta \langle k \rangle}}{1 + e^{2\alpha + 2\beta \langle k \rangle}} \tag{4.42}$$

We can now conclude that within this mean-field approach the links are distributed independently and identically, so we are in effect approximating the original graph ensemble by an Erdös–Rényi one (3.1), with the standard link probabilities $p = \langle k \rangle / (N-1)$, but where $\langle k \rangle$ is now chosen to be the solution of (4.42). The variance of the degree distribution can now be calculated in a similar way

$$\langle k^2 \rangle - \langle k \rangle^2 = \frac{1}{N} \sum_i \sum_{j \neq i} \sum_{k \neq i} \langle A_{ij} A_{ik} \rangle - \langle k \rangle^2$$

$$= \frac{1}{N} \sum_i \sum_{j \neq i} \sum_{k \neq i} \Big(\delta_{jk} \langle A_{ij} \rangle + (1 - \delta_{jk}) \langle A_{ij} \rangle \langle A_{ik} \rangle \Big) - \langle k \rangle^2$$

$$= \langle k \rangle - \frac{1}{N-1} \langle k \rangle^2 = (N-1)p(1-p) \tag{4.43}$$

Rewriting (4.42) in terms of p gives us the equivalent equation

$$\left(1-\frac{1}{N}\right)p = \frac{e^{2\alpha+2\beta N(1-1/N)p}}{1+e^{2\alpha+2\beta N(1-1/N)p}} = \frac{e^{\alpha+\beta N(1-1/N)p}}{2\cosh[\alpha+\beta N(1-1/N)p]} \tag{4.44}$$

Using the identity $e^x/(2\cosh x) = \frac{1}{2}[1+\tanh(x)]$ and upon limiting ourselves to the regime of large N, we may write this as

$$p = \frac{1}{2}\left[1+\tanh(\alpha+\beta Np)\right] \tag{4.45}$$

In the dense graph regime, where $p = \mathcal{O}(1)$, one has to take the limit $N \to \infty$ at constant βN to get non-trivial results, i.e. the ensemble parameter β has to be of order $\mathcal{O}(N^{-1})$ to produce dense graphs. Setting $\beta = 2J/N$ leads us to the equation

$$p = \frac{1}{2}\left[\tanh(\alpha+2Jp)+1\right] \tag{4.46}$$

which was derived in [141] for large and dense graphs. The crucial question now is: how many solutions p does (4.46) have? If more than one, we cannot uniquely control the link density in the ensemble, let alone the number of two-stars, but for given values of the control parameters there are multiple macroscopic states. The latter would be called *phases* of the system, and the values of the control parameters where multiple states first appear would be called *bifurcation* or *phase transition* points. Upon plotting both sides of (4.46) for some typical values of (α, J), as in Fig. 4.1, we can already see that the number of solutions can be either 1, 2 or 3, and that for $\alpha \to \pm\infty$ we have always one solution. New solution(s) of (4.46) appear or vanish when we satisfy (4.46) and at the same time have identical p-derivatives on both sides of (4.46):

$$\text{transition:}\quad p = \frac{1}{2}\left[\tanh(2Jp+\alpha)+1\right], \quad 1 = J - J\tanh^2(2Jp+\alpha) \tag{4.47}$$

Since $0 \le \tanh^2(x) < 1$, the second equation has no solution (so there cannot be a bifurcation) if $J \le 1$. Hence for $J \le 1$ there is a unique p. Moreover, from the shape of the right-hand side of (4.46) (see Fig. 4.1), we deduce that newly created solutions for $J > 1$ represent discontinuities (i.e. jumps) in the value of p, if *at* the solution of (4.47) the right-hand side of (4.46) is concave in p, i.e. if $d^2[\tanh(2Jp+\alpha)+1]/dp^2 < 0$, or

$$\tanh(2Jp+\alpha) > 0 \tag{4.48}$$

We can now answer our questions. We use the first equation of (4.47) to eliminate α from the second, via $\tanh(2Jp+\alpha) = 2p-1$, or, upon using the formula $\tanh^{-1}(x) = \frac{1}{2}\log[(1+x)/(1-x)]$ for the inverse of $\tanh(x)$, via $\alpha = \frac{1}{2}\log[p/(1-p)]-2Jp$. The second equation of (4.47) then gives us two potential solution branches $p_\pm = \frac{1}{2} \pm \frac{1}{2}\sqrt{1-1/J}$. The p_+ branch corresponds to a discontinuous transition, and the p_- branch to a continuous one. Combination with our formula for α in terms of p then gives us, after some simple rewriting, explicit simple formulae for the two curves in (α, J) parameter space where, at least for $J > 1$, transitions occur:

$$\text{discontinuous transition:}\quad \alpha_+(J) = \frac{1}{2}\log\left[\frac{\sqrt{J}+\sqrt{J-1}}{\sqrt{J}-\sqrt{J-1}}\right] - \frac{\sqrt{J}}{\sqrt{J}-\sqrt{J-1}} \tag{4.49}$$

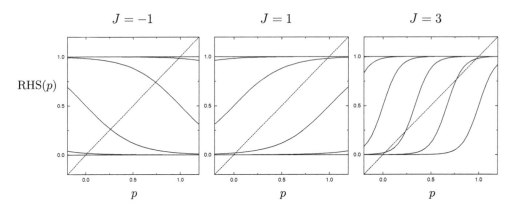

Fig. 4.1 Graphical illustration of the solution of (4.46) for the parameter p, for three values of J. Each panel shows with solid curves the right-hand side of (4.46), for the choices $\alpha \in \{-6, -4, -2, 0, 2, 4, 6\}$. The solutions of (4.46) are the locations of the intersections of these curves with the diagonal (shown as a dashed line).

$$\text{continuous transition:} \quad \alpha_-(J) = \frac{1}{2}\log\left[\frac{\sqrt{J}-\sqrt{J-1}}{\sqrt{J}+\sqrt{J-1}}\right] - \frac{\sqrt{J}}{\sqrt{J}+\sqrt{J-1}} \quad (4.50)$$

These curves are shown in Fig. 4.2. In the parameter regime between the two curves (bounded from above by the discontinuous transition line α_+, and from below by the continuous transition line α_-), the solution p is no longer unique: there are three solutions, of which the smallest and the largest are stable (proving this requires more work, and will not be done here). This is the dangerous area, where which graph is in practice being sampled from our ERGM will depend on the initial conditions of one's algorithm. Moreover, in view of the discontinuous jump in the value of p at the region's upper boundary, there will be combinations of our target observables L and S that cannot be achieved by *any* combination of the ensemble's control parameters[5].

The bi-stability in the solutions of (4.46) leads to a degeneracy in the degree statistics: in the dangerous region of the phase diagram, our ensemble will mainly produce graphs with either a high or a low average degree $\langle k \rangle = p$, and will only very rarely give intermediate values. Further details can be found in [140].

Exact solution of the two-star model in the sparse regime. We now discuss the exact solution for the two-star model in the sparse regime, where $\langle k \rangle = \mathcal{O}(1)$ as $N \to \infty$, and where the mean-field approximation does not apply. This is the realistic regime for most practical applications, but it requires more advanced techniques from statistical mechanics such as path integrals; see, e.g. [135, 72, 10]. First, however, it will be useful

---

[5]Note that the emergence of 'strictly impossible' (L, S) combinations is mathematically a consequence of having taken the limit $N \to \infty$. For large but finite N the presently discontinuous jumps will be smoothened into continuous but steep jumps. Nevertheless, we will find in practice even for graphs with only a few hundred nodes that, whatever algorithm we choose to work with, it is nearly impossible to steer our ensemble accurately towards certain (L, S) combinations. This is a fundamental statistical feature of the ERGM ensemble itself, of which we need to be aware.

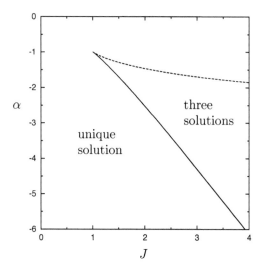

Fig. 4.2 The phase diagram of the two-star ERGM, calculated for the dense regime within the mean-field approximation. The upper (dashed) curve marks the location of continuous bifurcations of solutions of (4.46); the lower (solid) curve marks discontinuous bifurcations. They touch in the point $(J, \alpha) = (1, -1)$. In the region between the two curves, (4.46) has three solutions p, of which two are stable. As one moves into this region from above, the two solutions that bifurcate continuously from the unique low-density state as one crosses the dashed line become increasingly distinct (one having low, the other having high link density). At the lower boundary (solid line) the low-density solution disappears as a jump discontinuity, leaving the high-density state as the unique solution.

to derive bounds on the possible number of stars that a network can exhibit for a fixed number of edges, i.e. a fixed value of $\langle k \rangle$. We had found that the total number of two-stars in a graph equals $S = \frac{1}{2}N(\langle k^2 \rangle - \langle k \rangle)$. This can also be written as

$$S = \frac{1}{2}N\langle k \rangle(\langle k \rangle - 1) + \frac{1}{2}N\langle(k - \langle k \rangle)^2\rangle \qquad (4.51)$$

Hence, $S/N\langle k \rangle \geq \frac{1}{2}(\langle k \rangle - 1)$. This lower bound is realized for regular graphs, i.e. $p(k) = \delta_{k,\langle k \rangle}$. To find an upper bound for S, we must increase the variance of $p(k)$ for fixed $\langle k \rangle$. We can use Lagrange's method to maximize S, by solving

$$p(k) = 0 \quad \text{or} \quad \frac{\partial}{\partial p(k)}\left[\sum_{k'=0}^{N} p(k')k'^2 - \lambda_0\left(1 - \sum_{k'=0}^{N} p(k')\right) - \lambda_1\left(1 - \sum_{k'=0}^{N} p(k')k'\right)\right] = 0 \qquad (4.52)$$

The solution is that either $p(k) = 0$, or $k^2 + \lambda_1 k + \lambda_0 = 0$. Hence, since the quadratic equation for k has at most two solutions, we have either a regular graph (which, as we know, does not give the maximum but the minimum S) or one with a bimodal degree distribution, which due to the constraint on $\langle k \rangle$ must then inevitably be

$$p(k) = \frac{k_2 - \langle k \rangle}{k_2 - k_1}\delta_{k,k_1} + \frac{\langle k \rangle - k_1}{k_2 - k_1}\delta_{k,k_2}, \qquad k_1 < \langle k \rangle < k_2 \tag{4.53}$$

For this degree distribution, one would obtain $\langle k^2 \rangle = k_2 \langle k \rangle - k_1(k_2 - \langle k \rangle)$, which for a given k_2 is maximal when $k_1 = 0$, giving $\langle k^2 \rangle / \langle k \rangle = k_2$. Therefore, $S/N\langle k \rangle \le \max\{k_2\}/2$. Thus, the graph consists of a fraction $1 - \langle k \rangle / k_2$ of disconnected nodes, plus a fraction $\langle k \rangle / k_2$ of nodes that are wired as a regular 'core'. Working out for this case the graphicality conditions of the Erdös–Gallai theorem (4.23) reveals that we must ensure that $k_2 \le N\langle k \rangle / k_2$, i.e. $k_2 \le \sqrt{N\langle k \rangle}$, so $S/N\langle k \rangle \le \sqrt{N\langle k \rangle}$. It follows that will we have the bound $S/N\langle k \rangle \le \mathcal{O}(\sqrt{N})$ if we impose our graphs' sparsity, i.e. the demand that the total number of links scales as $\mathcal{O}(N)$, by just keeping $\langle k \rangle$ fixed. Only if we insist explicitly on also keeping $\langle k^2 \rangle$ finite will we find $S/N\langle k \rangle = \mathcal{O}(1)$.

Let us now solve the two-star model in the sparse regime exactly, for $N \to \infty$. We start from expression (4.40) and focus on the rescaled generating function or free-energy density $f(\alpha, \beta) = N^{-1}\log Z(\alpha, \beta)$, which can be written as

$$f(\alpha, \beta) = \frac{1}{N} \log \sum_{\mathbf{A} \in G} e^{\sum_{i<j} A_{ij}\left(2\alpha + \beta(k_i(\mathbf{A}) + k_j(\mathbf{A}))\right)} \tag{4.54}$$

According to (4.38), we can obtain from $f(\alpha, \beta)$ the *densities* L/N and S/N of links and two-stars via

$$2L/N = \frac{\partial}{\partial \alpha} f(\alpha, \beta), \qquad 2(S + L)/N = \frac{\partial}{\partial \beta} f(\alpha, \beta) \tag{4.55}$$

In the sparse regime, we know by definition that $L/N = \mathcal{O}(1)$. Our previous analysis of bounds on S, however, showed that whether also $S/N = \mathcal{O}(1)$ will depend on whether not only the average degree but also the degree variance remains finite.

In Appendix C, we calculate $f(\alpha, \beta)$ and work out the equations for the two constraints $L/N = \frac{1}{2}\langle k \rangle$ and $(S+L)/N \to \frac{1}{2}\langle k^2 \rangle$ in the limit $N \to \infty$ in the sparse scaling regime. One finds that for the ERGM to generate sparse graphs, we must choose $\alpha = \tilde{\alpha} - \frac{1}{2}\log N$, and solve $\tilde{\alpha}$ and β (which will then both be of order one as $N \to \infty$) from the following two equations:

$$\langle k \rangle = \sum_{k \ge 0} k p(k), \qquad \langle k^2 \rangle = \sum_{k \ge 0} k^2 p(k) \tag{4.56}$$

with
$$p(k) = \frac{\langle k \rangle^{k/2} e^{\tilde{\alpha} k + \beta k^2} / k!}{\sum_{\ell \ge 0} \langle k \rangle^{\ell/2} e^{\tilde{\alpha}\ell + \beta \ell^2} / \ell!} \tag{4.57}$$

As a simple test one verifies that for the simple choice $\beta = 0$ expression (4.36) describes an Erdös–Rènyi ensemble with average degree $\langle k \rangle = (N-1)e^{2\alpha}/(1 + e^{2\alpha})$:

$$p(\mathbf{A}) = \prod_{i<j} \left[\frac{e^{2\alpha}}{1 + e^{2\alpha}}\delta_{A_{ij},1} + \frac{1}{1 + e^{2\alpha}}\delta_{A_{ij},0} \right] \tag{4.58}$$

Choosing $\alpha = \tilde{\alpha} - \frac{1}{2}\log N$, we find the average degree $\langle k \rangle = e^{2\tilde{\alpha}} + \mathcal{O}(N^{-1})$. Formula (4.57) reduces for $\beta = 0$ to a Poisson distribution, as is indeed expected for an Erdös–Rènyi ensemble, with the correct value $\sqrt{\langle k \rangle}e^{\tilde{\alpha}} = \langle k \rangle$ for the average degree.

The important question now is whether solutions for $\tilde{\alpha}$ and β can in the sparse regime be found for *any* values of $\langle k \rangle$ and $\langle k^2 \rangle$, or whether (as in the dense regime) the system will exhibit phase transitions and certain values of the moments can never be attained. Deriving phase transition conditions from (4.56) is still relatively easy. One calculates the infinitesimal changes in $\langle k \rangle$ and $\langle k^2 \rangle$ resulting from infinitesimal changes in the control parameters $(\tilde{\alpha}, \beta)$; bifurcations are then marked by the non-invertibility of this relation, giving

$$\left(\langle k^2 \rangle - \langle k \rangle^2\right)\left(\sum_{k \geq 0} k^4 p(k) - \langle k^2 \rangle^2\right) = \left(\sum_{k \geq 0} k^3 p(k) - \langle k^2 \rangle \langle k \rangle\right)^2 \tag{4.59}$$

Unfortunately, if $\beta \neq 0$ we cannot solve (4.56) or (4.59) analytically, and need to rely generally on numerical methods to find the parameters $(\tilde{\alpha}, \beta)$ or the transition lines (if any). Nevertheless, one can still see via a simple argument that also in the sparse regime there will be forbidden values of degree moments. We note that for $\beta = 0$ the degree distribution (4.57) will be Poissonian, so there we must have $\langle k^2 \rangle = \langle k \rangle^2 + \langle k \rangle$ (this relation between moments is a fundamental property of all Poissonian distributions). For $\beta < 0$ and fixed $\langle k \rangle$ we see that large k-values will be suppressed, relative to the case $\beta = 0$, leading to $\langle k^2 \rangle < \langle k \rangle^2 + \langle k \rangle$. Hence, to generate sparse graphs with $\langle k^2 \rangle > \langle k \rangle^2 + \langle k \rangle$, we would require $\beta > 0$, but for such values the denominator of (4.57) does not converge. We conclude that it is impossible to generate sparse graphs with $\langle k^2 \rangle > \langle k \rangle^2 + \langle k \rangle$, i.e. with numbers of two-stars significantly above the value one would find in ensembles where only the link density is constrained.

Numerical solutions of (4.56) for $\beta < 0$ are shown in Figure 4.3, and are compared to results from Markov Chain Monte–Carlo (MCMC) simulations (a numerical sampling protocol for ERGMs, which is discussed in more detail in subsequent chapters) and to the predictions from the mean-field approximation. The latter predicts that $\langle k \rangle = Np(\alpha, \beta)$ and $\langle k^2 \rangle = N^2 p^2(\alpha, \beta)$, with $p(\alpha, \beta)$ being the solution of (4.46) with $J = \frac{1}{2}N\beta$. The plots in Figure 4.3 show the average degree $\langle k \rangle$ and the width $\sigma_k = (\langle k^2 \rangle - \langle k \rangle^2)^{\frac{1}{2}}$ of the degree distribution, as functions of β, for $\alpha = \frac{1}{2}\log(4/N)$, for graphs with $N = 300$. As expected, the mean-field predictions are seen to be quite acceptable at high connectivity, but become inaccurate at low connectivity. It will be clear from the figure that the predicted phase transitions, although derived for $N \to \infty$, are not red herrings: even for modest graph sizes they are very prominent phenomena that severely limit our ability in practice to control observables by tuning the ensemble's parameters.

4.4   ERGMs with phase transitions: the Strauss (triangle) model

Overview: The so-called Strauss (or random triangle) model is another notorious example of a simple ERGM with complex phase transitions, in which the tuning of control parameters is problematic. In this model, one seeks to control the average number of links and the average number of triangles in the graph.

Definition of the random triangle model. In the *random triangle model* [94] or *Strauss model* [169], one considers again the set G of nondirected simple N-node graphs, and

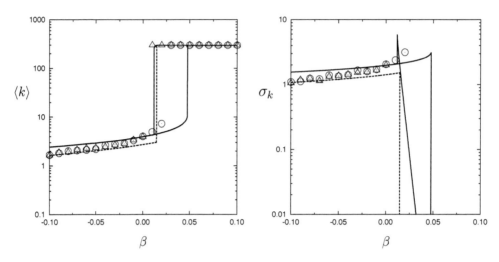

Fig. 4.3 Comparison of the predicted and observed values of the average degree $\langle k \rangle$ (left) and the width $\sigma_k = (\langle k^2 \rangle - \langle k \rangle^2)^{\frac{1}{2}}$ of the degree distribution (right) of the two-star model, as a function of β. Markers: MCMC sampling results (see Chapter 5 for details) for ensembles with $N = 300$ and $\alpha = \frac{1}{2}\log(4/N)$, starting either from a high-density (triangles) or a low-density initialization (circles). Solid lines: Prediction of the mean-field equations (giving two solutions for certain parameter values, one with high link density and the other with low link density). Dashed line: Exact solution describing the sparse scaling regime. The density of two-stars equals $S/N = \frac{1}{2}(\sigma_k^2 + \langle k^2 \rangle - \langle k \rangle)$, so the explosion of $\langle k \rangle$ at the transition (note the logarithmic vertical scale) is accompanied by a similar explosion in the number of two-stars. Although the theoretical predictions were obtained for $N \to \infty$, and for finite N the size of the region of two coexistent states is reduced, even for graphs as small as $N = 300$ the phase transition is prominent. This illustrates the difficulties in controlling observables by tuning of the parameters (α, β) even for modest graph sizes.

one chooses as soft-constrained observables the total number of links and the total number of triangles in the graph. A triangle is a triplet of distinct nodes (i, j, k) such that $A_{ij} = A_{jk} = A_{ki} = 1$. Thus, upon ordering the nodes of each triangle such that $i < j < k$ (which is always possible), we have

$$\Omega_1(\mathbf{A}) = \sum_{i<j} A_{ij}, \qquad \Omega_2(\mathbf{A}) = \sum_{i<j<k} A_{ij} A_{jk} A_{ki} \qquad (4.60)$$

and hence

$$p(\mathbf{A}) = \frac{e^{\lambda_1 \sum_{i<j} A_{ij} + \lambda_2 \sum_{i<j<k} A_{ij} A_{jk} A_{ki}}}{Z(\lambda_1, \lambda_2)} \qquad (4.61)$$

$$Z(\lambda_1, \lambda_2) = \sum_{\mathbf{A} \in G} e^{\lambda_1 \sum_{i<j} A_{ij} + \lambda_2 \sum_{i<j<k} A_{ij} A_{jk} A_{ki}} \qquad (4.62)$$

It will be clear from the definition (4.61) that for $\lambda_2 = 0$ we return to the Erdös–Rényi ensemble (3.1), whereas for $\lambda_2 > 0$ we promote graphs with larger numbers of triangles. Solving the model means finding values for (λ_1, λ_2) such that we achieve specified average numbers $L = \langle \sum_{i<j} A_{ij} \rangle$ of links and $T = \langle \sum_{i<j<k} A_{ij} A_{jk} A_{ki} \rangle$ of triangles, i.e. solving the following coupled equations:

$$L = \frac{\partial}{\partial \lambda_1} \log Z(\lambda_1, \lambda_2), \qquad T = \frac{\partial}{\partial \lambda_2} \log Z(\lambda_1, \lambda_2) \tag{4.63}$$

Pseudocode for the Strauss model is provided in Algorithm 5. As with the two-star model, this ERGM will also exhibit phase transitions: we will find transitions between states with low link density and states with high link density. Moreover, here too the ensemble averages cannot be forced to take any value we wish. It turns out that the number of triangles T cannot be increased significantly without inducing a similar increase in the link density L.

Mean-field approximation with identically distributed links. Following our previous mean-field approach for the two-star model and using the symmetry of $\mathbf{A}$, we start rewriting $Z(\lambda_1, \lambda_2)$ in the pseudo-factorized form

$$Z(\lambda_1, \lambda_2) = \sum_{\mathbf{A} \in G} e^{\lambda_1 \sum_{i<j} A_{ij} + \frac{1}{3}\lambda_2 \sum_{i<j} \sum_k A_{ij} A_{jk} A_{ki}}$$

$$= \sum_{\mathbf{A} \in G} \prod_{i<j} e^{A_{ij}\left(\lambda_1 + \frac{1}{3}\lambda_2 \sum_k A_{ik} A_{jk}\right)} \tag{4.64}$$

Since our graphs have no self-interactions, the sum $\sum_k A_{ik} A_{jk}$ cannot contain the entry A_{ij}. Similar to the two-star model, in the mean-field approximation we replace this sum by its ensemble average, i.e. we approximate our expression for $Z(\lambda_1, \lambda_2)$ by

$$Z(\lambda_1, \lambda_2) = \sum_{\mathbf{A} \in G} \prod_{i<j} e^{A_{ij}\left(\lambda_1 + \frac{1}{3}\lambda_2 \sum_{k \notin \{i,j\}} \langle A_{ik} A_{jk}\rangle\right)} \tag{4.65}$$

But this implies having statistically independent links, each distributed according to $p(A_{ij}) = p_{ij}\delta_{A_{ij},1} + (1-p_{ij})\delta_{A_{ij},0}$, in which the link probabilities $\{p_{ij}\}$ must be solved self-consistently from

$$p_{ij} = \frac{e^{\lambda_1 + \frac{1}{3}\lambda_2 \sum_{k \notin \{i,j\}} p_{ik} p_{jk}}}{1 + e^{\lambda_1 + \frac{1}{3}\lambda_2 \sum_{k \notin \{i,j\}} p_{ik} p_{jk}}} \tag{4.66}$$

The simplest solution of (4.66) has all p_{ij} independent of (i,j). This means approximating our ensemble by an Erdös–Rényi one, with a link probability p that satisfies

$$p = \frac{e^{\lambda_1 + \frac{1}{3}\lambda_2(N-2)p^2}}{1 + e^{\lambda_1 + \frac{1}{3}\lambda_2(N-2)p^2}}$$

$$= \frac{1}{2}\left[\tanh\left(\frac{1}{2}\lambda_1 + \frac{1}{6}\lambda_2(N-2)p^2\right) + 1\right] \tag{4.67}$$

Now we can work out (4.65) in full, giving

$$Z(\lambda_1, \lambda_2) = \left(1 + e^{\lambda_1 + \frac{1}{3}\lambda_2(N-2)p^2}\right)^{\frac{1}{2}N(N-1)} \tag{4.68}$$

Our equations (4.63) from which to solve the control parameters (λ_1, λ_2) then give $L = \frac{1}{2}N(N-1)p$ and $T = \frac{1}{6}N(N-1)(N-2)p^3$. Since T and L are both functions of the

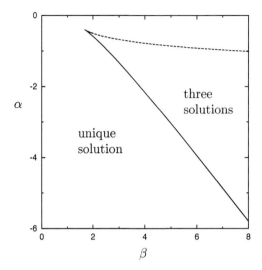

Fig. 4.4 The phase diagram of the random triangle or Strauss model, calculated for the dense regime within the mean-field approximation. The upper (dashed) curve marks the location of continuous bifurcations of solutions of (4.67); the lower (solid) curve marks discontinuous bifurcations. In the region between the two curves, (4.67) has three solutions p, of which two are stable. As one moves into this region from above, the two solutions that bifurcate continuously from the unique low-density state as one crosses the dashed line become increasingly distinct (one having low, the other having high link density). At the lower boundary (solid line), the low-density solution disappears as a jump discontinuity, leaving the high-density state as the unique solution.

solution of (4.67), they cannot be controlled independently. The mean-field approach makes sense only in the dense regime, so we seek solutions of (4.67) that scale for $N \to \infty$ as $p = \mathcal{O}(1)$. We are thus led to choose $\lambda_1 = 2\alpha$ (for notation convenience) and $\lambda_2 = 6\beta/N$ (to have the correct scaling of p), whereupon we obtain for $N \to \infty$:

$$p = \frac{1}{2}\big[\tanh(\alpha+\beta p^2) + 1\big] \tag{4.69}$$

This equation differs from the mean-field equation of the two-star model in the dense regime only in the replacement $2Jp \to \beta p^2$, and since the bifurcation analysis is therefore nearly identical, we will only give the results. Phase transitions happen when in addition to (4.67) we also satisfy $1 = \beta p[1 - \tanh^2(\alpha+\beta p^2)]$. The transitions are discontinuous if $\tanh(\alpha+\beta p^2) > \frac{1}{2}$ and continuous otherwise. This leads directly to the phase diagram shown in Figure 4.4, which is indeed qualitatively identical to that of the two-star model in the dense regime.

This mean-field route is identical to the one we applied to the two-star model, but here it is too crude. We saw that it implies replacing $\langle A_{ik}A_{jk}\rangle \to \langle A_{ik}\rangle\langle A_{jk}\rangle$ in the exponent of $Z(\lambda_1, \lambda_2)$, so even those links that share a node (i.e. that are adjacent in the graph) will be distributed independently. However, in the present ensemble, boosting the number of triangles via $\lambda_2 > 0$ will cause precisely those adjacent links to be correlated, and we must therefore demand of any acceptable approximation that it can accommodate solutions with $\sum_{k \notin \{i,j\}} \langle A_{ik} A_{jk} \rangle > \sum_{k \notin \{i,j\}} \langle A_{ik} \rangle \langle A_{jk} \rangle$.

Mean-field approximation with non-identically distributed links. Since in the present model approximations apparently have to be postponed to a later (and more harmless) stage in the argument, we are forced to return to the original probabilities (4.61) and ask whether we can perhaps work out alternative exact expressions for the key averages $\langle A_{ij} \rangle$ and $\langle A_{ik} A_{jk} \rangle$. To save ink and trees, we first introduce the shorthand

$$F(\mathbf{A}) = \lambda_1 \sum_{r<s} A_{rs} + \lambda_2 \sum_{r<\ell<s} A_{r\ell} A_{\ell s} A_{sr} \tag{4.70}$$

and the shorthand $\sum_{\mathbf{A}/A_{ij}}$ to denote summation over all entries of $\mathbf{A}$ except A_{ij}. Following [140], we now try to write the averages $\langle A_{ij} \rangle$ and $\langle A_{ik} A_{jk} \rangle$ such that the sums over A_{ij} or $\{A_{ik}, A_{jk}\}$ are done explicitly, but the results are written once more as ensemble averages over the remaining entries of $\mathbf{A}$. First we evaluate $\langle A_{ij} \rangle$:

$$\langle A_{ij} \rangle = \frac{1}{Z(\lambda_1, \lambda_2)} \sum_{\mathbf{A} \in G} A_{ij} \, e^{F(\mathbf{A})} = \frac{1}{Z(\lambda_1, \lambda_2)} \sum_{\mathbf{A}/A_{ij}} e^{F(\mathbf{A})|_{A_{ij}=1}}$$

$$= \frac{1}{Z(\lambda_1, \lambda_2)} \sum_{\mathbf{A}/A_{ij}} e^{F(\mathbf{A})|_{A_{ij}=1}} \left(\frac{\sum_{A_{ij}} e^{F(\mathbf{A})}}{e^{F(\mathbf{A})|_{A_{ij}=1}} + e^{F(\mathbf{A})|_{A_{ij}=0}}} \right)$$

$$= \frac{1}{Z(\lambda_1, \lambda_2)} \sum_{\mathbf{A} \in G} e^{F(\mathbf{A})} \left(\frac{e^{F(\mathbf{A})|_{A_{ij}=1}}}{e^{F(\mathbf{A})|_{A_{ij}=1}} + e^{F(\mathbf{A})|_{A_{ij}=0}}} \right)$$

$$= \left\langle \frac{1}{1 + e^{F(\mathbf{A})|_{A_{ij}=0} - F(\mathbf{A})|_{A_{ij}=1}}} \right\rangle \tag{4.71}$$

Using $F(\mathbf{A})|_{A_{ij}=0} - F(\mathbf{A})|_{A_{ij}=1} = -\lambda_1 - \lambda_2 \sum_\ell A_{i\ell} A_{\ell j}$, which follows from the fact that the total number of triangles $T(\mathbf{A}) = \sum_{r<\ell<s} A_{r\ell} A_{\ell s} A_{sr}$ is reduced by $\sum_\ell A_{i\ell} A_{j\ell}$ upon removal of link A_{ij}, we obtain

$$\langle A_{ij} \rangle = \left\langle \frac{1}{1 + e^{-\lambda_1 - \lambda_2 \sum_\ell A_{i\ell} A_{\ell j}}} \right\rangle \tag{4.72}$$

This average indeed involves only entries of $\mathbf{A}$ that are unequal to A_{ij}. The same approach can be applied to our second ensemble average, which is slightly more involved:

$$\langle A_{ik} A_{jk} \rangle = \frac{1}{Z(\lambda_1, \lambda_2)} \sum_{\mathbf{A} \in G} A_{ik} A_{jk} \, e^{F(\mathbf{A})} = \frac{1}{Z(\lambda_1, \lambda_2)} \sum_{\mathbf{A}/A_{ik}, A_{jk}} e^{F(\mathbf{A})|_{A_{ik}=A_{jk}=1}}$$

$$= \frac{1}{Z(\lambda_1, \lambda_2)} \sum_{\mathbf{A}/A_{ik}, A_{jk}} e^{F(\mathbf{A})|_{A_{ik}=A_{jk}=1}} \times$$

$$\frac{\sum_{A_{ik}, A_{jk}} e^{F(\mathbf{A})}}{e^{F(\mathbf{A})|_{A_{ik}=A_{jk}=1}} + e^{F(\mathbf{A})|_{A_{ik}=A_{jk}=0}} + e^{F(\mathbf{A})|_{A_{ik}=1, A_{jk}=0}} + e^{F(\mathbf{A})|_{A_{ik}=0, A_{jk}=1}}}$$

$$= \frac{1}{Z(\lambda_1, \lambda_2)} \sum_{\mathbf{A}} e^{F(\mathbf{A})} \times$$

$$\frac{e^{F(\mathbf{A})|_{A_{ik}=A_{jk}=1}}}{e^{F(\mathbf{A})|_{A_{ik}=A_{jk}=1}} + e^{F(\mathbf{A})|_{A_{ik}=A_{jk}=0}} + e^{F(\mathbf{A})|_{A_{ik}=1,A_{jk}=0}} + e^{F(\mathbf{A})|_{A_{ik}=0,A_{jk}=1}}}$$

$$= \left\langle \frac{e^{F(\mathbf{A})|_{A_{ik}=A_{jk}=1}}}{e^{F(\mathbf{A})|_{A_{ik}=A_{jk}=1}} + e^{F(\mathbf{A})|_{A_{ik}=A_{jk}=0}} + e^{F(\mathbf{A})|_{A_{ik}=1,A_{jk}=0}} + e^{F(\mathbf{A})|_{A_{ik}=0,A_{jk}=1}}} \right\rangle$$

$$(4.73)$$

Here we need to work out the following three quantities, with $i < j$ and the total number of triangles $T(\mathbf{A}) = \sum_{r<\ell<s} A_{r\ell} A_{\ell s} A_{sr}$:

$$F(\mathbf{A})|_{A_{ik}=A_{kj}=0} - F(\mathbf{A})|_{A_{ik}=A_{kj}=1} = \lambda_2 \Big(T(\mathbf{A})|_{A_{ik}=A_{kj}=0} - T(\mathbf{A})|_{A_{ik}=A_{kj}=1} \Big) - 2\lambda_1$$

$$F(\mathbf{A})|_{A_{ik}=1,A_{kj}=0} - F(\mathbf{A})|_{A_{ik}=A_{kj}=1} = \lambda_2 \Big(T(\mathbf{A})|_{A_{ik}=1,A_{kj}=0} - T(\mathbf{A})|_{A_{ik}=A_{kj}=1} \Big) - \lambda_1$$

$$F(\mathbf{A})|_{A_{ik}=0,A_{kj}=1} - F(\mathbf{A})|_{A_{ik}=A_{kj}=1} = \lambda_2 \Big(T(\mathbf{A})|_{A_{ik}=0,A_{kj}=1} - T(\mathbf{A})|_{A_{ik}=A_{kj}=1} \Big) - \lambda_1$$

The tricky terms, those proportional to λ_2, are best evaluated graphically. In all three cases we simply need to assess how the number of triangles $T(\mathbf{A})$ is affected by specific local link removals of the form $(1,1) = (A_{ik}, A_{kj}) \to (A'_{ik}, A'_{kj}) \neq (1,1)$:

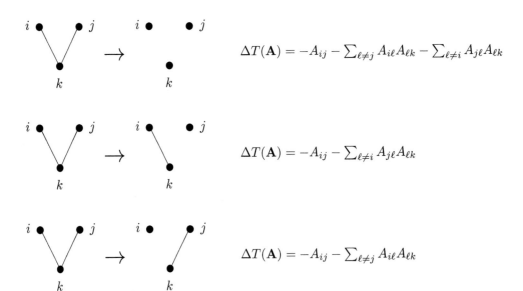

$$\Delta T(\mathbf{A}) = -A_{ij} - \sum_{\ell \neq j} A_{i\ell} A_{\ell k} - \sum_{\ell \neq i} A_{j\ell} A_{\ell k}$$

$$\Delta T(\mathbf{A}) = -A_{ij} - \sum_{\ell \neq i} A_{j\ell} A_{\ell k}$$

$$\Delta T(\mathbf{A}) = -A_{ij} - \sum_{\ell \neq j} A_{i\ell} A_{\ell k}$$

Upon inserting these three expressions, we obtain, after some rearranging of terms,

$$\langle A_{ik} A_{jk} \rangle = \left\langle \frac{e^{\lambda_2 A_{ij}}}{e^{\lambda_2 A_{ij}} - 1 + \big(1 + e^{-\lambda_1 - \lambda_2 \sum_{\ell \neq i} A_{j\ell} A_{\ell k}}\big)\big(1 + e^{-\lambda_1 - \lambda_2 \sum_{\ell \neq j} A_{i\ell} A_{\ell k}}\big)} \right\rangle \quad (4.74)$$

At this stage, we make a mean-field approximation in (4.72) and (4.74). In both expressions, we replace sums of the form $\sum_\ell A_{i\ell} A_{\ell j}$ by $\sum_{\ell \notin \{i,j\}} \langle A_{i\ell} A_{\ell j} \rangle$. We also assume that $\langle A_{ij} \rangle = p$ for all (i, j), and that $\langle A_{i\ell} A_{\ell j} \rangle = q$ for all distinct (i, ℓ, j). This leads us to two coupled equations for p and q. First, (4.72) gives us

$$p = \frac{1}{1 + e^{-\lambda_1 - \lambda_2 (N-2)q}} = \frac{1}{2}\left[\tanh\left(\frac{1}{2}\lambda_1 + \frac{1}{2}\lambda_2 (N-2)q \right) + 1 \right] \qquad (4.75)$$

Equation (4.74) just needs minor manipulations[6] to deal with the remaining entry A_{ij}:

$$q = \left\langle \frac{e^{\lambda_2 A_{ij}}}{e^{\lambda_2 A_{ij}} - 1 + \left(1 + e^{-\lambda_1 - \lambda_2 (N-3)q}\right)^2} \right\rangle$$

$$= \frac{\langle A_{ij} \rangle e^{\lambda_2}}{e^{\lambda_2} - 1 + \left(1 + e^{-\lambda_1 - \lambda_2 (N-3)q}\right)^2} + \frac{1 - \langle A_{ij} \rangle}{\left(1 + e^{-\lambda_1 - \lambda_2 (N-3)q}\right)^2}$$

$$= \frac{p e^{\lambda_2}}{e^{\lambda_2} - 1 + \left(1 + e^{-\lambda_1 - \lambda_2 (N-3)q}\right)^2} + \frac{1 - p}{\left(1 + e^{-\lambda_1 - \lambda_2 (N-3)q}\right)^2} \qquad (4.76)$$

Substituting (4.75) into the latter equation then gives us finally a self-consistent equation for q only, of the form $q = \mathcal{F}(q)$, with

$$\mathcal{F}(q) = \frac{1}{\left(1 + e^{-\lambda_1 - \lambda_2 (N-3)q}\right)^2}\left\{ 1 + \frac{e^{\lambda_2} - 1}{1 + e^{-\lambda_1 - \lambda_2 (N-2)q}} \frac{\left(1 + e^{-\lambda_1 - \lambda_2 (N-3)q}\right)^2 - 1}{e^{\lambda_2} - 1 + \left(1 + e^{-\lambda_1 - \lambda_2 (N-3)q}\right)^2} \right\}$$

$$(4.77)$$

If $\lambda_2 = 0$ the function $\mathcal{F}(q)$ is independent of q, and we find the trivial result $q = p^2 = (1 + e^{-\lambda_1})^{-2}$, describing the Erdös–Rényi ensemble (as it should). If $\lim_{N \to \infty} \lambda_2 \neq 0$, the function $\mathcal{F}(q)$ acquires a non-trivial shape; see [140]. Now we will no longer have $q = p^2$, not all values of q will be accessible, and the equation $\mathcal{F}(q) = q$ will exhibit discontinuous bifurcations (phase transitions). However, if $\lambda_2 \neq 0$ and N is large, we must ensure that $p = \mathcal{O}(1)$ for $N \to \infty$, in order for mean-field arguments to be acceptable. This is seen to require $\lambda_1 = \mathcal{O}(1)$ and $\lambda_2 = \mathcal{O}(N^{-1})$, but upon setting $\lambda_2 = 2\beta/N$ with $\beta = \mathcal{O}(1)$ and sending $N \to \infty$, we find expression (4.77) collapsing to $\mathcal{F}(q) = p^2 = \left(1 + e^{-\lambda_1 - 2\beta q}\right)^{-2}$. Hence, we can find the solution of the mean-field calculation in this limit simply by inserting $q = p^2$ into the right-hand side of (4.75). If we also choose $\lambda_1 = 2\alpha$ for notational convenience, we then recover (4.67), which we derived first with simpler arguments:

$$p = \frac{1}{2}\left[\tanh\left(\alpha + \beta p^2 \right) + 1 \right] \qquad (4.78)$$

The present (more precise) mean-field derivation is reflected only in a different relation between β and the original control parameter λ_2. In spite of this derivation having allowed explicitly for non-trivial correlations between adjacent links, the model appears

---

[6]Note that at this stage the authors of [140] make a further (unnecessary) approximation, in that they make the replacement $A_{ij} \to \langle A_{ij} \rangle$ in the denominator of (4.74). Here we continue with the more precise evaluation.

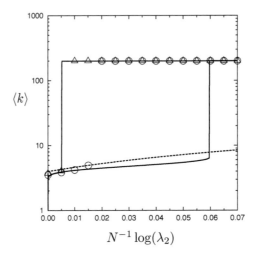

$$N^{-1}\log(\lambda_2)$$

Fig. 4.5 Comparison of the predicted and observed values of the average degree $\langle k \rangle$ in the Strauss (or random triangle) model, as a function of $N^{-1}\exp(\lambda_2)$. Markers: MCMC sampling results (see Chapter 5 for details) for ensembles with $N = 200$ and $\lambda_1 = \log(4/N)$, starting either from a high-density (triangles) or a low-density initialization (circles). Solid lines: Prediction of the mean-field equations (giving two solutions for certain parameter values, one with high link density and the other with low link density). Dashed line: Exact solution describing the sparse scaling regime. As with the two-star model, although for finite N the region of coexistent states is reduced (compared to the predictions for $N \to \infty$), even for $N = 200$ the phase transition and the existence of multiple stable states (which cause graph samples to depend on algorithm initialization) are prominent.

unable to exhibit such correlations; in the dense regime, it is apparently impossible to boost the number of triangles without boosting simultaneously the average connectivity. Depending on the values chosen for the ensemble parameters, and if N is large, we may have a unique solution or two distinct solutions, and there will be intermediate values of the link and triangle densities that cannot be reached for any value of the ensemble's control parameters.

Random triangle model in the sparse regime. Unfortunately, and in contrast to the two-star model, there is no exact solution yet for the random triangle model in the sparse scaling regime, where $p = \mathcal{O}(N^{-1})$. The cyclic appearance of the node indices in $T(\mathbf{A})$ prevents us from achieving factorization over links in (4.62). Various approximation and expansion strategies have been followed to study the critical behaviour of graphs from the Strauss ensemble in the sparse regime, and all lead to similar results. See papers such as [94, 35, 34]. Here, we follow a simple approximate approach that starts by rewriting (4.62) as an average over an Erdös–Rényi (3.1) ensemble and introduce a Kronecker delta to isolate the number of triangles $T(\mathbf{A}) = \sum_{i<j<k} A_{ij}A_{jk}A_{ki}$:

$$Z(\lambda_1, \lambda_2) = \sum_{\ell \geq 0} e^{\lambda_2 \ell} \sum_{\mathbf{A} \in G} e^{\lambda_1 \sum_{i<j} A_{ij}} \delta_{\ell, T(\mathbf{A})}$$

$$= \sum_{\ell \geq 0} e^{\lambda_2 \ell} \sum_{\mathbf{A} \in G} \delta_{\ell, T(\mathbf{A})} \prod_{i<j} \left(e^{\lambda_1} \delta_{A_{ij}, 1} + \delta_{A_{ij}, 0} \right)$$

$$= (1 + e^{\lambda_1})^{\frac{1}{2} N(N-1)} \sum_{\ell \geq 0} e^{\lambda_2 \ell} \langle \delta_{\ell, T(\mathbf{A})} \rangle_{\mathrm{ER}} \qquad (4.79)$$

Here $\langle \ldots \rangle_{\mathrm{ER}}$ denotes averaging over an Erdös–Rényi ensemble with link probability $\tilde{p} = e^{\lambda_1}/(1 + e^{\lambda_1})$, and the average $\langle \delta_{\ell, T(\mathbf{A})} \rangle_{\mathrm{ER}}$ is the probability of finding exactly ℓ triangles in such an ensemble. Upon writing $P(\ell | \lambda_1) = \langle \delta_{\ell, T(\mathbf{A})} \rangle_{\mathrm{ER}}$, we may apparently write equations (4.63) for our ensemble's control parameters as

$$L = \frac{1}{2} N(N-1) \frac{e^{\lambda_1}}{1 + e^{\lambda_1}} + \frac{\sum_{\ell \geq 0} e^{\lambda_2 \ell} \frac{\partial}{\partial \lambda_1} P(\ell | \lambda_1)}{\sum_{\ell \geq 0} e^{\lambda_2 \ell} P(\ell | \lambda_1)} \qquad (4.80)$$

$$T = \frac{\sum_{\ell \geq 0} \ell e^{\lambda_2 \ell} P(\ell | \lambda_1)}{\sum_{\ell \geq 0} e^{\lambda_2 \ell} P(\ell | \lambda_1)} \qquad (4.81)$$

These transparent equations show how the introduction of $\lambda_2 \neq 0$ changes the average number of links away from the naive value $\frac{1}{2} N(N-1)\tilde{p}$ of the Erdös–Rényi ensemble, and how it deforms the triangle distribution from $P(\ell | \lambda_1)$ into $P(\ell | \lambda_1, \lambda_2) = e^{\lambda_2 \ell} P(\ell | \lambda_1) / \sum_{\ell' \geq 0} e^{\lambda_2 \ell'} P(\ell' | \lambda_1)$. All of this is still fully exact. We have so far simply changed the problem from calculating a simple quantity in a complex ensemble to calculating a complex quantity, namely $P(\ell | \lambda_1)$ with $\ell \in \mathbb{N}$, in a simple ensemble.

Our approximation now consists in making a sensible assumption regarding the form of $P(\ell | \lambda_1)$, which we take to be a Poisson distribution, i.e. $P(\ell | \lambda_1) = e^{-\tau} \tau^\ell / \ell!$. The unknown parameter τ of this distribution can then be calculated explicitly, since

$$\tau = \sum_{\ell \geq 0} \ell P(\ell | \lambda_1) = \sum_{i<j<k} \langle A_{ij} A_{jk} A_{ki} \rangle_{\mathrm{ER}}$$

$$= \frac{1}{6} N(N-1)(N-2)(1 + e^{-\lambda_1})^{-3} \qquad (4.82)$$

This enables us to calculate the required sums over ℓ analytically, and work out equations (4.80, 4.81) in detail. Using $d\tau/d\lambda_1 = \frac{1}{2} N(N-1)(N-2) e^{-\lambda_1} (1 + e^{-\lambda_1})^{-4}$, we find

$$L = \frac{1}{2} N(N-1) \left[\frac{1}{1 + e^{-\lambda_1}} + (N-2) \frac{(e^{\lambda_2} - 1) e^{-\lambda_1}}{(1 + e^{-\lambda_1})^4} \right] \qquad (4.83)$$

$$T = \frac{1}{6} N(N-1)(N-2) \frac{e^{\lambda_2}}{(1 + e^{-\lambda_1})^3} \qquad (4.84)$$

Upon eliminating the parameter λ_2 via (4.84), we then obtain a single equation from which λ_1 is to be solved in terms of the desired ensemble averages L and T:

$$\frac{2L}{N} = \frac{N-1}{1 + e^{-\lambda_1}} \left[1 + \frac{(N-2) e^{-\lambda_1}}{(1 + e^{-\lambda_1})^3} \left(\frac{6T(1 + e^{-\lambda_1})^3}{N(N-1)(N-2)} - 1 \right) \right] \qquad (4.85)$$

In the sparse regime, we must have $L = \frac{1}{2}N\langle k \rangle$ with $\langle k \rangle = \mathcal{O}(1)$, and $T/N = \mathcal{O}(1)$. Hence, the appropriate scaling of λ_1 is $e^{-\lambda_1} = \mathcal{O}(N)$. We find that all our scaling requirements will be met upon setting

$$\lambda_1 = \log\left(\langle k \rangle - 6T/N\right) - \log N, \qquad \lambda_2 = \log N + \log\left(\frac{6T/N}{(\langle k \rangle - 6T/N)^3}\right) \qquad (4.86)$$

(modulo $\mathcal{O}(1/N)$ corrections). Once more we find also in the sparse regime the usual problems of ERGMs with non-trivial observables: not all values of L/N and T/N are achievable, only those combinations with $T/N < \frac{1}{6}\langle k \rangle$, and for large N we require infinitesimal relative changes in the control parameters to tune our ERGMs into the production of graphs with the required statistics. Numerical solutions of (4.86) are shown in Figure 4.5, and are compared to results from MCMC simulations and to the predictions from the mean-field approximation (4.78). The plot shows the average degree $\langle k \rangle$ as a function of $N^{-1}\log(\lambda_2)$, for a fixed value of λ_1. Again, the mean-field predictions are seen to be quite acceptable at high connectivity, but become inaccurate at low connectivity. Note that it is in principle perfectly possible to construct large graphs $\mathbf{A}$ with $\bar{k}(\mathbf{A}) = \mathcal{O}(1)$ and $T(\mathbf{A})/N > \bar{k}(\mathbf{A})/6$, as shown in the exercise below, so the observed inability to produce such graphs in the present model does not reflect any fundamental topological bound, similar to, e.g. the graphicality conditions (4.23), but simply a limitation of ERGMs. In fact, the Strauss model is even more tricky in that it exhibits further condensation transitions, from states where the triangles are distributed more or less homogeneously over the graph to states where they are concentrated onto just a small number of nodes.

Exercise 4.2 Calculate the average degree $\bar{k}(\mathbf{A}) = N^{-1}\sum_{ij} A_{ij}$ and the number of triangles $T(\mathbf{A})$ for the two example nondirected N-node graphs shown below, and verify that for sufficiently large N one will in both cases have $T(\mathbf{A})/N > \bar{k}(\mathbf{A})/6$.

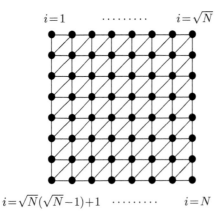

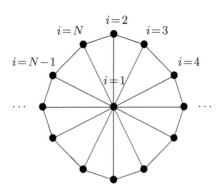

4.5 Stochastic block models for graphs with community structure

Overview: Stochastic block models (SBM) are a generalization of the Erdös–Rényi ensemble where nodes are divided into communities, and given different connection probabilities within (usually higher) versus between (usually lower) communities.

Standard definition of SBMs. A graph is said to have community structure if its nodes can be divided into subsets (or *modules* or *communities*), in such a way that the density of links *within* a subset, the so-called intra-modular links, is significantly different from the density of links *between* subsets, the inter-modular links. A simple and popular ERGM for generating such graphs is the SBM [87, 179, 56]. It is defined as follows. We first choose the number Q of node groups, which will be labelled by $q = 1 \ldots Q$, and we define the *a priori* probability $p(q) \in [0,1]$ for a node to be in group q, so $\sum_{q \leq Q} p(q) = 1$. This will define the required number and relative sizes of the communities. We then specify the so-called *affinity matrix* $\boldsymbol{P}$, with entries $P_{q,q'} \in [0,1]$ that give the probability for any two nodes in groups q and q', respectively, to be connected. Note that the latter entries are not subject to any normalization condition; we could have $P_{q,q'} = 0$ for all (q, q') (no links at all) or $P_{q,q'} = 1$ for all (q, q') (giving a fully connected graph). Note also that the affinity matrix need not be symmetric either, so that we can produce directed random graphs.

We now generate an N-node directed graph $\boldsymbol{A}$ in the following way. Each node $i = 1 \ldots N$ is given a group label $q_i \in \{1, \ldots, Q\}$. We use the shorthand $\boldsymbol{q}$ for the sequence $(q_1, \ldots, q_N)$ of group labels of all nodes. Each q_i is generated randomly and independently according to the *a priori* group probabilities $p(q)$, hence the probability for finding a given sequence $\boldsymbol{q}$ will be $p(\boldsymbol{q}) = \prod_{i \leq N} p(q_i)$. We then generate all N^2 entries $\{A_{ij}\}$ randomly and independently according to the relevant entry of the affinity matrix, i.e. $p(A_{ij}=1) = P_{q_i,q_j}$. This recipe will generate random directed N-node graphs $\boldsymbol{A}$ with the following probabilities:

$$p(\boldsymbol{A}) = \sum_{\boldsymbol{q}} p(\boldsymbol{A}|\boldsymbol{q})p(\boldsymbol{q}), \qquad p(\boldsymbol{A}|\boldsymbol{q}) = \prod_{ij} \left[P_{q_i,q_j} \delta_{A_{ij},1} + \left(1 - P_{q_i,q_j}\right) \delta_{A_{ij},0} \right] \quad (4.87)$$

One usually chooses the entries $P_{q,q'}$ to be largest for $q = q'$, so that intra-modular links in $\boldsymbol{A}$ will have the highest density, i.e. nodes tend to connect preferably to other nodes within their community. An example is shown in Figure 4.6. Adaptating this protocol to the generation of simple nondirected graphs is straightforward: one chooses $\boldsymbol{P}$ to be symmetric, and one generates the random links A_{ij} only for $i < j$, so that

$$p(\boldsymbol{A}) = \sum_{\boldsymbol{q}} p(\boldsymbol{A}|\boldsymbol{q})p(\boldsymbol{q}), \qquad p(\boldsymbol{A}|\boldsymbol{q}) = \prod_{i<j} \left[P_{q_i,q_j} \delta_{A_{ij},1} + \left(1 - P_{q_i,q_j}\right) \delta_{A_{ij},0} \right] \quad (4.88)$$

Rather than generating the group label sequence $\boldsymbol{q} = (q_1, \ldots, q_N)$ randomly, we could also set it by hand. This would be called a *planted partition*, and the graph probabilities would simply become $p(\boldsymbol{A}|\boldsymbol{q})$ as given in (4.87, 4.88) above.

SBMs with planted partitions as ERGMs. We now show that SBMs with planted partitions, i.e. those with fixed group labels $\boldsymbol{q} = (q_1, \ldots, q_N)$ and described by the probabilities $P(\boldsymbol{A}|\boldsymbol{q})$ given in (4.87) (for directed graphs) or (4.88) (for nondirected simple

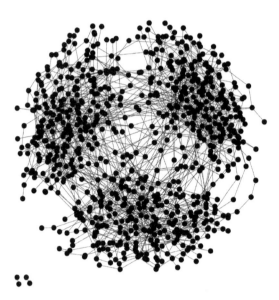

Fig. 4.6 Typical example of a simple nondirected graph generated by the stochastic block model. Here $N = 800$ and $Q = 3$ (so there are three communities), and the affinity matrix assigns probabilities $p_{\text{intra}} \approx 0.01355$ to all intra-modular links and $p_{\text{extra}} \approx 0.00075$ to all extra-modular links: $P_{q,q'} = p_{\text{intra}}\delta_{q,q'} + p_{\text{extra}}(1-\delta_{q,q'})$. The average connectivity is $\bar{k}(\mathbf{A}) \approx 4$.

graphs), are ERGMs. This will allow us to interpret SBMs as soft-constrained maximum entropy ensembles, and hence provide them with a solid information-theoretic basis. It will also enable us to assign appropriate values to the parameters of SBMs, should we wish to tailor them to resemble real-world networks. We start with directed graphs, so $G = \{0,1\}^{N^2}$, and we define an ERGM by applying the general formalism to the following Q^2 observables:

$$\forall q, q' = 1 \dots Q : \qquad \Omega_{q,q'}(\mathbf{A}) = \sum_{i,j=1}^{N} A_{ij}\delta_{q,q_i}\delta_{q',q_j} \tag{4.89}$$

The generic ERGM equations (4.1) then take the following form:

$$p(\mathbf{A}) = \frac{e^{\sum_{i,j=1}^{N} A_{ij}\lambda_{q_i,q_j}}}{Z(\boldsymbol{\lambda})}, \qquad Z(\boldsymbol{\lambda}) = \sum_{\mathbf{A} \in G} e^{\sum_{i,j=1}^{N} A_{ij}\lambda_{q_i,q_j}} \tag{4.90}$$

We note that the probabilities in (4.90) can be written as

$$p(\mathbf{A}) = \frac{1}{Z(\boldsymbol{\lambda})} \prod_{ij} \left[e^{\lambda_{q_i,q_j}}\delta_{A_{ij},1} + \delta_{A_{ij},0} \right]$$

$$= \prod_{ij} \left[P_{q_i,q_j}\delta_{A_{ij},1} + (1-P_{q_i,q_j})\delta_{A_{ij},0} \right] \tag{4.91}$$

with

$$P_{q,q'} = (1+e^{-\lambda_{q,q'}})^{-1} \tag{4.92}$$

This confirms that SBMs with planted partitions are indeed ERGMs, in which one enforces the values of the Q^2 quantities (4.89) as soft constraints. Within the ERGM framework, the entries $P_{q,q'}$ of the affinity matrix are expressed via the control parameters $\{\lambda_{q,q'}\}$ in terms of the averages $\langle \Omega_{q,q'}(\mathbf{A}) \rangle$, via

$$\langle \Omega_{q,q'}(\mathbf{A}) \rangle = \Big\langle \sum_{i,j=1}^{N} A_{ij} \delta_{q,q_i} \delta_{q',q_j} \Big\rangle = \frac{\partial}{\partial \lambda_{q,q'}} \log Z(\boldsymbol{\lambda})$$

$$= \sum_{ij} \frac{\partial}{\partial \lambda_{q,q'}} \log \left(1+e^{\lambda_{q_i,q_j}}\right) = \frac{\left(\sum_i \delta_{q,q_i}\right)\left(\sum_j \delta_{q',q_j}\right)}{1+e^{-\lambda_{q,q'}}} \tag{4.93}$$

which can also be written as

$$\lambda_{q,q'} = \log \left[\frac{\langle \sum_{ij} A_{ij} \delta_{q,q_i} \delta_{q',q_j} \rangle}{\langle \sum_{ij} (1-A_{ij}) \delta_{q,q_i} \delta_{q',q_j} \rangle} \right] \tag{4.94}$$

Combining this with (4.92) then immediately gives us the transparent prescription

$$P_{q,q'} = \sum_{ij} \langle A_{ij} \rangle \left(\frac{\delta_{q,q_i}}{\sum_\ell \delta_{q,q_\ell}} \right) \left(\frac{\delta_{q',q_j}}{\sum_\ell \delta_{q',q_\ell}} \right) \tag{4.95}$$

The corresponding derivation for nondirected graphs is nearly identical. We conclude that SBMs with planted partitions are ERGMs, and that SBMs with randomly generated partitions can be viewed as mixtures of ERGMs. SBMs can in fact be seen as a special case of a wider class of models, where first an underlying geometric structure of the desired networks is specified, followed by generating the detailed patterns of connectivity based on this geometric structure. In Chapter 10, we will revisit such generalizations of the SBM in greater detail.

Exercise 4.3 For which choice(s) of parameters will an SBM of simple nondirected graphs with planted partition become an Erdös–Rényi ensemble?

4.6   Strengths and weaknesses of ERGMs as null models

Advantages of working with ERGMs. In the context of defining and generating null models, i.e. random graphs against which to compare and interpret observations made in real-world networks, in such a way that the random graphs share specific topological features with the observed networks but are otherwise random, ERGMs have the great advantages of simplicity and probabilistic rigour. It is perfectly clear what we do, and what is the statistical basis for what we do. We make sure that our random null model graphs inherit *on average* from the observed real-world network the values $\Omega_\mu(\mathbf{A})$ of certain features that we select (so they are similar to their real-world counterparts in a well-defined way), and apart from imposing this condition, they are maximally random in an information-theoretic sense. The decision to impose such values as soft constraints, as opposed to insisting that *each individual* randomly generated $\mathbf{A}$ must have the specified values of the $\Omega_\mu(\mathbf{A})$, leads to the appealing simple exponential form of the probabilities in ERGMs. This form makes it significantly easier to calculate the values of the ensembles' control parameters analytically, as well as other quantities of interest, and will also (as we shall see) lead to relatively simple numerical sampling algorithms. The freedom to choose the observables $\Omega_\mu(\mathbf{A})$ as one wishes allows us to tailor our models to our specific application domain. This freedom is important since it will generally vary from field to field which are the important characteristics to be imposed upon a graph. Last but not least, in using ERGMs we can draw upon a large body of established results. For instance, Table 4.1 gives a list of the ERGMs that we have discussed so far in this chapter.

Disadvantages of working with ERGMs. However, the same core feature that leads to the convenient mathematical properties of ERGMs, namely the 'soft' implementation of the imposed constraints, is also their Achilles heel. As soon as our observables $\Omega_\mu(\mathbf{A})$ become more sophisticated than simply linear functions of the entries $\{A_{ij}\}$, and our graphs become large, we encounter serious complications. Phase transitions will make it nearly impossible in certain parameter regions to tune our control parameters effectively; the degeneracy of stable system states (or bi-stability) introduces a dependence on initializations of one's numerical sampling algorithm, and we find ourselves at times unable to control the values of the imposed quantities $\langle \Omega_\mu(\mathbf{A}) \rangle$ independently. Some combinations of values, while in principle realizable and indeed observed in our real-world network, will never be seen in our random graphs. We must conclude that for certain types of constraints a satisfactory and practical ERG formulation of graph ensembles is not always possible, which forces us to turn to ensembles with hard constraints instead. We are also led to consider the general and difficult question of which observables are in fact suitable to be used in ERGMs, if we want to tailor random graphs to mimic observed networks. We will return to such questions in Section 6.3, in the context of random graph ensembles with hard constraints.

In practice, ERGMs are found to be best suited to implement constraints on *local* observables, such as degrees. We have seen that if our observables depend linearly on the entries of $\mathbf{A}$, then we can always calculate the normalization factor $Z(\boldsymbol{\lambda})$ and its logarithm (which is the generator of our equations for the ensemble parameters) exactly. If this cannot be done, one may have to solve numerically the control parameters from a high-dimensional set of coupled equations. Techniques to accomplish this have been the subject of intense investigation over the last few decades and range from pseudo-likelihood estimation [18, 170, 7, 144] to Markov Chain Monte Carlo Maximum Likelihood techniques [166, 85, 167, 182], to Bayesian inference [11, 36, 37]. The worst-case scenario is the one where we cannot even find the latter equations in analytical form, and have to calculate ensemble averages by actually sampling from the probability $p(\mathbf{A})$. Since even in the sparse regime the number of graphs scales with the size

ERGM	OBSERVABLES $\{\Omega_\mu(\mathbf{A})\}$	PARAMETERS $\{\lambda_\mu\}$
Erdös–Rényi ensemble (simple, nondirected)	$\sum_{ij} A_{ij}$	$\lambda = \log\left[L/(\frac{1}{2}N(N-1)-L)\right]$ $L = \langle \sum_{ij} A_{ij} \rangle$
Constrained degrees (simple, nondirected)	$\sum_j A_{ij}$ for $i = 1 \ldots N$	$\lambda_i = \log\left[k_i/\sqrt{N\langle k\rangle} + \ldots\right]$ $k_i = \langle \sum_j A_{ij} \rangle$
Constrained degrees (directed)	$\sum_j A_{ij}$ and $\sum_j A_{ji}$ for $i = 1 \ldots N$	$\lambda_i^{\mathrm{in}} = \log\left[k_i^{\mathrm{in}}/\sqrt{N\langle k\rangle} + \ldots\right]$ $\lambda_j^{\mathrm{out}} = \log\left[k_j^{\mathrm{out}}/\sqrt{N\langle k\rangle} + \ldots\right]$ $k_i^{\mathrm{in}} = \langle \sum_j A_{ij} \rangle, \ \ k_j^{\mathrm{out}} = \langle \sum_j A_{ji} \rangle$
Reciprocity model (directed)	$\sum_{i<j}(A_{ij}+A_{ji})$ and $2\sum_{i<j} A_{ij} A_{ji}$	To be solved numerically from $(4.29, 4.30)$
Two-star model (simple, nondirected)	$\sum_{i<j} A_{ij}$ and $\frac{1}{2}\sum_{ijk} A_{ij} A_{jk}$	To be solved numerically from (4.56)
Strauss model (simple, nondirected)	$\sum_{i<j} A_{ij}$ and $\sum_{i<j<k} A_{ij} A_{jk} A_{ki}$	To be solved numerically from (4.63)
SBM (directed)	$\sum_{ij} A_{ij}\delta_{q,q_i}\delta_{q',q_j}$ for $q, q' = 1 \ldots Q$	$\lambda_{q,q'} = \log\langle \sum_{ij} A_{ij}\delta_{q,q_i}\delta_{q',q_j} \rangle$ $- \log\langle \sum_{ij}(1-A_{ij})\delta_{q,q_i}\delta_{q',q_j} \rangle$
SBM (nondirected)	$\sum_{i<j} A_{ij}\delta_{q,q_i}\delta_{q',q_j}$ for $q, q' = 1 \ldots Q$	$\lambda_{q,q'} = \log\langle \sum_{i<j} A_{ij}\delta_{q,q_i}\delta_{q',q_j} \rangle$ $- \log\langle \sum_{i<j}(1-A_{ij})\delta_{q,q_i}\delta_{q',q_j} \rangle$

Table 4.1 This table lists various standard exponential random graph models and their defining observables $\{\Omega_\mu(\mathbf{A})\}$ and control parameters $\{\lambda_\mu\}$. Their graph probabilities are given by $p(\mathbf{A}) = Z^{-1}(\boldsymbol{\lambda}) \exp[\sum_\mu \lambda_\mu \Omega_\mu(\mathbf{A})]$, where $Z(\boldsymbol{\lambda})$ is a normalization constant.

N as $\exp(N \log N)$ (which pretty soon becomes an astronomical number), that should be avoided at all costs.

Shortcut: *ERGMs are transparent and simple random graph ensembles in which one tunes the values of control parameters $\{\lambda_\mu\}$ to achieve desired ensemble averages for associated observables $\{\Omega_\mu(\mathbf{A})\}$. However, their simplicity comes at a price. Unless the observables depend in a relative simple way on $\mathbf{A}$ or the graph size N is small, one will observe (i) parameter regions with phase transitions, i.e. drastic changes in ensemble averages in response to small changes in control parameters, which make parameter tuning nearly impossible, and (ii) that certain combinations of ensemble averages can never be achieved. Both are fundamental consequences of the form of the probabilities* *in ERGMs and are independent of the numerical algorithm(s) used.*

4.7   Solutions of exercises

Exercise 4.1

Now $G = \{0,1\}^{N^2}$ and we prescribe $k_i^{\mathrm{in}} = \langle \sum_j A_{ij} \rangle$ and $k_i^{\mathrm{out}} = \langle \sum_j A_{ji} \rangle$ for all i, giving $2N$ parameters $\boldsymbol{\lambda}^{\mathrm{in}} = \{\lambda_1^{\mathrm{in}}, \ldots, \lambda_N^{\mathrm{in}}\}$ and $\boldsymbol{\lambda}^{\mathrm{out}} = \{\lambda_1^{\mathrm{out}}, \ldots, \lambda_N^{\mathrm{out}}\}$. The graph probabilities become

$$p(\mathbf{A}) = \frac{e^{\sum_{i=1}^N \left(\lambda_i^{\mathrm{in}} \sum_j A_{ij} + \lambda_i^{\mathrm{out}} \sum_j A_{ji} \right)}}{Z(\boldsymbol{\lambda})} = \frac{e^{\sum_{ij} A_{ij}(\lambda_i^{\mathrm{in}} + \lambda_j^{\mathrm{out}})}}{Z(\boldsymbol{\lambda}^{\mathrm{in}}, \boldsymbol{\lambda}^{\mathrm{out}})} \tag{4.96}$$

The generating function $F(\boldsymbol{\lambda}^{\mathrm{in}}, \boldsymbol{\lambda}^{\mathrm{out}}) = \log Z(\boldsymbol{\lambda}^{\mathrm{in}}, \boldsymbol{\lambda}^{\mathrm{out}})$ becomes

$$F(\boldsymbol{\lambda}^{\mathrm{in}}, \boldsymbol{\lambda}^{\mathrm{out}}) = \log \sum_{\mathbf{A} \in G} e^{\sum_{ij} A_{ij}(\lambda_i^{\mathrm{in}} + \lambda_j^{\mathrm{out}})} = \sum_{ij} \log(1 + e^{\lambda_i^{\mathrm{in}} + \lambda_j^{\mathrm{out}}}) \tag{4.97}$$

Working out (4.3) leads us to the equations from which to solve the control parameters:

$$k_i^{\mathrm{in}} = \frac{\partial F(\boldsymbol{\lambda}^{\mathrm{in}}, \boldsymbol{\lambda}^{\mathrm{out}})}{\partial \lambda_i^{\mathrm{in}}} = \sum_j \frac{1}{1 + e^{-\lambda_i^{\mathrm{in}} - \lambda_j^{\mathrm{out}}}} \tag{4.98}$$

$$k_j^{\mathrm{out}} = \frac{\partial F(\boldsymbol{\lambda}^{\mathrm{in}}, \boldsymbol{\lambda}^{\mathrm{out}})}{\partial \lambda_j^{\mathrm{out}}} = \sum_i \frac{1}{1 + e^{-\lambda_i^{\mathrm{in}} - \lambda_j^{\mathrm{out}}}} \tag{4.99}$$

We rewrite expression (4.96) to emphasize the mutual statistical independence of all N^2 links:

$$p(\mathbf{A}) = \frac{1}{Z(\boldsymbol{\lambda}^{\mathrm{in}}, \boldsymbol{\lambda}^{\mathrm{out}})} \prod_{ij} \left(e^{\lambda_i^{\mathrm{in}} + \lambda_j^{\mathrm{out}}} \delta_{A_{ij},1} + \delta_{A_{ij},0} \right)$$

$$= \prod_{ij} \left(\frac{e^{\lambda_i^{\mathrm{in}} + \lambda_j^{\mathrm{out}}}}{1 + e^{\lambda_i^{\mathrm{in}} + \lambda_j^{\mathrm{out}}}} \delta_{A_{ij},1} + \frac{1}{1 + e^{\lambda_i^{\mathrm{in}} + \lambda_j^{\mathrm{out}}}} \delta_{A_{ij},0} \right) \tag{4.100}$$

So the probability to have $A_{ij} = 1$ is $p_{ij} = (1 + e^{-\lambda_i^{\mathrm{in}} - \lambda_j^{\mathrm{out}}})^{-1}$. In the sparse regime, we require $\langle N^{-1} \sum_{ij} A_{ij} \rangle = N^{-1} \sum_{ij} p_{ij} = \mathcal{O}(1)$, which implies that $e^{\lambda_i^{\mathrm{in}} + \lambda_j^{\mathrm{out}}} = \mathcal{O}(1/N)$. We can then expand (4.98, 4.99) for large N:

$$k_i^{\mathrm{in}} = \sum_j \left(e^{\lambda_i^{\mathrm{in}} + \lambda_j^{\mathrm{out}}} + \mathcal{O}(N^{-2}) \right) = e^{\lambda_i^{\mathrm{in}}} \sum_j e^{\lambda_j^{\mathrm{out}}} + \mathcal{O}(N^{-1}) \tag{4.101}$$

$$k_j^{\mathrm{out}} = \sum_i \left(e^{-\lambda_i^{\mathrm{in}} - \lambda_j^{\mathrm{out}}} + \mathcal{O}(N^{-2}) \right) = e^{\lambda_j^{\mathrm{out}}} \sum_i e^{\lambda_i^{\mathrm{in}}} + \mathcal{O}(N^{-1}) \tag{4.102}$$

Summing these over i and j gives in both cases $N^{-1}[\sum_i e^{\lambda_i^{\mathrm{in}}}][\sum_j e^{\lambda_j^{\mathrm{out}}}] = \langle k \rangle + \mathcal{O}(N^{-1})$. These results enable us to write for the individual link probabilities in (4.100):

$$p_{ij} = \frac{e^{\lambda_i^{\mathrm{in}} + \lambda_j^{\mathrm{out}}}}{1 + e^{\lambda_i^{\mathrm{in}} + \lambda_j^{\mathrm{out}}}} = e^{\lambda_i^{\mathrm{in}} + \lambda_j^{\mathrm{out}}} + \mathcal{O}(N^{-2})$$

$$= \left[\frac{k_i^{\text{in}}}{\sum_j e^{\lambda_j^{\text{out}}}}(1+\mathcal{O}(N^{-1}))\right]\left[\frac{k_j^{\text{out}}}{\sum_i e^{\lambda_i^{\text{in}}}}(1+\mathcal{O}(N^{-1}))\right]+\mathcal{O}(N^{-2})$$

$$= \frac{k_i^{\text{in}}k_j^{\text{out}}(1+\mathcal{O}(N^{-1}))}{\left(\sum_i e^{\lambda_i^{\text{in}}}\right)\left(\sum_j e^{\lambda_j^{\text{out}}}\right)}+\mathcal{O}(N^{-2}) = \frac{k_i^{\text{in}}k_j^{\text{out}}}{N\langle k\rangle}+\mathcal{O}(N^{-2}) \qquad (4.103)$$

This then leads us directly to (4.24). ◻

Exercise 4.2

We start with the graph on the left. It has $(\sqrt{N}-2)^2$ nodes with degree $k_i = 6$ (the non-boundary ones), four with degree $k_i = 3$ (the corners), and $4(\sqrt{N}-2)$ with degree $k_i = 4$ (the boundary nodes that are not corners). This gives the average degree

$$\bar{k}(\mathbf{A}) = \frac{1}{N}\left[6(\sqrt{N}-2)^2 + 12 + 16(\sqrt{N}-2)\right] = 6 - 8N^{-\frac{1}{2}} + 4N^{-1} \qquad (4.104)$$

The number of triangles in $\mathbf{A}$ is simply twice the number of elementary squares in the picture, so the triangle density $T(\mathbf{A})/N$ is

$$T(\mathbf{A})/N = \frac{2}{N}(\sqrt{N}-1)^2 = 2 - 4N^{-\frac{1}{2}} + 2N^{-1} \qquad (4.105)$$

For $N \to \infty$ our formulae tell us that $\bar{k}(\mathbf{A})/6 \to 1$ and $T(\mathbf{A})/N \to 2$, so indeed $T(\mathbf{A})/N > \bar{k}(\mathbf{A})/6$. We now turn to the graph on the right. Here we have a central node with degree $k_1 = N-1$ and $N-1$ peripheral nodes with degree $k_i = 3$. This gives the average degree

$$\bar{k}(\mathbf{A}) = \frac{1}{N}\left[N-1 + 3(N-1)\right] = 4 - 4N^{-1} \qquad (4.106)$$

The number of triangles in $\mathbf{A}$ can be read off directly form the picture. There is precisely one triangle for each of the $N-1$ peripheral nodes, so the triangle density is

$$T(\mathbf{A})/N = (N-1)/N = 1 - N^{-1} \qquad (4.107)$$

In this second example graph, we observe for $N \to \infty$ that $\bar{k}(\mathbf{A})/6 \to 2/3$ and $T(\mathbf{A})/N \to 1$, so also here we find $T(\mathbf{A})/N > \bar{k}(\mathbf{A})/6$. ◻

Exercise 4.3

In the Erdös–Rényi ensemble (3.1), the link probabilities are identical for any pair (i,j) of distinct nodes. There are two ways to achieve this in the SBM. If we choose the entries of the affinity matrix to be uniform, i.e. $P_{q,q'} = p$ for all (q,q'), then the graph probabilities $p(\mathbf{A}|\mathbf{q})$ in (4.88) indeed reduce to those in (3.1), irrespective of the group label sequence $\mathbf{q}$. Alternatively we could assign all nodes to the same group, i.e. $q_i = q$ for all i, in which case we again obtain the formula for $p(\mathbf{A})$ in (3.1), with link probabilities $p = P_{q,q}$. ◻

5

Ensembles with hard constraints

Introduction. In exponential random graph models (ERGMs) there is no need for individual graphs to have specific values for chosen observables; we only demand that these values are recovered as ensemble averages. As a consequence, when working with conventional sets such as the set of all simple nondirected N-node graphs, or the set of all directed N-node graphs, we will be able to use Markov chain sampling processes in which the moves are simple operations that create or remove individual links. This is convenient, since we can then work with simple and standard formulae. However, ERGMs were found to have limitations. The simple exponential form of the probabilities in (4.1) restricts the possible (combinations of) values of observables that graphs from these ensembles can exhibit. Certain measurements will almost never be observed in large graphs from ERGMs, even if graphs do exist that would have such values. This fundamental limitation of ERGMs often forces us to use maximum entropy ensembles (3.15) with hard constraints. By using hard constraints, we are able to impose *any* combination of values that is physically achievable in a graph (i.e. that is graphical), including those which ERGMs struggle to generate.

5.1 Basic properties and tools

Overview: In this section, we introduce maximum entropy ensembles involving hard constraints as well as maximum entropy ensembles involving a combination of hard and soft constraints. One can also introduce ensembles where some observables are hard (and/or soft) constrained and some others are drawn from a prescribed distribution. These can be written as mixtures of maximum entropy ensembles. The number of graphs in a given ensemble can be calculated by using importance sampling.

Definitions. In random graph ensembles of the form (3.15), i.e. with hard topological constraints, we first choose a set G of graphs, and then assign equal probabilities to all graphs $\mathbf{A} \in G$ that have prescribed values for a set of K functions $\Omega_\mu(\mathbf{A})$. We assign probability zero to all other graphs. If N is finite, there will be a finite number $|G|$ of graphs in G (although this number is normally very large), so the functions $\Omega_\mu(\mathbf{A})$ take values from a discrete set, and we can therefore write

$$\forall \mathbf{A} \in G : \quad p(\mathbf{A}|\mathbf{\Omega}) = \frac{\delta_{\mathbf{\Omega}(\mathbf{A}),\mathbf{\Omega}}}{Z(\mathbf{\Omega})}, \qquad Z(\mathbf{\Omega}) = \sum_{\mathbf{A} \in G} \delta_{\mathbf{\Omega}(\mathbf{A}),\mathbf{\Omega}} \tag{5.1}$$

Here $\mathbf{\Omega}(\mathbf{A}) = \{\Omega_1(\mathbf{A}), \dots, \Omega_K(\mathbf{A})\}$, and $\delta_{\mathbf{\Omega},\mathbf{\Omega}'} = \prod_{\mu \leq K} \delta_{\Omega_\mu,\Omega'_\mu}$.

Generating Random Networks and Graphs. First Edition. A.C.C. Coolen, A. Annibale, and E.S. Roberts. © A.C.C. Coolen, A. Annibale, and E.S. Roberts 2017. Published in 2017 by Oxford University Press. DOI: 10.1093/acprof:oso/9780198709893.001.0001

In addition to the above ensembles, in which all constraints are hard, we may sometimes also wish to combine hard constraints with soft ones. Redoing the maximum entropy derivation then tells us that such ensembles will be of the following form, in which we now have K hard constraints for the observables $\Omega_\mu(\mathbf{A})$ and R further soft constraints for the functions $\hat{\Omega}_\nu(\mathbf{A})$:

$$\forall \mathbf{A} \in G: \quad p(\mathbf{A}|\boldsymbol{\lambda}, \boldsymbol{\Omega}) = \frac{e^{\sum_{\nu=1}^{R} \lambda_\nu \hat{\Omega}_\nu(\mathbf{A})} \delta_{\boldsymbol{\Omega}(\mathbf{A}),\boldsymbol{\Omega}}}{Z(\boldsymbol{\lambda}, \boldsymbol{\Omega})} \tag{5.2}$$

$$Z(\boldsymbol{\lambda}, \boldsymbol{\Omega}) = \sum_{\mathbf{A} \in G} e^{\sum_{\nu=1}^{R} \lambda_\nu \hat{\Omega}_\nu(\mathbf{A})} \delta_{\boldsymbol{\Omega}(\mathbf{A}),\boldsymbol{\Omega}} \tag{5.3}$$

The R control parameters $\boldsymbol{\lambda} = (\lambda_1, \ldots, \lambda_R)$ are as always solved from the soft constraint equations $\langle \hat{\Omega}_\nu(\mathbf{A}) \rangle = \hat{\Omega}_\nu$ for all $\nu = 1 \ldots R$.

As a further generalization of the above, we may also work with *mixture* models, where $p(\mathbf{A})$ is not itself a maximum entropy ensemble, but a mixture of maximum entropy ensembles involving hard constraints. For instance, we may wish to draw $\hat{\Omega}$ from a distribution $g(\hat{\Omega})$ *and* impose a hard constraint on Ω. This leads to

$$\forall \mathbf{A} \in G: \quad p(\mathbf{A}|\boldsymbol{\Omega}, g) = \sum_{\hat{\Omega}} g(\hat{\Omega}) p(\mathbf{A}|\boldsymbol{\Omega}, \hat{\Omega}) \tag{5.4}$$

with

$$\forall \mathbf{A} \in G: \quad p(\mathbf{A}|\boldsymbol{\Omega}, \hat{\Omega}) = \frac{\delta_{\boldsymbol{\Omega}(\mathbf{A}),\boldsymbol{\Omega}} \delta_{\hat{\Omega}(\mathbf{A}),\hat{\Omega}}}{Z(\boldsymbol{\Omega}, \hat{\Omega})}, \tag{5.5}$$

$$Z(\boldsymbol{\Omega}, \hat{\Omega}) = \sum_{\mathbf{A} \in G} \delta_{\boldsymbol{\Omega}(\mathbf{A}),\boldsymbol{\Omega}} \delta_{\hat{\Omega}(\mathbf{A}),\hat{\Omega}} \tag{5.6}$$

and $\sum_{\hat{\Omega}} g(\hat{\Omega}) = 1$. This can be rewritten, by performing the sum over $\hat{\Omega}$, as

$$\forall \mathbf{A} \in G: \quad p(\mathbf{A}|\boldsymbol{\Omega}, g) = \frac{\delta_{\boldsymbol{\Omega}(\mathbf{A}),\boldsymbol{\Omega}} g(\hat{\Omega}(\mathbf{A}))}{Z(\boldsymbol{\Omega}, \hat{\Omega}(\mathbf{A}))} \tag{5.7}$$

If the $\{\hat{\Omega}_\nu\}$ are drawn randomly and independently from the distribution g, one has

$$\forall \mathbf{A} \in G: \quad p(\mathbf{A}|\boldsymbol{\Omega}, g) = \frac{\delta_{\boldsymbol{\Omega}(\mathbf{A}),\boldsymbol{\Omega}} [\prod_{\nu=1}^{R} g(\hat{\Omega}_\nu(\mathbf{A}))]}{Z(\boldsymbol{\Omega}, \hat{\Omega}(\mathbf{A}))}. \tag{5.8}$$

This differs from the maximum entropy ensemble (5.2), where Ω is implemented as a hard constraint and $\hat{\Omega}$ as a soft constraint, unless g is itself a maximum entropy distribution conditioned on its expectation. One could generalize expression (5.4) further and introduce a mixture of maximum entropy ensembles, each involving a combination of K hard and S soft constraints

$$\forall \mathbf{A} \in G: \quad p(\mathbf{A}|\boldsymbol{\Omega}, \boldsymbol{\lambda}, g) = \sum_{\hat{\Omega}} g(\hat{\Omega}) p(\mathbf{A}|\boldsymbol{\Omega}, \boldsymbol{\lambda}, \hat{\Omega}) \tag{5.9}$$

with

$$\forall \mathbf{A} \in G: \quad p(\mathbf{A}|\mathbf{\Omega}, \boldsymbol{\lambda}, \hat{\mathbf{\Omega}}) = \frac{\mathrm{e}^{\sum_{\rho=1}^{S} \lambda_\rho \tilde{\Omega}_\rho(\mathbf{A})} \delta_{\mathbf{\Omega}(\mathbf{A}), \mathbf{\Omega}} \delta_{\hat{\mathbf{\Omega}}(\mathbf{A}), \hat{\mathbf{\Omega}}}}{Z(\mathbf{\Omega}, \boldsymbol{\lambda}, \hat{\mathbf{\Omega}})} \tag{5.10}$$

$$Z(\mathbf{\Omega}, \boldsymbol{\lambda}, \hat{\mathbf{\Omega}}) = \sum_{\mathbf{A} \in G} \mathrm{e}^{\sum_{\rho=1}^{S} \lambda_\rho \tilde{\Omega}_\rho(\mathbf{A})} \delta_{\mathbf{\Omega}(\mathbf{A}), \mathbf{\Omega}} \delta_{\hat{\mathbf{\Omega}}(\mathbf{A}), \hat{\mathbf{\Omega}}} \tag{5.11}$$

Exercise 5.1 Show that for large and finitely connected graphs, the ensemble

$$p(\mathbf{A}|\mathbf{k}, L) = \frac{1}{Z(\mathbf{k}, L)} \prod_{i<j} \left[\frac{k_i k_j}{N\langle k \rangle} \delta_{A_{ij}, 1} + \left(1 - \frac{k_i k_j}{N\langle k \rangle} \right) \delta_{A_{ij}, 0} \right] \delta_{L, \sum_{i<j} A_{ij}} \tag{5.12}$$

with soft-constrained degree sequence $\mathbf{k}$ and hard-constrained number of links L takes the maximum entropy form (5.2).

Exercise 5.2 Give an expression for the probabilities $p(\mathbf{A})$ describing an ensemble of simple nondirected graphs with hard-constrained number of links L, and with degrees drawn randomly and independently from a prescribed distribution $g(k)$.

The number of graphs that meet the constraints. As in the case of ERGMs, where the normalization constant was found to be useful as a generating function of averages, the normalizer $Z(\mathbf{\Omega})$ of (5.1) also carries important information: it gives us the number of graphs in G that satisfy the K imposed constraints. The latter quantity relates directly to the difficulty of generating graphs from (5.1); the larger $Z(\mathbf{\Omega})$, the faster and simpler we expect our algorithms to be. Since the adjacency matrix $\mathbf{A}$ of an N-node graph has N^2 binary entries, we know from $|G| \leq 2^{N^2}$ that $\log Z(\mathbf{\Omega}) \leq N^2 \log 2$, but especially if we work in the sparse (or finite connectivity) scaling regime, where all degrees scale as $k_i(\mathbf{A}) = \mathcal{O}(1)$, we must expect the number of admissible graphs to scale less explosively with N. If the K functions $\Omega_\mu(\mathbf{A})$ take integer values, we can use the integral representation $\delta_{nm} = (2\pi)^{-1} \int_{-\pi}^{\pi} \mathrm{d}\omega \, \mathrm{e}^{\mathrm{i}\omega(n-m)}$ of the Kronecker delta symbol, and write

$$\frac{1}{N} \log Z(\mathbf{\Omega}) = \frac{1}{N} \log \int_{-\pi}^{\pi} \frac{\mathrm{d}\hat{\mathbf{\Omega}}}{(2\pi)^K} \mathrm{e}^{\mathrm{i}\hat{\mathbf{\Omega}} \cdot \mathbf{\Omega}} \sum_{\mathbf{A} \in G} \mathrm{e}^{-\mathrm{i} \sum_{\mu=1}^{K} \hat{\Omega}_\mu \Omega_\mu(\mathbf{A})}$$

$$= -\frac{K}{N} \log(2\pi) + \frac{1}{N} \log \int_{-\pi}^{\pi} \mathrm{d}\hat{\mathbf{\Omega}} \, \mathrm{e}^{\mathrm{i}\hat{\mathbf{\Omega}} \cdot \mathbf{\Omega}} \sum_{\mathbf{A} \in G} \mathrm{e}^{-\mathrm{i} \sum_{\mu=1}^{K} \hat{\Omega}_\mu \Omega_\mu(\mathbf{A})} \tag{5.13}$$

with $\hat{\mathbf{\Omega}} \cdot \mathbf{x} = \sum_{\mu=1}^{K} \hat{\Omega}_\mu x_\mu$. We observe upon comparing this with the formula for $Z(\boldsymbol{\lambda})$ in (4.1), that, apart from the extra integral and a simple constant, the sum over all $\mathbf{A} \in G$ in the normalization constant for the present hard-constrained ensembles has the same form of the sum that we would be calculating for the normalization constant in the soft-constrained ERGMs, albeit with the replacement $\lambda_\mu \to -\mathrm{i}\hat{\Omega}_\mu$.

Importance sampling. To evaluate expressions such as (5.13) further, we will often use the following simple idea from statistics and computer science. Any sum can always be written as an average, with probabilities that for some (mathematical or numerical) reason we prefer to work with. If in our present formulae we choose any graph probabilities $w(\mathbf{A}) \in [0,1]$ with $\sum_{\mathbf{A} \in G} w(\mathbf{A}) = 1$ and abbreviate averages over these probabilities as $\sum_{\mathbf{A} \in G} w(\mathbf{A}) f(\mathbf{A}) = \langle f \rangle_w$, we may always write $\sum_{\mathbf{A} \in G} f(\mathbf{A}) = \langle (f/w) \rangle_w$. As it stands, this may seem a pointless step, but it becomes particularly useful if we choose probabilities $w(\mathbf{A})$ that can be expressed in terms of the quantities $\Omega_\mu(\mathbf{A})$.

Suppose, for example, that we work with simple nondirected graphs, and that our constraining functions are chosen such that the total number of links $L(\mathbf{A}) = \sum_{i<j} A_{ij}$ can be expressed in terms of these functions, i.e. $L(\mathbf{A}) = L[\Omega(\mathbf{A})]$ for some function $L[\Omega]$. We can then use the alternative expression (3.4) for the graph probabilities in the Erdös–Rényi ensemble (3.1), which we will now write as $p_{\mathrm{ER}}(\mathbf{A})$ and with the individual link probabilities $\langle k \rangle / (N-1) = 2L[\Omega]/N(N-1)$, to rewrite (5.13) as

$$
\frac{1}{N} \log Z(\Omega) = \frac{1}{N} \log \left\{ \sum_{\mathbf{A} \in G} \frac{p_{\mathrm{ER}}(\mathbf{A}) \, \delta_{\Omega(\mathbf{A}),\Omega}}{\left(\frac{2L[\Omega]}{N(N-1)}\right)^{L(\mathbf{A})} \left(1 - \frac{2L[\Omega]}{N(N-1)}\right)^{N(N-1)/2 - L(\mathbf{A})}} \right\}
$$

$$
= \frac{1}{N} \log \left\langle \delta_{\Omega(\mathbf{A}),\Omega} \right\rangle_{\mathrm{ER}}
$$
$$
- \frac{1}{N} \log \left[\left(\frac{2L[\Omega]}{N(N-1)}\right)^{L[\Omega]} \left(1 - \frac{2L[\Omega]}{N(N-1)}\right)^{\frac{1}{2}N(N-1) - L[\Omega]} \right]
$$

$$
= \frac{L[\Omega]}{N} \log \left(\frac{N(N-1)}{2L[\Omega]} \right) - \left(\frac{1}{2}(N-1) - \frac{1}{N} L[\Omega]\right) \log \left(1 - \frac{2L[\Omega]}{N(N-1)}\right)
$$
$$
+ \frac{1}{N} \log \int_{-\pi}^{\pi} \mathrm{d}\hat{\Omega} \; e^{\mathrm{i}\hat{\Omega} \cdot \Omega} \left\langle e^{-\mathrm{i} \sum_{\mu=1}^{K} \hat{\Omega}_\mu \Omega_\mu(\mathbf{A})} \right\rangle_{\mathrm{ER}} - \frac{K}{N} \log(2\pi) \qquad (5.14)
$$

Expression (5.14) did not require any approximations or expansions, it is still fully equivalent to its predecessor (5.13). We have thereby converted a sum over all $\mathbf{A} \in G$ into an average over an Erdös–Rényi ensemble. This is particularly useful in the sparse regime, where $\langle k \rangle$ remains finite and N becomes large, since now we know that $L[\Omega] = \mathcal{O}(N)$, and we can expand and simplify various terms. If we express the result fully in terms of $\langle k \rangle$, which is then understood to be defined, via $L[\Omega]$, in terms of the prescribed values Ω, we obtain

$$
\frac{1}{N} \log Z(\Omega) = \frac{1}{2} \langle k \rangle \left[1 + \log \left(\frac{N}{\langle k \rangle} \right) \right] - \frac{K}{N} \log(2\pi)
$$
$$
+ \frac{1}{N} \log \int_{-\pi}^{\pi} \mathrm{d}\hat{\Omega} \; e^{\mathrm{i}\hat{\Omega} \cdot \Omega} \left\langle e^{-\mathrm{i} \sum_{\mu=1}^{K} \hat{\Omega}_\mu \Omega_\mu(\mathbf{A})} \right\rangle_{\mathrm{ER}} + \mathcal{O}\left(\frac{1}{N}\right) \qquad (5.15)
$$

This final expression suggests (as it will turn out, correctly) that the typical number of *sparse* N-node graphs will for large N scale as $\mathcal{O}(\exp(N \log N))$. The main advantage of (5.15) is that it will make it easier for us to expand the formula in the limit of large N, using the fact that in the Erdös–Rényi average $\langle \ldots \rangle_{\mathrm{ER}}$ the individual link probabilities scale as $\mathcal{O}(N^{-1})$.

5.2   Nondirected graphs with hard-constrained number of links

Overview: In this section, we consider the hard-constrained version of the Erdös–Rényi ensemble, where the total number of links is hard constrained, and illustrate how to calculate the number of graphs in this ensemble.

Definition. As our first example of an ensemble with hard topological constraints, we consider simple nondirected N-node graphs in which we prescribe the total number of links, i.e. we insist that $\sum_{i<j} A_{ij} = L$ for all $\mathbf{A} \in G$, for some integer number L. This is the hard-constrained equivalent of the Erdös–Rényi model (3.1), in which the number of links was prescribed as a soft constraint. Here we have

$$G = \{0,1\}^{N(N-1)/2}, \qquad p(\mathbf{A}) = \frac{\delta_{L,\sum_{i<j} A_{ij}}}{Z(L)}, \qquad Z(L) = \sum_{\mathbf{A} \in G} \delta_{L,\sum_{i<j} A_{ij}} \qquad (5.16)$$

Using the identity $\langle k \rangle = 2L/N$, we can always write the ensemble probabilities of (5.16) in the alternative but mathematically equivalent form

$$p(\mathbf{A}) = \frac{1}{Z(\langle k \rangle)} \delta_{N\langle k \rangle, 2\sum_{i<j} A_{ij}}, \qquad Z(\langle k \rangle) = \sum_{\mathbf{A} \in G} \delta_{N\langle k \rangle, 2\sum_{i<j} A_{ij}} \qquad (5.17)$$

This will be useful in the sparse regime, where $N \to \infty$ while keeping $\langle k \rangle$ finite.

Number of graphs in the ensemble – the short route. The ensemble (5.17) is sufficiently simple to allow us to calculate the number $Z(\langle k \rangle)$ of N-node simple nondirected graphs that have average degree $\langle k \rangle$ directly: it is just the number of possible ways in which one can draw $L = \frac{1}{2}N\langle k \rangle$ links from a set of $N(N-1)/2$ candidates, giving

$$Z(\langle k \rangle) = \binom{N(N-1)/2}{N\langle k \rangle/2} \qquad (5.18)$$

In the finite connectivity regime, we can inspect the leading orders in N, using Stirling's formula $\log n! = n \log n - n + \mathcal{O}(\log n)$ for $n \to \infty$. If both n and m are large (with $m \ll n$), this formula, together with $\log(1+x) = x + \mathcal{O}(x^2)$, allows us to expand

$$\log\left(\frac{n!}{(n-m)!\, m!}\right) = n \log n - (n-m) \log(n-m) - m \log m + \mathcal{O}(\log n, \log m)$$

$$= n\left\{\log n - (1-\frac{m}{n}) \log(n-m) - \frac{m}{n} \log m\right\} + \mathcal{O}(\log n, \log m)$$

$$= n\left\{\frac{m}{n} \log \frac{n}{m} + \frac{m}{n}\right\} + \mathcal{O}(\log n, \log m, m^2/n^3)$$

$$= m \log \frac{n}{m} + m + \mathcal{O}(\log n, \log m, m^2/n^3) \qquad (5.19)$$

Application of this expansion to $n = \frac{1}{2}N(N-1)$ and $m = \frac{1}{2}N\langle k \rangle$ then gives us

$$\frac{1}{N} \log Z(\langle k \rangle) = \frac{1}{2}\langle k \rangle \log(N/\langle k \rangle) + \frac{1}{2}\langle k \rangle + \mathcal{O}\left(\frac{\log N}{N}\right) \qquad (5.20)$$

Thus, even in the sparse regime, for large N the number of graphs with finite average degree $\langle k \rangle$ still grows super-exponentially as $Z(\langle k \rangle) \sim \exp(\frac{1}{2}\langle k \rangle N \log(N) + \ldots)$. This means that one can never hope to sample numerically the space of all such graphs, even for relatively modest sizes N. For instance, working out the latter formula for small graphs already gives

$$N = 32, \ \langle k \rangle = 2: \qquad Z(\langle k \rangle) \approx 2.7 \ 10^{52} \qquad (5.21)$$

$$N = 50, \ \langle k \rangle = 2: \qquad Z(\langle k \rangle) \approx 4.1 \ 10^{91} \qquad (5.22)$$

These are *big* numbers. For comparison: the total number of atoms in the observable universe is estimated to be somewhere between 10^{78} and 10^{82}.

Exercise 5.3 Calculate the number $Z(L)$ of directed N-node graphs with exactly L links. Use Stirling's formula $\log n! = n \log n - n + \mathcal{O}(\log n)$ to find the leading orders in N of $N^{-1} \log Z(L)$ as $N \to \infty$, in the sparse scaling regime.

Number of graphs in the ensemble – the longer route. The above simple calculation is unfortunately only feasible for very simple ensembles; more interesting and realistic cases require alternative and more generic tools. However, as was already appreciated by Hannibal, there indeed exists more than one road that leads to Rome. Let us now see how we could have found $Z(\langle k \rangle)$ using expression (5.15), which for the ensemble (5.17) takes the form

$$\frac{1}{N} \log Z(\langle k \rangle) = \frac{\langle k \rangle}{2} \left[1 + \log \left(\frac{N}{\langle k \rangle} \right) \right] + \frac{1}{N} \log \int_{-\pi}^{\pi} d\omega \ e^{\frac{1}{2} i \omega N \langle k \rangle} \left\langle e^{-i\omega \sum_{i<j} A_{ij}} \right\rangle_{\mathrm{ER}} + \mathcal{O}\left(\frac{1}{N} \right)$$

$$= \frac{\langle k \rangle}{2} \left[1 + \log \left(\frac{N}{\langle k \rangle} \right) \right]$$

$$\qquad + \frac{1}{N} \log \int_{-\pi}^{\pi} d\omega \ e^{\frac{1}{2} i \omega N \langle k \rangle + \frac{1}{2} N(N-1) \log \left[1 + \frac{\langle k \rangle}{N-1} (e^{-i\omega} - 1) \right]} + \mathcal{O}\left(\frac{1}{N} \right)$$

$$= \frac{1}{2} \langle k \rangle \log \left(\frac{N}{\langle k \rangle} \right) + \frac{1}{N} \log \int_{-\pi}^{\pi} d\omega \ e^{\frac{1}{2} N \langle k \rangle (i\omega + e^{-i\omega})} + \mathcal{O}\left(\frac{1}{N} \right) \qquad (5.23)$$

We can now use steepest descent integration (see Appendix D) to work out the leading order in N of the term with the integral, and find

$$\frac{1}{N} \log Z(\langle k \rangle) = \frac{1}{2} \langle k \rangle \log \left(\frac{N}{\langle k \rangle} \right) + \mathrm{extr}_{\omega \in [-\pi, \pi]} \frac{1}{2} \langle k \rangle (i\omega + e^{-i\omega}) + \epsilon_N \qquad (5.24)$$

with $\lim_{N \to \infty} \epsilon_N = 0$. The extremum in this last expression is found by simply working out $d(i\omega + e^{-i\omega})/d\omega = 0$, giving $\omega = 0$ at the saddle point. Upon inserting this value for ω, we obtain the extremum $\mathrm{extr}_{\omega \in [-\pi, \pi]} \frac{1}{2} \langle k \rangle (i\omega + e^{-i\omega}) = \frac{1}{2} \langle k \rangle$, and hence we indeed recover correctly from (5.23) the leading orders of (5.20). For the present model, where we have a simple alternative for finding $N^{-1} \log Z$, following the latter route is clearly overkill. For more complex ensembles, it is often found to be the most effective and feasible option.

5.3 Prescribed degree statistics and hard-constrained number of links

Overview: In this section, we introduce a mixture ensemble where the number of links is hard constrained and the degrees are drawn randomly and independently from a distribution $g(k)$. We show that the degree distribution of the resulting network is equal to $g(k)$ only if the average connectivity imposed by the hard constraint on the links matches the average of $g(k)$. Otherwise, the degree distribution of the resulting graphs is drastically different from $g(k)$, showing either an exponential decay or a condensation behaviour.

Definitions. As our second example, we inspect a mixture ensemble (5.2) defined on the set G of simple nondirected N-node graphs, in which we first draw N degrees $k_1, \ldots, k_N$ randomly from a degree distribution $g(k)$, and impose a further hard constraint on the number of links. This construction is described by the probability distribution

$$p(\mathbf{A}|L, g) = \sum_{k_1 \ldots k_N} \left[\prod_{i=1}^{N} g(k_i) \right] \frac{1}{Z(L, \mathbf{k})} \delta_{L, \sum_{i<j} A_{ij}} \prod_i \delta_{k_i, \sum_j A_{ij}} \tag{5.25}$$

$$Z(L, \mathbf{k}) = \sum_{\mathbf{A} \in G} \delta_{L, \sum_{i<j} A_{ij}} \prod_i \delta_{k_i, \sum_j A_{ij}} \tag{5.26}$$

We use this example to illustrate the non-trivial interplay in mixture models between local constraints (here imposed upon on the N degrees) and global constraints (imposed upon the number of links). The crucial issue is how we should choose the distribution $g(k)$ such that graphs generated from (5.25) will for large N exhibit a desired target degree distribution $p(k) = \langle N^{-1} \sum_i \delta_{k,k_i(\mathbf{A})} \rangle$. Without the global constraint $\sum_{i<j} A_{ij} = L$, we can simply choose $g(k) = p(k)$, but in its presence this is generally no longer true.

Interplay between local and global degree constraints. The present model is sufficiently simple to allow us to understand the new phenomenology induced by competition between local and global constraints, which touches the field of large deviation theory [174]. We note that we may write the probabilities (5.25) alternatively as

$$p(\mathbf{A}|L, g) = \sum_{k_1 \ldots k_N} p(\mathbf{A}|\mathbf{k}) p(\mathbf{k}|L, g) \tag{5.27}$$

with the individually normalized distributions

$$p(\mathbf{k}|L, g) = \frac{\delta_{L, \frac{1}{2} \sum_i k_i}}{\tilde{Z}(L)} \left[\prod_{i=1}^{N} g(k_i) \right], \qquad \tilde{Z}(L, g) = \sum_{k_1 \ldots k_N} \delta_{L, \frac{1}{2} \sum_i k_i} \left[\prod_{i=1}^{N} g(k_i) \right] \tag{5.28}$$

$$p(\mathbf{A}|\mathbf{k}) = \frac{\prod_i \delta_{k_i, \sum_j A_{ij}}}{Z(\mathbf{k})}, \qquad Z(\mathbf{k}) = \sum_{\mathbf{A} \in G} \prod_i \delta_{k_i, \sum_j A_{ij}} \tag{5.29}$$

The degree statistics of the graphs generated by (5.27) are determined completely by the distribution $p(\mathbf{k}|L)$, so we no longer need worry[1] about graph statistics to find the ensemble's average degree distribution. Using the invariance of $p(\mathbf{k}|L)$ under permutations of the node indices $\{1,\dots,N\}$, we can infer from (5.27) that

$$
\begin{aligned}
p(k) &= \sum_{\mathbf{A}\in G}\sum_{k_1\dots k_N} p(\mathbf{A}|\mathbf{k})p(\mathbf{k}|L,g)\left(\frac{1}{N}\sum_i \delta_{k,k_i(\mathbf{A})}\right)\\
&= \sum_{k_1\dots k_N} p(\mathbf{k}|L)\frac{1}{N}\sum_i \delta_{k,k_i} = \sum_{k_1\dots k_N} p(\mathbf{k}|L,g)\delta_{k,k_N}\\
&= \frac{g(k)}{\tilde{Z}(L)}\sum_{k_1\dots k_{N-1}}\left[\prod_{i<N} g(k_i)\right]\delta_{2L,k+\sum_{i<N}k_i}
\end{aligned}
\tag{5.30}
$$

As in the previous model, we write the Kronecker delta of the global constraint in integral form. Anticipating that our main interest will be in the sparse scaling regime, we also put $L = \frac{1}{2}N\langle k\rangle$, where $\langle k\rangle$ is now the imposed average degree. We also write the normalization factor $\tilde{Z}(L)$, which follows simply from normalization of $p(k)$, in explicit form. This gives us

$$
\begin{aligned}
p(k) &= \frac{g(k)\int_{-\pi}^{\pi}d\omega\; e^{i\omega(N\langle k\rangle-k)}\left(\prod_{i<N}\sum_{k_i} g(k_i)e^{-i\omega k_i}\right)}{\sum_{k'\ge 0} g(k')\int_{-\pi}^{\pi}d\omega\; e^{i\omega(N\langle k\rangle-k')}\left(\prod_{i<N}\sum_{k_i} g(k_i)e^{-i\omega k_i}\right)}\\
&= \frac{\int_{-\pi}^{\pi}d\omega\left(\frac{g(k)e^{-i\omega k}}{\sum_{k'} g(k')e^{-i\omega k'}}\right)e^{N\left[i\langle k\rangle\omega+\log\sum_{k'} g(k')\exp[-i\omega k']\right]}}{\sum_{k''}\int_{-\pi}^{\pi}d\omega\left(\frac{g(k'')e^{-i\omega k''}}{\sum_{k'} g(k')e^{-i\omega k'}}\right)e^{N\left[i\langle k\rangle\omega+\log\sum_{k'} g(k')\exp[-i\omega k']\right]}}
\end{aligned}
\tag{5.31}
$$

Once more we arrive at an expression that for large N can be evaluated via steepest descent integration (see Appendix D). The two integrals over $\omega \in [-\pi,\pi]$ are for $N\to\infty$ dominated by their saddle-point value. The location ω of the saddle point is the solution of

$$
\frac{d}{d\omega}\left[i\langle k\rangle\omega + \log\sum_{k'} g(k')\exp[-i\omega k']\right] = 0
\tag{5.32}
$$

After differentiation with respect to ω, we recognize that the simplest solution of the resulting equation is purely imaginary, $\omega = -i\alpha$, where $\alpha \in \mathbb{R}$ is the solution of

$$
\langle k\rangle = \frac{\sum_{k\ge 0} kg(k)e^{-\alpha k}}{\sum_{k\ge 0} g(k)e^{-\alpha k}}
\tag{5.33}
$$

and the degree distribution $p(k)$ of the ensemble will for $N\to\infty$ approach

---

[1] This is a special feature, of course, of the present ensemble (5.25). As soon as the global hard constraint can no longer be written in terms of the degrees, we will no longer be able to disentangle graph statistics from degree statistics, and understanding quantitatively the competition between global and local constraints becomes much harder.

$$p(k) = \frac{g(k)e^{-\alpha k}}{\sum_{k' \geq 0} g(k')e^{-\alpha k'}} \tag{5.34}$$

We observe that if the average of $g(k)$ coincides with the value $\langle k \rangle$ imposed by the hard global constraint, i.e. if $\sum_k g(k)k = \langle k \rangle$, then $\alpha = 0$ and the so-called 'dressed' distribution $p(k)$ will coincide with the 'bare' distribution $g(k)$. Hence, in the special case of ensemble (5.27), we are indeed allowed to choose $g(k) = p(k)$, if we seek to generate graphs that for large N will on average have the degree distribution $p(k)$, provided the latter distribution is indeed compatible with the global constraint, i.e. it obeys $\sum_k p(k)k = \langle k \rangle = 2L/N$.

We saw in the analysis above that an imposed global constraint has the ability to modify the observed degree statistics $p(k)$, such that it becomes different from the distribution $g(k)$ from which the degrees were generated in the mixed ensemble. Because in the present ensemble the global constraint can be expressed fully in terms of the quantities that are constrained at individual sites (the degrees), we could see explicitly the impact of the global constraint in the language of the local ones, and consequently see how to avoid such modification of the degree statistics. For ensembles where a global constraint takes a more general form, this will no longer be possible, and we should therefore anticipate that in mixture ensembles of the form (5.2) the observed degree distribution $p(k)$ will generally differ from $g(k)$.

Exercise 5.4 Choose in ensemble (5.27) the individual degrees from the 'bare' Poisson distribution $g(k) = e^{-q}q^k/k!$, with arbitrary $q > 0$. Calculate the 'dressed' average degree distribution $p(k)$ that will be exhibited by graphs generated in this ensemble for $N \to \infty$, in terms of the average degree q of $g(k)$ and the value $\langle k \rangle$ imposed by the global constraint.

Condensation transitions. As with the soft-constrained ensembles, we can also in hard-constrained ensembles find phenomena associated with phase transitions, i.e. abrupt changes in ensemble averages as external parameters are tuned. In particular, systems that can be described by equations of the type (5.28) are known to undergo condensation transitions, when the 'bare' distribution $g(k)$ decays more slowly than an exponential, but faster than k^{-2} (see, e.g. [68]). In particular, for heavy-tailed distributions $g(k) \sim k^{-\gamma}$ and $\gamma > 2$, a solution to (5.33) can only be found for $\alpha \in [0, \infty)$. Since increasing α in (5.33) will always *decrease* the value of the right-hand side, there is a maximum value of $\langle k \rangle$ beyond which no solutions will exist. This maximum value is obtained for $\alpha = 0$, which defines a critical average degree value

$$\mu_c = \sum_k kg(k) \tag{5.35}$$

A solution $\alpha > 0$ of (5.33) exists only when $\langle k \rangle < \mu_c$. At $\langle k \rangle = \mu_c$, a transition takes place. For $\langle k \rangle > \mu_c$, one can observe that the excess number of links $N(\langle k \rangle - \mu_c)$ will all attach to a single node, forming a *condensate*. This new behaviour cannot be described by the current equations, since they have been derived by assuming that all degrees remain finite.

Such phase transition phenomena will obviously have important implications for the formulation of graph generation algorithms. As always, they appear strictly speaking only for $N \to \infty$ (since for finite N all sums over degrees will run over $k = 0 \dots N{-}1$, we can never have truly diverging sums), but in practice, graphs of a few hundred nodes already behave as infinitely large ones. While in the hard-constrained ensembles we can indeed (by force) always make sure that we produce graphs that exhibit the topological features we require (unlike in the soft-constrained ones), we see that if our constraints are difficult to reconcile with each other (i.e. if we demand too much), the graphs being generated may well have unexpected and peculiar features.

5.4 Ensembles with prescribed numbers of links and two-stars

Definitions. In this section, we investigate the hard-constrained twin ensemble of the model (4.36). Here our graphs are taken from the set G of simple nondirected N-node graphs, and we prescribe the numbers of links and two-stars, or, equivalently, the average and the width of the degree distribution. With hard constraints this implies

$$\forall \mathbf{A} \in G: \quad p(\mathbf{A}) = \frac{1}{Z(L,S)}\, \delta_{\sum_i k_i(\mathbf{A}),2L}\, \delta_{\sum_i k_i^2(\mathbf{A}),2(S+L)} \tag{5.36}$$

$$Z(L,S) = \sum_{\mathbf{A} \in G} \delta_{\sum_i k_i(\mathbf{A}),2L}\, \delta_{\sum_i k_i^2(\mathbf{A}),2(S+L)} \tag{5.37}$$

In contrast to the soft-constrained ensemble (4.36), where L and S represented the *average* numbers of links and two-stars of graphs $\mathbf{A}$ generated by the ensemble, here *each individual graph* of the ensemble will have exactly L links and exactly S two-stars. In the sparse scaling regime, we will as before switch from parametrization in terms of the pair (L, S), both of which scale as $\mathcal{O}(N)$, to the language of the two quantities $\langle k \rangle = 2L/N$ and $\sigma_k^2 = \langle k^2 \rangle - \langle k \rangle^2 = 2(S+L)/N - 4L^2/N^2$, which in this regime both remain of order one as $N \to \infty$. Compared to the previous ensemble it is now less clear whether and how the two hard ensemble constraints can be reconciled, and how this depends on the values (L, S) that we choose to enforce. To shed light on this question, we again turn to the number of admissible graphs, measured by the normalization constant (5.37).

Number of graphs in the ensemble. Since the most relevant scaling regime is usually the sparse (i.e. finite connectivity) one, we will work with the pair $\langle k \rangle$ and $\langle k^2 \rangle$ instead of (L, S), and write the constant (5.37) consequently as $Z(\langle k \rangle, \langle k^2 \rangle)$. In the sparse regime, application of (5.15) to the model (5.36) tells us that

$$\frac{1}{N} \log Z(\langle k \rangle, \langle k^2 \rangle) = \frac{1}{2}\langle k \rangle \left[1 + \log\left(\frac{N}{\langle k \rangle}\right)\right] + \mathcal{O}\left(\frac{1}{N}\right)$$
$$+ \frac{1}{N} \log \int_{-\pi}^{\pi} \mathrm{d}\phi\mathrm{d}\phi'\; \mathrm{e}^{\mathrm{i}N(\phi\langle k \rangle + \phi'\langle k^2 \rangle)} \left\langle \mathrm{e}^{-\mathrm{i}\sum_i \left[\phi k_i(\mathbf{A}) + \phi' k_i^2(\mathbf{A})\right]} \right\rangle_{\mathrm{ER}} \tag{5.38}$$

In Appendix E, we calculate the leading orders in N of this expression, and find the result (E.11):

$$\frac{1}{N} \log Z(\langle k \rangle, \langle k^2 \rangle) = \frac{1}{2} \langle k \rangle \log \left(N \langle k \rangle \right) - \frac{1}{2} \langle k \rangle + \epsilon_N$$

$$+ \, \text{extr}_{\alpha, \beta} \Big\{ \alpha \langle k \rangle + \beta \langle k^2 \rangle + \log \sum_{k \geq 0} \frac{1}{k!} e^{-\alpha k - \beta k^2} \Big\} \quad (5.39)$$

The remaining extremization over (α, β) is easy, and shows that the equations from which to solve these parameters in terms of $\langle k \rangle$ and $\langle k^2 \rangle$ take the following transparent form: $\langle k \rangle = \sum_k k p(k)$ and $\langle k^2 \rangle = \sum_k k p(k)$, in which

$$p(k) = \frac{e^{-\alpha k - \beta k^2} / k!}{\sum_{\ell \geq 0} e^{-\alpha \ell - \beta \ell^2} / \ell!} \quad (5.40)$$

If $\langle k^2 \rangle = \langle k \rangle^2 + \langle k \rangle$, we observe that our equations are solved by $\alpha = -\log \langle k \rangle$ and $\beta = 0$, with the Poissonian degree distribution $p(k) = e^{-\langle k \rangle} \langle k \rangle^k / k!$, and in that case we find (5.39) reducing to expression (5.20). This is as expected, since for $\beta = 0$ we remove the constraint on the degree variance, and our present ensemble reduces to the one where only the number of links is constrained.

Exercise 5.5 Show that, in the limit where we impose values of the first and second moment of the degree distribution such that $\langle k \rangle^2 = \langle k \rangle$, for some integer $\langle k \rangle = k^\star$, the parameters (α, β) will tend to values for which the degree distribution (5.40) becomes that of a regular graph, i.e. $p(k) = \delta_{kk^\star}$.

Condensation transitions. It will be clear that for fixed $\langle k \rangle$ the second moment of the degree distribution (5.40) will decrease monotonically with β. Since for $\beta = 0$ we have $\langle k^2 \rangle = \langle k \rangle^2 + \langle k \rangle$, we must deduce that larger imposed values of the degree variance, i.e. $\langle k^2 \rangle - \langle k \rangle^2 > \langle k \rangle$, cannot be achieved for any $\beta \in [0, \infty)$. However, Stirling's formula [124] tells us that we cannot allow for parameter values $\beta < 0$ either, since for large k we will have

$$\frac{1}{k!} e^{-\alpha k - \beta k^2} \sim e^{-(\alpha - 1) k - \beta k^2 - (k + \frac{1}{2}) \log k} \quad (5.41)$$

Hence, the denominator of (5.40) diverges as soon as β is real and negative. We conclude that precisely when $\langle k^2 \rangle = \langle k \rangle^2 + \langle k \rangle$ there must be a phase transition in the system. As in the previous model, the new graph realizations that take over in the regime $\langle k^2 \rangle > \langle k \rangle^2 + \langle k \rangle$, where the current analysis breaks down, involve condensation of a diverging number of links onto a single node. This situation could not have been described by the present equations, since these rely on the assumption that all degrees remain finite as $N \to \infty$.

Simple arguments suffice to get some intuition for the type of graphs that will be produced by our ensemble in the regime of large degree variance, for large N. On the basis of the above, we expect to see $N - 1$ nodes with Poissonian degree statistics, with average μ, plus a single 'special' node with degree ξN^γ, where $\xi, \gamma \in (0, 1)$. The corresponding degree distribution would be

$$p(k) = \left(1 - \frac{1}{N}\right) \frac{e^{-\mu} \mu^k}{k!} + \frac{1}{N} \delta_{k, \xi N^\gamma} \tag{5.42}$$

One can now readily calculate the first two moments:

$$\langle k \rangle = \left(1 - \frac{1}{N}\right) \mu + \xi N^{\gamma - 1}, \qquad \langle k^2 \rangle = \left(1 - \frac{1}{N}\right) \mu(\mu + 1) + \xi^2 N^{2\gamma - 1} \tag{5.43}$$

If we impose finite values for $\langle k \rangle$ and $\langle k \rangle^2$, with $\langle k^2 \rangle - \langle k \rangle^2 > \langle k \rangle$, we must find for large N that $\gamma \to \frac{1}{2}$, $\mu \to \langle k \rangle$, and $\xi \to \sqrt{\langle k^2 \rangle - \langle k \rangle^2 - \langle k \rangle}$. Hence, we expect to find that in the high-variance regime, our ensemble will generate graphs that have a bulk of $N - 1$ nodes with Poissonian distributed degrees of average $\langle k \rangle$, plus condensation of a diverging number of links onto a single node, whose degree will be approximately

$$k_c \approx \sqrt{N[\langle k^2 \rangle - \langle k \rangle^2 - \langle k \rangle]} \tag{5.44}$$

Again, the conclusion is that while graph ensembles with hard constraints do allow us to enforce 'extreme' or 'graphically unnatural' values for the chosen topological observables, we may not always find that the graphs that are being generated under such conditions are what we are looking for. In the present ensemble, for instance, we found that, while well-behaved graphs with large degree variances $\langle k^2 \rangle - \langle k \rangle^2 > \langle k \rangle$ certainly do exist, we will only observe those specific realizations $\mathbf{A} \in G$ in which the required large variance is achieved via condensation, i.e. by having a bimodal degree distribution in which there is one special node with a degree that scales as $\mathcal{O}(\sqrt{N})$. It is important to realize that this is a fundamental property of the ensemble itself, independent of which algorithm is used to sample graphs from it.

Shortcut: *Graph ensembles with hard constraints can remedy some of the problems of ERGMs, in that they allow us to achieve any desired graphical combination of values for our observables. However, we again encounter phase transitions. In the presently studied ensembles, with controlled statistical characteristics of degrees, these tend to be condensation transitions, where for demanding combinations of the imposed values of our observables, the ensembles start focusing on graphs in which one node acquires a diverging number of links. While such extremely inhomogeneous graphs satisfy our quantitative constraints, they are often not the types we seek to generate.*

5.5   Ensembles with constrained degrees and short loops

Controlling degrees, triangles and squares. The previous section showed that in the case of hard constraints one may well find graph realizations in which the imposed values of observables are achieved in unexpected or even undesirable ways. More extreme examples of this are found in ensembles of simple nondirected graphs of the following form, which share with the random triangle model [169,94] the soft-constrained number of triangles (complemented further with a soft-constrained number of squares), but where now also all N degrees $k_i(\mathbf{A}) = \sum_j A_{ij}$ are prescribed via hard constraints:

$$p(\mathbf{A}) = \frac{\delta_{\mathbf{k},\mathbf{k}(\mathbf{A})}}{Z(\alpha_3, \alpha_4; \mathbf{k})} \, e^{\alpha_3 \operatorname{Tr}(\mathbf{A}^3) + \alpha_4 \operatorname{Tr}(\mathbf{A}^4)} \tag{5.45}$$

with $\operatorname{Tr}(\mathbf{A}^\ell) = \sum_{i=1}^{N} (\mathbf{A}^\ell)_{ii}$, and where $Z(\alpha_3, \alpha_4; \mathbf{k})$ is a normalization constant. The mathematical analysis of these ensembles is highly non-trivial, and beyond the scope of this book[2]. Here we will only illustrate the phase transition phenomenology of random graphs generated from this ensemble, which has important practical implications and is entirely independent of which numerical algorithm one uses. We will see in Chapter 6 that generating random graphs according to the recipe (5.45) numerically is relatively straightforward.

Phenomenology monitored via eigenvalue spectra. It turns out that in ensembles such as (5.45) the eigenvalue spectra $\varrho(\lambda|\mathbf{A})$ of the adjacency matrices, as defined in (2.17), are effective markers of phase transitions. For $\alpha_3 = \alpha_4 = 0$, and upon choosing $k_i(\mathbf{A}) = q$ for all i, the ensemble (5.45) simplifies to a maximum entropy ensemble of q-regular random graphs. The ensemble-averaged spectrum $\varrho(\lambda) = \sum_{\mathbf{A}} p(\mathbf{A})\varrho(\lambda|\mathbf{A})$ can now be calculated analytically in the limit $N \to \infty$, leading to McKay's [119] formula

$$|\lambda| \geq 2\sqrt{q-1}: \quad \varrho(\lambda) = 0 \tag{5.46}$$

$$|\lambda| < 2\sqrt{q-1}: \quad \varrho(\lambda) = \frac{q}{2\pi} \frac{\sqrt{4(q-1) - \lambda^2}}{q^2 - \lambda^2} \tag{5.47}$$

Individual graphs will typically exhibit spectra $\varrho(\lambda|\mathbf{A})$ that differ from (5.47) only via $\mathcal{O}(N^{-1/2})$ finite size corrections. In contrast, as one moves away from $(\alpha_3, \alpha_4) = (0,0)$ and boosts the number of triangles or squares, one observes in numerical simulations discontinuous transitions towards graphs that have very different spectra.

This is illustrated in Figure 5.1. For instance, for regular graphs with $q = 3$, $\alpha_4 = 0$ and $\alpha_3 > 0$ (see middle row in the figure) one observes in numerical simulations that for large enough α_3, modulo finite size effects, the spectrum is of the form

$$\varrho(\lambda) \approx (1 - \gamma)\tilde{\varrho}(\lambda) + \gamma\left\{\frac{3}{4}\delta(\lambda+1) + \frac{1}{4}\delta(\lambda-3)\right\} \tag{5.48}$$

in which $\gamma \in (0,1)$, and $\tilde{\varrho}(\lambda)$ is McKay's formula (5.47) with $q = 3$. One can confirm that expression (5.48) is compatible with $\langle\lambda\rangle = 0$ and $\langle\lambda^2\rangle = 3$ for any γ, as required[3]. A simple calculation confirms that the second term in (5.48) is the eigenvalue spectrum of the adjacency matrix of a fully connected simple graphlet of size four. Hence, (5.48) describes graphical phase separation. By increasing α_3, we tell the system to increase the number of triangles, and it is found to do so by disconnecting a fraction γ of the

---

[2] The reason is the *cyclic* occurrence of the link variables $\{A_{ij}\}$ in $\operatorname{Tr}(\mathbf{A}^3) = \sum_{ijk} A_{ij} A_{jk} A_{ki}$ and in $\operatorname{Tr}(\mathbf{A}^4) = \sum_{ijk\ell} A_{ij} A_{jk} A_{k\ell} A_{\ell i}$, which prevents us from summing over all graphs, which is required for the evaluation of ensemble averages or ensemble entropies.

[3] These two requirements follow from the following general identities for simple nondirected graphs: $\int d\lambda\, \varrho(\lambda|\mathbf{A})\lambda = \frac{1}{N}\sum_i A_{ii} = 0$ and $\int d\lambda\, \varrho(\lambda|\mathbf{A})\lambda^2 = \frac{1}{N}\sum_i(\mathbf{A}^2)_{ii} = \frac{1}{N}\sum_{ij} A_{ij} = \frac{1}{N}\sum_i k_i(\mathbf{A})$.

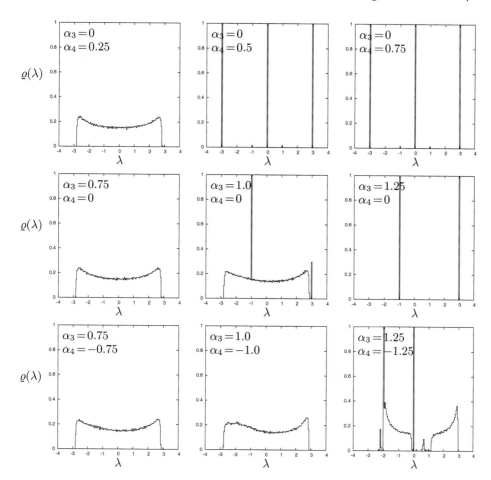

Fig. 5.1 Eigenvalue spectra of the adjacency matrices **A** of simple nondirected random graphs, generated numerically according to the ensemble probabilities (5.45), with $N = 5000$ and $k_i(\mathbf{A}) = 3$ for all $i = 1 \ldots N$. Top row: $\alpha_3 = 0$ and $\alpha_4 > 0$ (where we encourage graphs with many squares). Middle row: $\alpha_4 = 0$ and $\alpha_3 > 0$ (where we encourage graphs with many triangles). Bottom row: $\alpha_4 = -\alpha_3$ and $\alpha_3 > 0$ (where we boost the number of triangles and suppress the number of squares). For small values of (α_3, α_4), one obtains the smooth spectrum of tree-like regular random graphs as predicted by [119], as given by formula (5.47) (with $q = 3$). Upon boosting the number of triangles or squares by increasingly positive values of α_3 or α_4, one observes discontinuous transitions towards non-smooth spectra, reflecting the breaking up of the generated graphs into a large number of disconnected components.

nodes from the initially tree-like bulk, to form $\gamma N/4$ disconnected four-node cliques. Similarly, for regular graphs with $q = 3$, $\alpha_3 = 0$ and $\alpha_4 > 0$ (see top row in the figure), one observes for large enough α_4 that

$$\varrho(\lambda) \approx (1-\gamma)\tilde{\varrho}(\lambda) + \gamma\left\{\frac{2}{3}\delta(\lambda) + \frac{1}{6}\delta(\lambda-3) + \frac{1}{6}\delta(\lambda+3)\right\} \qquad (5.49)$$

$$\alpha_3 = 1.00 \qquad \alpha_4 = 0.0 \qquad\qquad\qquad \alpha_3 = 0.0 \qquad \alpha_4 = 0.5$$

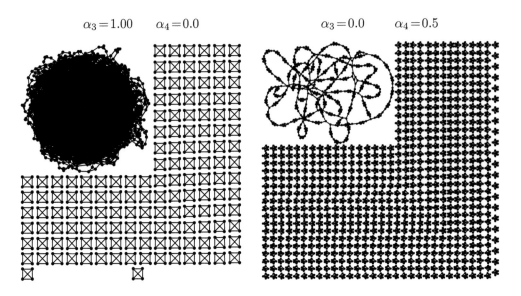

Fig. 5.2 Random graphs typically generated from ensemble (5.45), with $N = 5000$ nodes. Left graph: $(\alpha_3, \alpha_4) = (1, 0)$, corresponding to the middle eigenvalue spectrum in Figure 5.1. Right graph: $(\alpha_3, \alpha_4) = (0, 0.5)$, corresponding to middle spectrum in the top row in Figure 5.1. These pictures confirm the interpretation in the main text of formulae (5.48, 5.49) for the eigenvalue spectra of the adjacency matrices, in terms of phase separation. The graph on the left has separated from the connected bulk an extensive number of isolated five-node cliques. The graph on the left has separated from the connected bulk an extensive number of isolated six-node graphlets without triangles. For larger values of α_3 and α_4, respectively, the bulk will vanish completely, and only the disconnected microscopic graphlets will remain.

in which again $\tilde{\varrho}(\lambda)$ is McKay's formula (5.47) for $q = 3$. Again $\langle \lambda \rangle = 0$ and $\langle \lambda^2 \rangle = 3$ for any γ, as required. Here the second term of (5.49) is confirmed to be the spectrum of the six-node graphlet with adjacency matrix $A_{ij} = \delta_{i,j+1} + \delta_{i,j-1} + \delta_{i,j+3} + \delta_{i,j-3}$ $(i, j$ mod 6), which is a six-node clique from which all triangles are removed. Apparently, the system now increases the number of squares, as requested by $\alpha_4 > 0$, by disconnecting a fraction γ of the nodes from the initially tree-like bulk, to form $\gamma N/6$ suitable disconnected six-node graphlets. This interpretation of the observed spectra is indeed confirmed upon producing images of the graphs generated according to (5.45), as shown in Figure 5.2.

Upon inspecting more complex non-trivial combinations of $\alpha_3, \alpha_4 \neq 0$, especially with α_3 and α_4 having opposite signs (so that some closed paths are boosted and other are suppressed), one will observe transitions to more complex spectra.

As with the ensemble in the previous section, we see again that in the case of multiple topological constraints, our ensemble may resolve the different demands in unexpected ways. In the ensemble (5.45) with fully regular degrees $k_i(\mathbf{A}) = q$ for all i, we might have expected that for large N the spectra $\varrho(\lambda|\mathbf{A})$ for $\alpha_3 > 0$ and/or $\alpha_4 > 0$

would be continuous deformations of the continuous $\alpha_3 = \alpha_4 = 0$ spectrum (5.47), but instead one finds transitions towards the production of graphs that achieve the required increase in triangles or squares by sacrificing overall connectivity.

5.6 Solutions of exercises

Exercise 5.1

We can write

$$p(\mathbf{A}|\mathbf{k}, L) = \frac{1}{Z(\mathbf{k}, L)} \prod_{i<j} \left[\frac{k_i k_j}{N\langle k \rangle} \delta_{A_{ij},1} + \left(1 - \frac{k_i k_j}{N\langle k \rangle}\right) \delta_{A_{ij},0} \right] \delta_{L,\sum_{i<j} A_{ij}}$$

$$= \frac{1}{Z(\mathbf{k}, L)} \left[\prod_{i<j} \left(\frac{k_i k_j}{N\langle k \rangle}\right)^{A_{ij}} \left(1 - \frac{k_i k_j}{N\langle k \rangle}\right)^{1-A_{ij}} \right] \delta_{L,\sum_{i<j} A_{ij}}$$

$$= \frac{1}{Z(\mathbf{k}, L)} \left[\prod_{i<j} \left(1 - \frac{k_i k_j}{N\langle k \rangle}\right) \right] \left[\prod_{i<j} \left(\frac{1}{N\langle k \rangle/k_i k_j - 1}\right)^{A_{ij}} \right] \delta_{L,\sum_{i<j} A_{ij}}$$

$$= \frac{1}{Z'(\mathbf{k}, L)} e^{\sum_{i<j} A_{ij} \log\left(\frac{1}{N\langle k \rangle/k_i k_j - 1}\right)} \delta_{L,\sum_{i<j} A_{ij}} \tag{5.50}$$

For large finitely connected graphs, one has $k_i \ll \sqrt{N\langle k \rangle}$, and the above can be approximated by the following formula, which has the desired form with $\lambda_i = \log(k_i/\sqrt{N\langle k \rangle})$:

$$p(\mathbf{A}|\mathbf{k}, L) = \frac{1}{Z'(\mathbf{k}, L)} e^{\sum_{i<j} A_{ij}\left(\log \frac{k_i}{\sqrt{N\langle k \rangle}} + \log \frac{k_j}{\sqrt{N\langle k \rangle}}\right)} \delta_{L,\sum_{i<j} A_{ij}}$$

$$= \frac{1}{Z'(\mathbf{k}, L)} e^{\sum_i k_i(\mathbf{A}) \log \frac{k_i}{\sqrt{N\langle k \rangle}}} \delta_{L,\sum_{i<j} A_{ij}} \tag{5.51}$$

□

Exercise 5.2

The requested probabilities are

$$p(\mathbf{A}|g, L) = \sum_{\mathbf{k}} [\prod_i g(k_i)] \frac{1}{Z(\mathbf{k}, L)} \left[\prod_i \delta_{k_i, \sum_j A_{ij}} \right] \delta_{L,\sum_{i<j} A_{ij}}$$

$$= [\prod_i g(k_i(\mathbf{A}))] \frac{1}{Z(\mathbf{k}(\mathbf{A}), L)} \delta_{L,\frac{1}{2}\sum_i k_i} \tag{5.52}$$

□

Exercise 5.3

For directed N-node graphs, $Z(L)$ equals the number of possible ways in which one can draw $L = \sum_{ij} A_{ij} = N\langle k \rangle$ links from a set of N^2 candidates, giving

$$Z(L) = \binom{N^2}{L} \tag{5.53}$$

In the finite connectivity regime, we put $L = N\langle k \rangle$ and keep $\langle k \rangle$ fixed, while letting $N \to \infty$. From Stirling's formula, we had derived in the main text expansion (5.19), which applies when $1 \ll m \ll n$. Applying this expansion to $n = N^2$ and $m = N\langle k \rangle$, which in the sparse regime indeed obey its conditions of validity, then gives

$$N^{-1} \log Z(N\langle k \rangle) = \langle k \rangle \log(N/\langle k \rangle) + \langle k \rangle + \mathcal{O}\left(\frac{\log N}{N}\right) \tag{5.54}$$

This is exactly twice the corresponding value found for simple nondirected graphs. $\square$

Exercise 5.4

Inserting $g(k) = e^{-q}q^k/k!$ into the numerator and denominator of expression (5.34), followed by summing over k in the denominator, gives

$$p(k) = \frac{(qe^{-\alpha})^k/k!}{\sum_{k' \geq 0}(qe^{-\alpha})^{k'}/k'!} = e^{-qe^{-\alpha}}(qe^{-\alpha})^k/k! \tag{5.55}$$

The value of α follows from the demand $\langle k \rangle = \sum_k kp(k)$. However, since the above 'dressed' degree distribution is of a Poissonian form, with average $qe^{-\alpha}$, we can immediately write the equation for α as $\langle k \rangle = qe^{-\alpha}$. Hence $p(k)$ becomes

$$p(k) = e^{-\langle k \rangle}\langle k \rangle^k/k! \tag{5.56}$$

Thus, the 'dressed' degree distribution converts any 'bare' Poissonian $g(k)$, with arbitrary average degree, into a new Poissonian degree distribution with the correct average degree $\langle k \rangle$. $\square$

Exercise 5.5

Upon using the standard identity $\langle k^2 \rangle - \langle k \rangle^2 = \langle (k - \langle k \rangle)^2 \rangle$, we can formulate our problem as solving (α, β) in the limit $\epsilon \downarrow 0$ from the two coupled equations

$$k^* = \frac{\sum_{k \geq 0} ke^{-\alpha k - \beta k^2}/k!}{\sum_{k \geq 0} e^{-\alpha k - \beta k^2}/k!}, \qquad \epsilon = \frac{\sum_{k \geq 0}(k - k^*)^2 e^{-\alpha k - \beta k^2}/k!}{\sum_{k \geq 0} e^{-\alpha k - \beta k^2}/k!} \tag{5.57}$$

We now complete the squares in the exponentials, via $\alpha k + \beta k^2 = \beta(k + \frac{1}{2}\alpha/\beta)^2 - \frac{1}{4}\alpha^2/\beta$, and choose $\alpha = -2k^*\beta$. Our equations thereby become

$$k^* = \frac{\sum_{k \geq 0} ke^{-\beta(k-k^*)^2}/k!}{\sum_{k \geq 0} e^{-\beta(k-k^*)^2}/k!}, \qquad \epsilon = \frac{\sum_{k \geq 0}(k - k^*)^2 e^{-\beta(k-k^*)^2}/k!}{\sum_{k \geq 0} e^{-\beta(k-k^*)^2}/k!} \tag{5.58}$$

Both equations will be satisfied for $\epsilon \downarrow 0$, and hence we will indeed find $p(k) = \delta_{kk^*}$ if $\lim_{\epsilon \downarrow 0} \beta^{-1} = 0$. We find that reducing the imposed degree variance to zero is achievable in the hard-constrained ensemble, but it requires diverging parameter values $\beta \to \infty$ and $\alpha \to -\infty$, in the specific ratio $-\frac{1}{2}\alpha/\beta = k^*$. $\square$

Part III

Generating graphs from graph ensembles

6

Markov Chain Monte Carlo sampling of graphs

Introduction. In the previous chapters, we have introduced and discussed random graph ensembles, which provide by construction a complete and explicit description of the statistical features of our random graphs[1]: we define the set of allowed graphs, together with a probability measure on this set. We have also discussed a systematic way of tailoring such ensembles in order to generate graphs that will have certain topological features that we may choose to build in, but that are otherwise strictly unbiased in an information-theoretic sense. In this chapter, we discuss numerical protocols for actually generating sample graphs from such ensembles, in the form of graph randomization algorithms of the Markov chain form (3.20) that will be simulated until we can trust the chain to have reached equilibrium. We first consider the simplest Markov chains that sample from ensembles with soft constraints, i.e. from exponential random graph models (ERGMs) (4.1). We then consider the more tricky situation of sampling from ensembles with hard constraints.

6.1 The Markov Chain Monte Carlo sampling method

Motivation and desiderata of Markov chains. Given a set G of N-node graphs, and given a probability distribution $p(\mathbf{A})$ on this set, we seek to construct a Markov chain, i.e. a process of the form (3.20), that will evolve in time such that $\lim_{t \to \infty} p_t(\mathbf{A}) = p(\mathbf{A})$. Moreover, we would ideally want a chain that will do so for *any* initialization of the form $p_0(\mathbf{A}) = \delta_{\mathbf{A},\mathbf{A}^*}$; i.e. we want an *ergodic* Markov chain[2]. There is a clear rationale for focusing on Markov chains: owing to their discrete definition of time, and the dependence of a new state on only the current state, such processes have the natural form of numerical algorithms, where we can use random number generators[3] to build in stochasticity. The core of our problem is then to construct appropriate transition probabilities $W[\mathbf{A}' \to \mathbf{A}] \in [0, 1]$, which define the likelihood of moving to state $\mathbf{A}$ at the next step, if the present state is $\mathbf{A}'$. They must be normalized according to $\sum_{\mathbf{A} \in G} W[\mathbf{A}' \to \mathbf{A}] = 1$ for all $\mathbf{A}' \in G$, because of probability conservation.

---

[1] Unlike growth algorithms, where one sets a process in motion and still needs to predict the statistical characteristics of the resulting graphs that will emerge after this process has terminated.

[2] The shorthand $\delta_{\mathbf{A},\mathbf{A}'}$ is defined in terms of ordinary Kronecker deltas via $\delta_{\mathbf{A},\mathbf{A}'} = \prod_{ij} \delta_{A_{ij},A'_{ij}}$.

[3] Numerical random number generators are never truly random, so we always need to be aware of possible periodicity problems if we make large numbers of random number calls.

Generating Random Networks and Graphs. First Edition. A.C.C. Coolen, A. Annibale, and E.S. Roberts. © A.C.C. Coolen, A. Annibale, and E.S. Roberts 2017. Published in 2017 by Oxford University Press. DOI: 10.1093/acprof:oso/9780198709893.001.0001

A naive way to construct such a stochastic process would be to draw at each time step a configuration $\mathbf{A} \in G$ fully randomly with uniform probabilities, and accept this as the new state with probability $p(\mathbf{A})$. This would in principle indeed be a Markov chain, defined by the trivial transition probabilities $\mathcal{W}[\mathbf{A}' \to \mathbf{A}] = p(\mathbf{A})$; successive graphs would all be generated with the correct probabilities, and the chain would even converge in a single iteration step. However, because of the enormous size of the configuration space G^4, the probabilities associated with each configuration are typically very small, so that nearly all our computation time would be spent on generating configurations which are subsequently rejected, which makes this primitive approach completely impractical.

We need a stochastic process that moves more purposefully from one state to another, such that after equilibration it visits all states $\mathbf{A}$ with frequencies proportional to their required statistical weight $p(\mathbf{A})$. Fortunately, much is known about Markov chains. Since (3.20) is linear, it can be interpreted as the multiplication of a $|G|$-dimensional vector with entries $p_t(\mathbf{A})$ by a (generally non-symmetric) $|G| \times |G|$ matrix (the *stochastic matrix*), resulting in a new $|G|$-dimensional vector whose entries are the state probabilities at the next step. Hence, upon defining matrix multiplication in the usual way, i.e. $\mathcal{W}^2[\mathbf{A}' \to \mathbf{A}] = \sum_{\mathbf{A}'' \in G} \mathcal{W}[\mathbf{A}' \to \mathbf{A}''] \mathcal{W}[\mathbf{A}'' \to \mathbf{A}']$ (and similarly for higher powers), we may write directly for any $t \in \mathbb{N}$:

$$\forall \mathbf{A} \in G: \quad p_t(\mathbf{A}) = \sum_{\mathbf{A}' \in G} \mathcal{W}^t[\mathbf{A}' \to \mathbf{A}] \, p_0(\mathbf{A}') \tag{6.1}$$

$\mathcal{W}^t[\mathbf{A}' \to \mathbf{A}] \in [0,1]$ is the likelihood of finding state $\mathbf{A}$ after t iterations of the chain, if the present state is $\mathbf{A}'$. A *regular* Markov chain has the following further property: there exists an $n \in \mathbb{N}$ such that $\mathcal{W}^t[\mathbf{A}' \to \mathbf{A}] > 0$ for all $t \geq n$ and for all $\mathbf{A}, \mathbf{A}' \in G$. All regular Markov chains are ergodic, but the converse is not true. In particular, an ergodic chain has a finite probability of visiting any configuration, for times sufficiently large, starting from any configuration. If there are periodicities in the dynamics, whereby certain configurations can only be visited at certain times, an ergodic chain will not be regular, as one cannot find any time large enough, for which *any* pair of configurations is dynamically connected. However, for each pair of configurations there may still be a time that is large enough such that they are dynamically accessible to each other. A Markov chain is regular if it is ergodic and there are no periodicities in the dynamics[5].

One of the core results of Markov chain theory, based on the Perron–Frobenius theorem, is that each regular Markov chain will evolve to a unique stationary measure $p(\mathbf{A})$, for any initialization, i.e. the limit $p(\mathbf{A}) = \lim_{t \to \infty} \mathcal{W}^t[\mathbf{A}^\star \to \mathbf{A}]$ exists and is independent of $p_0(\mathbf{A})$. This stationary state is the unique solution of the following equation, which simply rephrases the stationarity condition:

$$\forall \mathbf{A} \in G: \quad p(\mathbf{A}) = \sum_{\mathbf{A}' \in G} \mathcal{W}[\mathbf{A}' \to \mathbf{A}] \, p(\mathbf{A}') \tag{6.2}$$

---

[4] As we have seen earlier, even in the sparse regime where $L = \mathcal{O}(N)$, the number of simple nondirected N-node graphs with L links grows super-exponentially as $\mathcal{O}(\exp(N \log N))$ for large N.

[5] A more rigorous definition of an aperiodic Markov chain is that the greatest common divisor of the return times to any configuration is one. If it is greater than one, the chain is periodic.

Since the stationary state of a regular Markov chain is unique[6], our problem can be reduced to this: construct a definition for $\mathcal{W}[\mathbf{A}' \to \mathbf{A}]$ such that (i) the corresponding Markov chain is regular, and (ii) our ensemble graph probabilities $p(\mathbf{A})$ solve equation (6.2). One can easily confirm that there will be infinitely many possible choices for the transition probabilities that meet our two requirements. For instance, the trivial (and impractical) recipe $\mathcal{W}[\mathbf{A}' \to \mathbf{A}] = \epsilon \delta_{\mathbf{A},\mathbf{A}'} + (1-\epsilon)p(\mathbf{A})$ corresponds for any $\epsilon \in [0,1)$ to a regular Markov chain that satisfies (6.2). Hence, the problem is well defined in the sense that we can be sure that solutions of (6.2) exist.

Move proposal and move acceptance. The most commonly used recipe in Markov Chain Monte Carlo sampling (MCMC) is of the following form. We define a set Φ of allowed auto-invertible moves F, each of which maps a state $\mathbf{A} \in G$ to a new state $F\mathbf{A} \in G$, such that $F^2\mathbf{A} = \mathbf{A}$ for all $\mathbf{A}$. We exclude the identity operation, so $F\mathbf{A} \neq \mathbf{A}$ for all $F \in \Phi$ and all $\mathbf{A} \in G$. The Markov chain dynamics is now at each time step decomposed into two stages. If we find ourselves in state $\mathbf{A}$, we first draw a move $F \in \Phi$ at random, with uniform probabilities $1/|\Phi|$, where $|\Phi|$ is the number of moves in the set Φ. This leads to the *move proposal* $\mathbf{A} \to F\mathbf{A}$. We accept this proposal with *move acceptance* probability $\mathcal{A}[\mathbf{A} \to F\mathbf{A}] \in [0,1]$. If the proposed move is rejected, which happens with probability $1 - \mathcal{A}[\mathbf{A} \to F\mathbf{A}]$, we stay at $\mathbf{A}$. This protocol corresponds to Markov chain transition probabilities of the form

$$\mathcal{W}[\mathbf{A}' \to \mathbf{A}] = \frac{1}{|\Phi|} \sum_{F \in \Phi} \left\{ \mathcal{A}[\mathbf{A}' \to F\mathbf{A}']\delta_{\mathbf{A},F\mathbf{A}'} + \left(1 - \mathcal{A}[\mathbf{A}' \to F\mathbf{A}']\right)\delta_{\mathbf{A},\mathbf{A}'} \right\} \quad (6.3)$$

It is clear that (6.3) obeys $\mathcal{W}[\mathbf{A}' \to \mathbf{A}] \in [0,1]$ for all $\mathbf{A}, \mathbf{A}' \in G$. It is also correctly normalized, since for any $\mathbf{A}' \in G$ we find:

$$\sum_{\mathbf{A} \in G} \mathcal{W}[\mathbf{A}' \to \mathbf{A}] = \frac{1}{|\Phi|} \sum_{F \in \Phi} \sum_{\mathbf{A} \in G} \left\{ \mathcal{A}[\mathbf{A}' \to F\mathbf{A}']\delta_{\mathbf{A},F\mathbf{A}'} + \left(1 - \mathcal{A}[\mathbf{A}' \to F\mathbf{A}']\right)\delta_{\mathbf{A},\mathbf{A}'} \right\}$$

$$= \frac{1}{|\Phi|} \sum_{F \in \Phi} \left\{ \mathcal{A}[\mathbf{A}' \to F\mathbf{A}'] + \left(1 - \mathcal{A}[\mathbf{A}' \to F\mathbf{A}']\right) \right\} = 1 \quad (6.4)$$

Expression (6.3) thus always defines a genuine Markov chain. Furthermore, it defines an aperiodic Markov chain, as there is a finite probability of remaining in the same state at any iteration, which removes any periodicity in the dynamics. Hence, to ensure convergence of the chain to a unique stationary state, we need only to define a set of moves that make the chain ergodic.

The detailed balance condition. We now show how a simple additional condition, called *detailed balance*, leads to a specific transparent and attractive recipe for constructing Markov chains (i.e. transition probabilities) that converge to any desired stationary state. The detailed balance condition states that

$$\forall \mathbf{A}, \mathbf{A}' \in G: \quad p(\mathbf{A}')\mathcal{W}[\mathbf{A}' \to \mathbf{A}] = p(\mathbf{A})\mathcal{W}[\mathbf{A} \to \mathbf{A}'] \quad (6.5)$$

---

[6]The stationary state is in fact unique in any ergodic Markov chain, although convergence to the unique steady state is only guaranteed in a regular Markov chain.

It represents microscopic reversibility of the stochastic dynamics, since the left-hand side and the right-hand of (6.5) are proportional to the expected average number of observed transitions $\mathbf{A}' \to \mathbf{A}$ and observed transitions $\mathbf{A} \to \mathbf{A}'$, respectively. If one chooses transition probabilities that satisfy (6.5), then these will automatically have $p(\mathbf{A})$ as their invariant measure. This is seen by summing both sides over $\mathbf{A}' \in G$, since owing to normalization of the transition probabilities this immediately leads us to (6.2). Intuitively, property (6.5) implies that if we record a movie of the Markov chain dynamics at stationarity, we will not be able to tell whether the film is running forward or backwards. A stationary state with this property is called an equilibrium state. If we choose the transition probabilities $W[\mathbf{A}' \to \mathbf{A}]$ in such a way that (6.5) is satisfied, and we can ensure that the process is ergodic, then the resulting Markov chain is guaranteed to converge to equilibrium, and will, after equilibration, visit the graphs in the space G with frequencies proportional to $p(\mathbf{A})$.

Inserting (6.3) into the condition (6.5) tells us that the process will obey detailed balance if and only if for all $\mathbf{A}, \mathbf{A}' \in G$ the following identity holds:

$$p(\mathbf{A}') \sum_{F \in \Phi} \mathcal{A}[\mathbf{A}' \to F\mathbf{A}'] \big[\delta_{\mathbf{A},F\mathbf{A}'} - \delta_{\mathbf{A},\mathbf{A}'}\big] = p(\mathbf{A}) \sum_{F \in \Phi} \mathcal{A}[\mathbf{A} \to F\mathbf{A}] \big[\delta_{\mathbf{A}',F\mathbf{A}} - \delta_{\mathbf{A},\mathbf{A}'}\big]$$

$$(6.6)$$

For $\mathbf{A} = \mathbf{A}'$ this identity is obviously true. It is also true if there exists no $F \in \Phi$ such that $\mathbf{A}' = F\mathbf{A}$, i.e. there is no move in the chosen arsenal Φ that can bring us in one iteration of the process from state $\mathbf{A}$ to state $\mathbf{A}'$, since in that case we simply find zeros on both sides of (6.6). This leaves us only with the case where $\mathbf{A}' = F\mathbf{A}$ for some $F \in \Phi$, and hence also $\mathbf{A} = F\mathbf{A}'$ (due to the assumed auto-invertibility of the moves). Here we obtain from (6.6) the following non-trivial condition on the acceptance probabilities:

$$\forall \mathbf{A} \in G, \ \forall F \in \Phi: \quad p(F\mathbf{A})\mathcal{A}[F\mathbf{A} \to \mathbf{A}] = p(\mathbf{A})\mathcal{A}[\mathbf{A} \to F\mathbf{A}] \qquad (6.7)$$

Although this condition seems nearly identical to (6.5), unlike (6.5), the present condition (6.7) refers to *acceptance* probabilities, rather than to *transition* probabilities. As a consequence, we no longer have to ensure normalization; this has already been built into the form (6.3). The acceptance probabilities in (6.7) just need to take values in the interval $[0, 1]$. It follows that while in a large configuration space the probability of any configuration is small, transition probabilities are generally not (they will be the *ratio* of small numbers). The two most popular solutions of the remaining problem (6.7) for the acceptance probabilities are the following:

$$\text{Metropolis-Hastings}: \quad \mathcal{A}[\mathbf{A}' \to \mathbf{A}] = \min\Big\{1, \frac{p(\mathbf{A})}{p(\mathbf{A}')}\Big\} \qquad (6.8)$$

$$\text{Glauber}: \quad \mathcal{A}[\mathbf{A}' \to \mathbf{A}] = \frac{p(\mathbf{A})}{p(\mathbf{A}) + p(\mathbf{A}')} \qquad (6.9)$$

The former is computationally more efficient, since fewer candidate moves will be rejected during simulations[7]. The latter is more convenient when proving convergence

---

[7] This follows from the inequality $p(\mathbf{A})/(p(\mathbf{A})+p(\mathbf{A}')) = (1+p(\mathbf{A}'/p(\mathbf{A}))^{-1} \le \min\big\{1, p(\mathbf{A})/p(\mathbf{A}')\big\}$.

properties, since its dependence on the probabilities $p(\mathbf{A})$ is more smooth, and can alternatively be written as $\mathcal{A}[\mathbf{A}' \to \mathbf{A}] = \frac{1}{2} + \frac{1}{2} \tanh\left[\frac{1}{2} \log[p(\mathbf{A})/p(\mathbf{A}')]\right]$.

The chosen set Φ of moves will generally depend on the type of measures $p(\mathbf{A})$ from which we seek to sample. It can include single link removals/additions of the form $A_{ij} \to 1 - A_{ij}$, pair-swaps and so on. The main condition on this set is that the Markov chain must be regular. For any pair $\mathbf{A}, \mathbf{A}' \in G$, there must exist a path of successive moves with non-zero probabilities via which we can go from $\mathbf{A}$ to $\mathbf{A}'$.

Exercise 6.1 Prove that the recipes (6.8,6.9) of Metropolis–Hastings and Glauber for the move acceptance probabilities $\mathcal{A}[\mathbf{A}' \to \mathbf{A}]$ both satisfy the detailed balance condition (6.7). Show that these two recipes are not the only choices that obey (6.7), by constructing explicitly alternative acceptance probabilities that also meet the detailed balance requirements.

Complications associated with having state-dependent move sets. The above construction of MCMC graph sampling algorithms is simple and practical. However, there are situations where it is too limited. For instance, we will be interested in situations where each individual move F can act only on a subset $G_F \subseteq G$ (see subsequent sections on sampling from ensembles with hard constraints), and conversely, where on each state $\mathbf{A}$ we can execute only a subset $\Phi_{\mathbf{A}} \subseteq \Phi$ of all possible moves. Because of the assumed auto-invertibility of our moves, we will require that if $\mathbf{A} \in G_F$ then also $F\mathbf{A} \in G_F$, and that if $F \in \Phi_{\mathbf{A}}$ then also $F \in \Phi_{F\mathbf{A}}$ (to ensure the microscopic reversibility of detailed balance).

There are two distinct ways to incorporate this type of situation. The first solution, called switch-and-hold, is to set all acceptance probabilities of forbidden moves to zero by hand, i.e. we continue to use the standard formalism (6.3), but we insist that

$$\mathcal{A}[\mathbf{A} \to F\mathbf{A}] = \mathcal{A}[\mathbf{A} \to F\mathbf{A}] = 0 \quad \text{if} \quad F \notin \Phi_{\mathbf{A}} \tag{6.10}$$

For those moves where $F \in \Phi_{\mathbf{A}}$, i.e. the allowed ones, we continue to use recipes such as (6.8,6.9). This is in principle a transparent and perfectly acceptable method for handling state-dependent restricted move sets. However, especially when the move sets $\Phi_{\mathbf{A}}$ tend to be relatively small, i.e. if $|\Phi_{\mathbf{A}}|/|\Phi| \ll 1$, we may find that most of the generated moves are rejected as impossible and switch-and-hold will spend most of its time on the same configurations, potentially leading to poor sampling.

An alternative route is to limit ourselves explicitly at each step of the MCMC process to those moves that we know can actually be executed on the instantaneous state $\mathbf{A}$, i.e. to propose only moves $F \in \Phi_{\mathbf{A}}$. The set of moves from which we will be selecting will now change over time, as the process runs its course, since generally $\Phi_{F\mathbf{A}} \neq \Phi_{\mathbf{A}}$. This implies that we will no longer have a process of the form (6.3), but must write instead

$$W[\mathbf{A}' \to \mathbf{A}] = \frac{1}{|\Phi_{\mathbf{A}'}|} \sum_{F \in \Phi_{\mathbf{A}'}} \left\{ \mathcal{A}[\mathbf{A}' \to F\mathbf{A}']\delta_{\mathbf{A}, F\mathbf{A}'} + \left(1 - \mathcal{A}[\mathbf{A}' \to F\mathbf{A}']\right)\delta_{\mathbf{A}, \mathbf{A}'} \right\}$$

$$\tag{6.11}$$

We can now once more demand that our Markov chain satisfies detailed balance, and will find upon inserting (6.11) into (6.5) that our previous condition (6.7) on the acceptance probabilities will be modified to

$$\forall \mathbf{A} \in G, \ \forall F \in \Phi: \quad \frac{p(F\mathbf{A})}{|\Phi_{F\mathbf{A}}|} \mathcal{A}[F\mathbf{A} \to \mathbf{A}] = \frac{p(\mathbf{A})}{|\Phi_{\mathbf{A}}|} \mathcal{A}[\mathbf{A} \to F\mathbf{A}] \tag{6.12}$$

The extra factors $|\Phi_{\mathbf{A}}|^{-1}$ and $|\Phi_{F\mathbf{A}}|^{-1}$, which were not present in condition (6.7), have non-trivial implications. Instead of being able to use simple prescriptions such as (6.8,6.9), we will now evolve to the measure $p(\mathbf{A})$ only if we adapt those recipes to

$$\text{Metropolis}-\text{Hastings}: \quad \mathcal{A}[\mathbf{A}' \to \mathbf{A}] = \min\left\{1, \frac{p(\mathbf{A})/|\Phi_{\mathbf{A}}|}{p(\mathbf{A}')/|\Phi_{\mathbf{A}'}|}\right\} \tag{6.13}$$

$$\text{Glauber}: \quad \mathcal{A}[\mathbf{A}' \to \mathbf{A}] = \frac{p(\mathbf{A})/|\Phi_{\mathbf{A}}|}{p(\mathbf{A})/|\Phi_{\mathbf{A}}| + p(\mathbf{A}')/|\Phi_{\mathbf{A}'}|} \tag{6.14}$$

If $|\Phi_{\mathbf{A}}|$ is independent of $\mathbf{A}$, we simply return to the previous standard equations. The quantity $|\Phi_{\mathbf{A}}|$, i.e. the number of moves $F \in \Phi$ that can be executed on graph $\mathbf{A}$, is called its *mobility* [52]. Compared to the switch-and-hold algorithms, we will here spend more computing time on calculating the acceptance probabilities, but we will no longer waste time on generating forbidden moves. Which of the two MCMC routes is fastest will depend on the ratio $|\Phi_{\mathbf{A}}|/|\Phi|$ and the computational complexity of calculating $|\Phi_{\mathbf{A}}|$.

Exercise 6.2 Prove that the modified recipes (6.13, 6.14) of Metropolis–Hastings and Glauber for the move acceptance probabilities $\mathcal{A}[\mathbf{A}' \to \mathbf{A}]$ both satisfy the modified detailed balance condition (6.12).

Common pitfalls. The preceding discussion underlines the importance of being meticulous in how we select candidate moves in our MCMC algorithms, since this will generally affect the final measure $p(\mathbf{A})$. If we choose to run the standard switch-and-hold protocol, with the standard acceptance probabilities (6.8) or (6.9), we must really generate *all* moves $F \in \Phi$ with uniform probabilities, even though we know beforehand that the vast majority of these moves will subsequently be rejected. The reason is that the extended periods in such processes where no state changes happen are effectively contributing to getting the overall state probabilities right. If we skip (some of) the forbidden moves during switch-and-hold, we mess up the graph sampling probabilities. Similarly, if we run the alternative protocol with state-dependent move sets and acceptance probabilities (6.13) or (6.14), we must make sure that all moves in $|\Phi_{\mathbf{A}}|$ are *equally likely* to be picked. Apparently innocent programming shortcuts can be very dangerous, as they can influence the probabilities $p(\mathbf{A})$.

To see this more clearly, let us imagine we want to generate graphs with uniform probabilities, i.e. we want to achieve sampling according to $p(\mathbf{A}) = 1/|G|$ for all $\mathbf{A} \in G$, and we propose at each MCMC step only one of the at-that-moment allowed moves $F \in \Phi_{\mathbf{A}}$. We have seen that we would target the measure $p(\mathbf{A})$ only by using the modified acceptance probabilities such as (6.13) or (6.14). If, however, instead of doing the right thing, we apply the conventional (non-corrected) formulae (6.8) or (6.9) to the uniform measure $p(\mathbf{A}) = 1/|G|$, we end up implementing the acceptance probabilities $\mathcal{A}[\mathbf{A}' \to \mathbf{A}] = 1$ or $\mathcal{A}[\mathbf{A}' \to \mathbf{A}] = \frac{1}{2}$, respectively. We see upon inspecting (6.13) and (6.14) that these choices will in fact target the following non-uniform measure:

$$p(\mathbf{A}) = \frac{|\Phi_{\mathbf{A}}|}{\sum_{\mathbf{A}' \in G} |\Phi_{\mathbf{A}'}|} \tag{6.15}$$

So instead of generating graphs uniformly, our sampler will be biased towards those graphs on which many moves can be applied, to the detriment of less mobile graphs. Especially in the context of null models, this can have disastrous consequences.

Shortcut: *MCMC algorithms are efficient tools for sampling graphs numerically from random graph ensembles with prescribed probabilities $p(\mathbf{A})$. They involve a set of moves $\mathbf{A} \to F\mathbf{A}$, defined by the user, from which at each step a candidate move is chosen at random, and a formula for the probability with which any such move is to be either accepted (and executed) or rejected. If the set of allowed moves is fixed, we have standard formulae (Metropolis–Hastings, Glauber) for these acceptance probabilities. If, in contrast, the set of allowed moves is state dependent, then in order to achieve the correct sampling probabilities, we must either generate all moves and reject the forbidden ones explicitly, or use suitably adapted move acceptance probabilities.*

6.2 MCMC sampling for exponential random graph models

MCMC based on individual link creation or removal. In the absence of hard and/or complicated topological constraints and for simple graph sets G, which is usually the situation with ERGMs (4.1), we can generally work with relatively simple definitions for the allowed set of moves. In particular, we can as a rule work with sets Φ that are independent of the current state $\mathbf{A}$, and we may hence use the standard move acceptance probabilities (6.8, 6.9). In the case of ERGMs, the probabilities $\mathcal{A}[\mathbf{A} \to F\mathbf{A}]$ with which we accept and execute a proposed transition $\mathbf{A} \to F\mathbf{A}$ take the form

$$\text{Metropolis–Hastings}: \quad \mathcal{A}[\mathbf{A} \to F\mathbf{A}] = \min\left\{1, e^{\sum_{\mu=1}^{K} \lambda_\mu \left[\Omega_\mu(F\mathbf{A}) - \Omega_\mu(\mathbf{A})\right]}\right\} \tag{6.16}$$

$$\text{Glauber}: \quad \mathcal{A}[\mathbf{A} \to F\mathbf{A}] = \left(1 + e^{-\sum_{\mu=1}^{K} \lambda_\mu \left[\Omega_\mu(F\mathbf{A}) - \Omega_\mu(\mathbf{A})\right]}\right)^{-1} \tag{6.17}$$

Here we work out the equations for our MCMC sampling process upon building the move set Φ from the operations $\{F_{ij}\}$, where each F_{ij} changes only entry A_{ij} of $\mathbf{A}$:

$$F_{ij} A_{ij} = 1 - A_{ij} \tag{6.18}$$
$$F_{ij} A_{rs} = A_{rs} \quad \text{if} \quad (r, s) \neq (i, j) \tag{6.19}$$

Thus, F_{ij} creates a link (i, j) if this link is absent from $\mathbf{A}$, and removes this link if it is present. It is clearly auto-invertible, i.e. $F_{ij}^2 = \mathbf{1}$ for all (i, j). It will be clear that one can go in principle from any initial state $\mathbf{A}$ to any final state $\mathbf{A}'$ by executing a sequence of elementary moves F_{ij}, and especially if we use the acceptance probabilities (6.17) (which are always non-zero), any such sequence of moves has a non-zero probability of being accepted. So the Markov chain is regular, and hence is guaranteed to converge to the desired probabilities $p(\mathbf{A})$ in (4.1).

If G is the set of simple nondirected graphs, all index pairs (i, j) are ordered according to $i < j$, and the action of F_{ij} affects both A_{ij} and A_{ji}, such that the symmetry of $\mathbf{A}$ is preserved. The chosen set Φ of moves is independent of $\mathbf{A}$, with $|\Omega| = \frac{1}{2}N(N-1)$. Hence, we can indeed use the standard move acceptance formulae (6.16, 6.17), which depend on the current state $\mathbf{A}$ only via the quantities

$$\Delta_{ij}(\mathbf{A}) = \sum_{\mu=1}^{K} \lambda_\mu \left[\Omega_\mu(F_{ij}\mathbf{A}) - \Omega_\mu(\mathbf{A}) \right] \tag{6.20}$$

Let us illustrate the sampling protocols we will find if we try to target some of the simple ERGMs in Chapter 4.

Sampling from an ERGM with prescribed average number of links. Sampling from the ERGM (4.9) means effectively sampling from the Erdös–Rényi (3.1) ensemble, so we could simply draw all links randomly and independently with the appropriate link probability. However, in anticipation of the more complicated ensembles to follow, it is instructive to see how we would approach this task via the MCMC route. Upon writing our ensemble probabilities in the form (4.9), with λ given for any desired number L of the average number of links by (4.13), we find that expression (6.20) becomes

$$\Delta_{ij}(\mathbf{A}) = \lambda \sum_{r<s}(F_{ij}A_{rs} - A_{rs}) = \lambda(1 - 2A_{ij}) \tag{6.21}$$

The acceptance probabilities (6.16, 6.17) for a move $\mathbf{A} \to F_{ij}\mathbf{A}$ thus simplify to

$$\text{Metropolis}-\text{Hastings}: \quad \mathcal{A}[\mathbf{A} \to F_{ij}\mathbf{A}] = \min\left\{1, e^{\lambda(1-2A_{ij})}\right\} \tag{6.22}$$

$$\text{Glauber}: \quad \mathcal{A}[\mathbf{A} \to F_{ij}\mathbf{A}] = \left(1 + e^{-\lambda(1-2A_{ij})}\right)^{-1} \tag{6.23}$$

We provide in Algorithm 1 (see Appendix J) the resulting Metropolis–Hastings version of the MCMC pseudocode to generate random graphs $\mathbf{A}$ from the ERGM, defined for the set G of simple nondirected N-node graphs with the prescribed average number of links $L = \langle \sum_{ij} A_{ij} \rangle$.

The termination condition should alert us to the expected achievement of equilibration of the Markov chain, as measured by, e.g. the stabilization of specific relevant observables that could be monitored during the process, or by the Hamming distance between the initial state $\mathbf{A}_0$ and the instantaneous state $\mathbf{A}$ (i.e. the numbers of non-identical entries) becoming equal to what would have been found for two completely uncorrelated graphs with similar densities. For the Glauber version, one makes the appropriate replacement for Prob(execute move) in Algorithm 1 of Appendix J.

Sampling from the two-star ERGM. Next we turn to one of the graph ensembles that were found to exhibit phase transitions, namely the two-star model (4.36). Here one effectively constrains the average values of the first and the second moments of the degree distribution. Since (4.36) depends on the state $\mathbf{A}$ only via the degree sequence $\mathbf{k}(\mathbf{A})$, we need to work out the effect of the operation F_{ij} on individual degrees:

$$k_r(F_{ij}\mathbf{A}) - k_r(\mathbf{A}) = (1-2A_{ij})(\delta_{ir}+\delta_{jr}) \tag{6.24}$$

We can then use $k_r(F_{ij}\mathbf{A}) = k_r(\mathbf{A}) + (1-2A_{ij})(\delta_{ir} + \delta_{jr})$ to work out the relevant quantity (6.20) that controls the move acceptance probabilities:

$$
\begin{aligned}
\Delta_{ij}(\mathbf{A}) &= \alpha\Big[\sum_r k_r(F_{ij}\mathbf{A}) - \sum_r k_r(\mathbf{A})\Big] + \beta\Big[\sum_r k_r^2(F_{ij}\mathbf{A}) - \sum_r k_r^2(\mathbf{A})\Big] \\
&= \alpha(1-2A_{ij})\sum_r(\delta_{ir}+\delta_{jr}) + \beta\sum_r\Big[\big(k_r(\mathbf{A})+(1-2A_{ij})(\delta_{ir}+\delta_{jr})\big)^2 - k_r^2(F_{ij}\mathbf{A})\Big] \\
&= 2\alpha(1-2A_{ij}) + \beta\sum_r\Big[2k_r(\mathbf{A})(1-2A_{ij})(\delta_{ir}+\delta_{jr}) + (\delta_{ir}+\delta_{jr})^2\Big] \\
&= 2(1-2A_{ij})\Big[\alpha + \beta[k_i(\mathbf{A})+k_j(\mathbf{A})]\Big] + 2\beta \tag{6.25}
\end{aligned}
$$

The acceptance probabilities (6.16, 6.17) for a move $\mathbf{A} \to F_{ij}\mathbf{A}$ thus become

Metropolis–Hastings : $\quad \mathcal{A}[\mathbf{A} \to F\mathbf{A}] = \min\Big\{1, e^{2(1-2A_{ij})\left[\alpha+\beta[k_i(\mathbf{A})+k_j(\mathbf{A})]\right]+2\beta}\Big\}$

$$\tag{6.26}$$

Glauber : $\quad \mathcal{A}[\mathbf{A} \to F\mathbf{A}] = \big(1 + e^{-2(1-2A_{ij})\left[\alpha-\beta[k_i(\mathbf{A})+k_j(\mathbf{A})]\right]-2\beta}\big)^{-1}$

$$\tag{6.27}$$

In this model, we have no exact formulae for either the average number of links and two-stars, or equivalently for the average of the first and second moments of the degree distribution, expressed in terms of the two control parameters (α, β). We only have asymptotic expressions in the dense and in the sparse regimes. Hence, we can retain the pair (α, β) as control parameters in the algorithm for the generation of graphs $\mathbf{A}$ randomly from (4.36), whose pseudocode, for the Metropolis–Hastings choice, is provided in Algorithm 4 (Appendix J).

As in the previous example, this algorithm includes an appropriate termination condition to tell us when we may expect the process to have equilibrated, and we can make a simple modification if we prefer to use Glauber's move acceptance recipe. The above algorithm was used to generate the earlier simulation data of Figure 4.3, starting from initialization graphs $\mathbf{A}_0$ of the Erdös–Rényi type (3.1) with different densities.

It is quite instructive to monitor the progress of the above MCMC algorithm, by measuring at regular intervals of the algorithmic time t (which counts the number of proposed moves F_{ij}) the values of the relevant observables that the algorithm is designed to tune. For the current ensemble (4.36), the relevant observables are the average degree $\langle k \rangle$ and the width $\sigma_k = (\langle k^2 \rangle - \langle k \rangle^2)^{1/2}$ of the degree distribution $p(k|\mathbf{A})$ of the instantaneous state $\mathbf{A}$. The results of carrying out this exercise are shown in Figure 6.1, for the Glauber recipe (6.17) of move acceptance probabilities. For each control parameter combination (α, β) the MCMC algorithm is run multiple times,

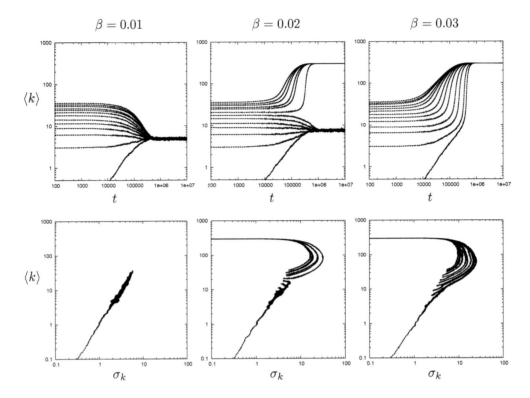

Fig. 6.1 Result of monitoring an MCMC graph sampling process constructed to generate simple nondirected graphs with $N = 300$ nodes from the two-star ensemble (4.36), using the Glauber prescription (6.17) for the move acceptance probabilities. Here $\alpha = 4$ and $\beta \in \{0.01, 0.02, 0.03\}$. All panels show simulation results (in log–log plots) following multiple initializations $\mathbf{A}_0$ consisting of Erdös–Rényi graphs (3.1), with link probabilities $p \in \{0, 0.01, 0.02, \ldots, 0.11, 0.12\}$ (and hence different initial average degrees). Top row: Average degree $\langle k \rangle$ plotted versus the number of iterations $t \geq 0$ of the algorithm. Bottom row: Average degree $\langle k \rangle$ plotted versus the width $\sigma_k = (\langle k^2 \rangle - \langle k \rangle^2)^{1/2}$. The control parameter values are such that the left diagrams ($\beta = 0.01$) refer to the regime with a unique (low density) state, the right diagrams ($\beta = 0.03$) to the regime with a unique (high density) state, and the middle diagrams ($\beta = 0.02$) to the regime where the state is not unique. In the middle two panels one observes two stable states, one with low and the other with high link density. To which of these the algorithm evolves depends on the chosen initialization. Tuning the control parameters to achieve states with intermediate average degree values will be hard.

following initializations $\mathbf{A}_0$ to sparse Erdös–Rényi graphs[8] (3.1) with different choices for their average degrees $\langle k \rangle_{t=0}$. In the complex parameter regime of the ensemble, where the mean-field approximation (which is valid, strictly speaking, only for $N \to \infty$)

---

[8] In large sparse Erdös–Rényi graphs, the degree distribution will approach $p(k) = e^{-\langle k \rangle} \langle k \rangle^k / k!$, giving $\sigma_k = (\langle k^2 \rangle - \langle k \rangle^2)^{1/2} = \langle k \rangle^{1/2}$, so in a log–log plot of $\langle k \rangle$ versus σ_k one expects the initial states to lie on a line. This is indeed visible in the bottom panels of Figure 6.1.

predicts two stable states, we can observe parameter combinations where even for $N = 300$ the MCMC process has two attractors, and where as a consequence the outcome of the algorithm depends crucially on its initialization. Here, for small values of $\langle k \rangle_{t=0}$ the algorithm will generate low-density graphs upon equilibration, whereas for larger values of $\langle k \rangle_{t=0}$ it will generate high-density graphs.

This observed behaviour of our MCMC algorithm, while confirming the predictions of the mean-field approximation in Chapter 4, appears to be in direct conflict with our assertion that, certainly for the Glauber move acceptance probabilities (where *all* paths based on repeated single link changes have a non-zero probability of occurring, unlike the Metropolis—Hastings version, where some move acceptances are compulsory), our Markov chain is regular, and our MCMC process must hence evolve to a state that is independent of the initial conditions. The explanation lies in the answer to a question that we have avoided so far: what is the timescale for the Markov chain to reach full equilibrium? Figure 6.1 suggests that the MCMC process converges on timescales $t \sim N^2$ to a fixed point. However, there turns out to be a further relevant timescale in the process, which is much larger than $\mathcal{O}(N^2)$. The two apparently stable states visible in the middle panels of Figure 6.1, with low $\langle k \rangle$ versus high $\langle k \rangle$, are in fact not 100% stable; there is always a small probability that a succession of improbable steps will at some point occur and make $\mathbf{A}$ switch from a high-density state to a low-density state (or vice versa). If we wait long enough, such switches will ultimately happen, and the algorithm's sampling probabilities will become a (unique) weighted average of the probabilities of low- and high-density states. The process is therefore ergodic on those larger timescales where such switches can come into play. However, as N becomes larger, the time required for ergodicity to be restored is astronomical, growing super-exponentially with N, so that for all practical purposes we must treat the process as non-ergodic. This phenomenon, which is very common in stochastic systems with many interacting particles, is called *ergodicity breaking*. It explains why one can have phase transitions and multiple macroscopic states in systems with ergodic dynamics. Ergodicity is only truly broken for $N \to \infty$; only then will we *never* see switches between the attractors no matter how long we wait. However, Figure 6.1 suggests clearly that, in the context of MCMC sampling, a random graph of just a few hundred nodes already behaves in practice like an infinitely large one.

Exercise 6.3 Construct an MCMC sampling algorithm of the detailed balance type, based on Metropolis–Hastings move acceptance probabilities, to generate N-node simple nondirected graphs from the ERGM (4.14), with prescribed average degrees $\mathbf{k} = \langle\langle k_1(\mathbf{A}), \ldots, k_N(\mathbf{A}) \rangle\rangle$.

Exercise 6.4 Construct an MCMC sampling algorithm of the detailed balance type, based on Glauber move acceptance probabilities, to generate N-node simple nondirected graphs with the following average degree distribution $p(k) = \langle p(k|\mathbf{A}) \rangle$: $p(0) = 0$ and $p(k > 0) = C_N^{-1} k^{-\gamma}$ (where $\gamma > 0$ and $C_N = \sum_{k>0} k^{-\gamma}$).

Exercise 6.5 Construct an MCMC sampling algorithm of the detailed balance type, based on Metropolis–Hastings move acceptance probabilities, to generate N-node simple nondirected graphs from the ERGM (4.61) with prescribed average numbers of links and triangles.

Shortcut: For ERGMs, where the values of observables are built in via soft constraints, one can always use MCMC sampling processes with simple move sets Φ, such as the operations F_{ij}, which only change the entry A_{ij} (and the entry A_{ji} for nondirected graphs) according to $A_{ij} \to 1 - A_{ij}$. Here the number $|\Phi|$ of elementary moves is independent of the state $\mathbf{A}$, and one can use the standard move acceptance probabilities (6.16, 6.17). In ERGMs with phase transitions, however, one observes even for relatively small graphs the phenomenon of ergodicity breaking: on $\mathcal{O}(N^2)$ timescales (in algorithmic time), the process will converge to one of its multiple attractors, and return a graph sample $\mathbf{A}$ with characteristics that are dependent upon the chosen initialization. Ergodicity is restored only on astronomical timescales. In practice, this implies that it is always necessary to run the algorithm from multiple distinct initializations.

6.3 MCMC sampling for graph ensembles with hard constraints

Switch-and-hold versus MCMC with state-dependent move sets. Unfortunately, when we turn to the maximum entropy ensembles (3.10) with hard constraints, the graph generation problem becomes more complicated. Now the only moves F that we are allowed to execute on a given state $\mathbf{A} \in G$ are those that satisfy the requirements $\Omega_\mu(F\mathbf{A}) = \Omega_\mu(\mathbf{A})$ for all $\mu = 1 \ldots K$. The simple single link operations of (6.18, 6.19) will generally violate one or more of these demands; indeed, the only observable that is manifestly invariant under the action of all $\{F_{ij}\}$ must by definition be independent of all entries $\{A_{ij}\}$ of the adjacency matrix, and is therefore a constant. The family of single link flips is thus ruled out, and we will generally have to work with more complicated move sets. This leaves us with the following two options.

Our first option is to rescue the desirable ability to work with state-independent move sets Φ as follows. We can enlarge the graph set G to include all states $\mathbf{A}$ (both the acceptable and the forbidden ones), but we assign $p(\mathbf{A}) = 0$ to those states that violate our hard constraints. We can then continue to use the standard MCMC formulae (6.3, 6.8, 6.9). As soon as a move $F \in \Phi$ is proposed that would lead to a forbidden new state $F\mathbf{A}$, (6.8, 6.9) will automatically assign acceptance probability zero to such moves, and our Metropolis–Hastings (6.8) MCMC algorithm for graph generation will take the simple standard form given in Algorithm 10, with G now including the forbidden states. Ideally, the initial state $\mathbf{A}_0$ should be chosen to meet all applicable constraints, if this is feasible[9]. The Glauber version of this pseudocode is obtained in the usual manner by simply replacing $\min\{1, p(F\mathbf{A})/p(\mathbf{A})\}$ by the recipe (6.9). We

[9]If we are seeking to generate random proxies for an observed real-world N-node network $\mathbf{A}^\star$, from which the constraint values were measured, we may as well use this $\mathbf{A}^\star$ as our algorithm initialization.

will pay a price for this simplicity[10]: many of our proposed moves $F \in \Phi$ might be rejected, i.e. we might find most of the time Prob(execute move) $= 0$ in step 3(b), as they would bring us from $\mathbf{A}$ to a forbidden new configuration $F\mathbf{A}$. We cannot tamper with the protocol by skipping forbidden moves to speed things up, since the very infrequency of accepting moves in the process and hence the long periods of stationarity caused by rejections contribute to ensuring the correct sampling probabilities. This is the *switch-and-hold* route.

Our second option is to construct more sophisticated move sets in which most (preferably all) moves preserve our hard constraints. Such sets will no longer contain the many forbidden moves that slow down the switch-and-hold protocol, but here also there is a price to be paid. Our move set will generally depend both on the chosen constrained observables and on the instantaneous state $\mathbf{A}$, so we write $\Phi_{\mathbf{A}}$. To achieve the correct sampling probabilities, we must now use the modified MCMC formulae (6.3, 6.8, 6.9), that take into account the impact of the mobility $|\Phi_{\mathbf{A}}|$ of a state, i.e. the fact that states $\mathbf{A}$ on which many moves F can be executed will (everything else being equal) be observed more often. Our modified Metropolis–Hastings (6.13) MCMC algorithm for graph generation will take the non-standard form given in Algorithm 11, with G now only including the states that satisfy our constraints. Depending on the complexity of the chosen moves, it will often be efficient to avoid having to calculate $|\Phi_{\mathbf{A}}|$ afresh after every state change, in the following way. We can calculate its value explicitly only at the start of the process, and update this value after every accepted and executed move F by only the change $\Delta_F|\Phi_{\mathbf{A}}| = |\Phi_{F\mathbf{A}}| - |\Phi_{\mathbf{A}}|$ that is triggered by this move. This modifies the Algorithm 11 into 12. Again, the Glauber version of both versions of the pseudocode are obtained in the usual manner upon replacing $\min\{1, p(F\mathbf{A})|\Phi_{\mathbf{A}}|/p(\mathbf{A})|\Phi_{F\mathbf{A}}|\}$ by the recipe (6.14). Whenever we use an MCMC algorithm of the type where we only propose admissible moves, and therefore have a state-dependent move set $\Phi_{\mathbf{A}}$, we will require additional computing time to evaluate the mobilities in step 4(b), but we will also waste less computing resource on proposing and rejecting forbidden moves.

Popular move sets and their properties. The MCMC equations demand that each move $F \in \Phi$ is its own inverse, and that always $F\mathbf{A} \neq \mathbf{A}$. Which precise move set Φ can be used in a given hard-constrained ensemble will depend crucially on the chosen functions $\Omega_\mu(\mathbf{A})$ in terms of which we formulated the constraints, since our moves have to be constructed such that they leave the values of these functions invariant. Unlike soft-constrained ensembles, where we could always work with the set of link flips $\{F_{ij}\}$, there can be no universal choice of move set that can be used in all hard-constrained ensembles. In the interest of efficiency and speed, we obviously want our moves to be as simple as possible. We can benefit from the second formulation of the algorithm discussed above (Algorihm 12), based on calculating after each executed move F only the change $\Delta_F|\Phi_{\mathbf{A}}| = |\Phi_{F\mathbf{A}}| - |\Phi_{\mathbf{A}}|$, only if our moves affect just a small number of nodes. In that case the changes $\Delta_F|\Phi_{\mathbf{A}}|$ can be computed with a relatively

---

[10]Note that even within the switch-and-hold protocol, the single link flips $\Phi = \{F_{ij}\}$ will still be ruled out as soon as our hard constraints include degrees or the overall number of links, since these moves would then *always* be rejected.

MOVE SET	INVARIANTS	ACTION
Link flips $\{F_{ij}\}$	none	
Hinge flips $\{F_{ijk}\}$	average degree $\bar{k}(\mathbf{A}) = \frac{1}{N}\sum_{rs} A_{rs}$	
Edge swaps $\{F_{ijk\ell}\}$	all individual degrees $k_i(\mathbf{A}) = \sum_j A_{ij}, \quad i = 1\ldots N$	

Table 6.1 Definitions of the most commonly used ergodic move sets Φ in MCMC algorithms for the generation of simple nondirected graphs, together with their invariants. The invariants are those functions $\Omega_\mu(\mathbf{A})$ whose values are preserved under the action of *all* operations $F \in \Phi$. Link flips are the simplest moves, but they have no invariants and can therefore be used only in soft-constrained ensembles. Hinge flips are the natural choice for ensembles where the average degree (or, equivalently, the total number of links) is prescribed as a hard constraint. Here, for F_{ijk} to be applicable to a state $\mathbf{A}$ we require that $A_{ij} = 1 - A_{ik}$. Edge swaps (also called *Seidel switches*) are the natural choice for ensembles with hard-constrained degree sequences. Here, for $F_{ijk\ell}$ to be applicable to a state $\mathbf{A}$, we require that $A_{ij} = A_{k\ell} = 1 - A_{jk} = 1 - A_{i\ell}$.

modest computational cost.

In Table 6.1, we list the most common choices of MCMC move sets Φ that are used in practice for the generation of simple nondirected graphs, together with the observables $\{\Omega_\mu(\mathbf{A})\}$ whose values are invariant under the action of all members $F \in \Phi$. It can be shown that all three families of moves in this table are ergodic on their respective graph sets G, i.e. that it is always possible to go from any initial state $\mathbf{A} \in G$ to any final state $\mathbf{A}' \in G$ via a succession of moves $F \in \Phi$. The question of ergodicity for hinge flips has been studied in [25] and for edge swaps a formal proof by induction was given by Taylor in [173]. Taylor's arguments focus on connected graphs[11], and this represents the harder case. Indeed, ergodicity for connected graphs would immediately transfer to generic graphs by a simple argument. Given a non-connected graph, add an auxiliary node connected to every other node. This extra node creates a new connected graph. However, the new links can never be switched with any of the pre-existing links.

---

[11] Actually, in [173] it is shown (Theorems 3.1 and 3.2) that ergodicity holds for generic connected pseudographs, a class of structures that allows for multiple links between edges and for loops of length 1, which is much broader than just the simple graphs we consider here. The present case is covered by Theorem 3.3 in [173].

Hence, they are not modified by the dynamics. Therefore, if ergodicity holds for the auxiliary connected graph, it is automatically valid also for the original non-connected graph. A more rigorous proof of the latter fact can be found in, e.g. [66].

Note that we can view each edge swap as a combination of two specific hinge flips, and each hinge flip as a combination of two link flips. One can confirm quite easily that there exists no family of moves that involves fewer than two nodes, that any move which preserves the average degree must affect at least three nodes, and that any move that preserves the values of all degrees must affect at least four nodes[12]. In this sense, we can regard the examples in Table 6.1 as minimal move sets.

If our hard constraints are indeed among those in Table 6.1, and if we choose to work with the switch-and-hold MCMC protocol, nothing further is required, and we can proceed to code and execute the relevant algorithm. If, however, we wish to use the second type of MCMC protocol (whether version 1 or version 2), which is based on proposing at each stage only allowed moves $F \in \Phi_{\mathbf{A}}$, then we must first find a formula for the number $|\Phi_{\mathbf{A}}|$ of such moves (the so-called mobility of state $\mathbf{A}$), since this information is required in our move acceptance probabilities.

Shortcut: *For hard-constrained ensembles, the MCMC method for generating random graphs becomes more complicated. Here we need to work with a set of moves that preserve the chosen constraints, and this set of acceptable moves will generally depend on the instantaneous state $\mathbf{A}$ of the process. The switch-and-hold approach is to pick randomly a candidate move $F \in \Phi$ with uniform probabilities at each iteration step from the full set Φ of moves, and then reject those that would violate a constraint if acting on the current state $\mathbf{A}$ (i.e. those F with $F \notin \Phi_{\mathbf{A}}$). Depending on the nature of the constraints and the ensemble, the vast majority of candidate moves may be rejected. The alternative route is to pick at each iteration step only a move with uniform probabilities directly from the set $\Phi_{\mathbf{A}}$ of presently acceptable moves. Here we pay the price of having to keep track of the number $|\Phi_{\mathbf{A}}|$ of such moves, which are required in the formulae for the move acceptance probabilities.*

6.4 Properties of move families – hinge flips and edge swaps

Formula for graph mobilities $|\Phi_{\mathbf{A}}|$ based on hinge flips. The precise enumeration and counting of all possible elementary moves is crucial for the correct sampling of graphs from ensembles. We will give examples later that illustrate the dangers associated with imprecision. Each hinge flip is parametrized by a triplet of indices (see Table 6.1) (i, j, k), in which i plays a special role. Since j and k are interchangeable in the applicability condition $A_{ij} = 1 - A_{jk}$, we will find all applicable hinge flip moves (and avoid overcounting) by limiting ourselves to

---

[12]It is not difficult to construct move sets that have the same invariants as the examples in Table 6.1, but involve a larger number of nodes. For instance, any closed path $i_1 \to i_2 \to \ldots \to i_L \to i_1$ of even length L in the graph, such that $A_{i_1 i_2} = A_{i_3 i_4} = \ldots = A_{i_{L-1} i_L} = 1$ and $A_{i_2 i_3} = A_{i_4 i_5} = \ldots = A_{i_L i_1} = 0$ (i.e. when walking along this path, links are alternately present and absent), corresponds to an auto-invertible operation F that preserves all degrees, where $FA_{ij} = 1 - A_{ij}$ for all A_{ij} along the path. The shortest such path has length $L = 4$, and these are exactly the edge swaps.

$$\Phi_{\mathbf{A}} = \Big\{ F_{ijk} \big| \ 1 \le i \le N, \ \ 1 \le j < k \le N, \ \ j,k \ne i$$

$$\text{and either } A_{ij}(1-A_{ik}) = 1 \text{ or } A_{ik}(1-A_{ij}) = 1 \Big\} \qquad (6.28)$$

Each combination (i,j,k) that meets the above requirements must in our algorithm have *equal probability* of being proposed. Deviation from this rule invalidates the applicability of our move acceptance formulae[13]. If operation F_{ijk} is admissible on graph **A**, i.e. if $F_{ijk} \in \Phi_{\mathbf{A}}$, then it acts according to

$$F_{ijk}A_{rs} = A_{rs} \qquad \text{if } (r,s) \notin \{(i,j),(j,i),(i,k),(k,i)\} \qquad (6.29)$$

$$F_{ijk}A_{rs} = 1-A_{rs} \quad \text{if } (r,s) \in \{(i,j),(j,i),(i,k),(k,i)\} \qquad (6.30)$$

Using simple identities such as $A_{ii} = 0$ for all i, and $A_{ij}(1-A_{ij}) = 0$ for all (i,j), it then follows that the number of such moves that can be executed on a simple nondirected graph **A** equals

$$|\Phi_{\mathbf{A}}| = \sum_{i=1}^{N} \sum_{j<k=1}^{N} (1-\delta_{ij})(1-\delta_{ik})\big[A_{ij}(1-A_{ik}) + A_{ik}(1-A_{ij})\big]$$

$$= \sum_{ijk=1}^{N} (1-\delta_{ij})(1-\delta_{ik})(1-\delta_{jk})A_{ij}(1-A_{ik})$$

$$= \sum_{ijk=1}^{N} (1-\delta_{ik}-\delta_{jk})A_{ij}(1-A_{ik})$$

$$= (N-1)\sum_{i=1}^{N} k_i(\mathbf{A}) - \sum_{i=1}^{N} k_i^2(\mathbf{A}) \qquad (6.31)$$

Using hinge flips as our move sets implies that the average degree (equivalently, the number of links) is conserved. Hence, the first term of (6.31) is invariant during the process, and needs to be calculated only once. The non-trivial term is the second one, which will generally *not* remain stationary during the execution of our algorithm.

We can now also work out the impact of the execution of an individual admissible hinge flip F_{ijk} on the mobility $|\Phi_{\mathbf{A}}|$. For individual degrees we find, using the facts that always $i \notin \{j,k\}$ and that the degree $k_i(\mathbf{A})$ of the pivotal node i is clearly not affected by this move:

$$k_r(F_{ijk}\mathbf{A}) - k_r(\mathbf{A}) = (1-\delta_{ir})\sum_{s} \big(F_{ijk}A_{rs} - A_{rs}\big)$$

$$= (1-\delta_{ir})\big[\delta_{rj}(1-2A_{ij}) + \delta_{rk}(1-2A_{ik})\big]$$

---

[13]This does not mean that it is impossible to weave computational shortcuts into our graph generation algorithms, aimed at picking up the admissible subset of moves more quickly. However, if we follow that route, we must be able to quantify precisely our adapted probabilities of picking moves, and change the definition of the move acceptance probabilities accordingly.

$$= (1-2A_{ij})\delta_{rj} + (1-2A_{ik})\delta_{rk} \qquad (6.32)$$

This obviously makes sense: it tells us that only the degrees of nodes j and k will change, such that $(k_j(\mathbf{A}), k_k(\mathbf{A})) \to (k_j(\mathbf{A})-1, k_k(\mathbf{A})+1)$ if $A_{ij} = 1$ and $A_{ik} = 0$, whereas $(k_j(\mathbf{A}), k_k(\mathbf{A})) \to (k_j(\mathbf{A})+1, k_k(\mathbf{A})-1)$ if $A_{ij} = 0$ and $A_{ik} = 1$. Hence

$$\Delta_{ijk}|\Phi_{\mathbf{A}}| = |\Phi_{F_{ijk}\mathbf{A}}| - |\Phi_{\mathbf{A}}| = \sum_{r=1}^{N} k_r^2(\mathbf{A}) - \sum_{r=1}^{N} k_r^2(F_{ijk}\mathbf{A})$$

$$= \sum_{r=1}^{N} k_r^2(\mathbf{A}) - \sum_{r=1}^{N}\left(k_r(\mathbf{A}) + (1-2A_{ij})\delta_{rj} + (1-2A_{ik})\delta_{rk}\right)^2$$

$$= -\sum_{r=1}^{N}(\delta_{rj}+\delta_{rk}) - 2(1-2A_{ij})k_j(\mathbf{A}) - 2(1-2A_{ik})k_k(\mathbf{A})$$

$$= 2(2A_{ij}-1)k_j(\mathbf{A}) + 2(2A_{ik}-1)k_k(\mathbf{A}) - 2 \qquad (6.33)$$

Since for every admissible move we have $A_{ik} = 1-A_{ij}$, this can also be written as

$$A_{ij} = 1: \quad \Delta_{ijk}|\Phi_{\mathbf{A}}| = 2\big[k_j(\mathbf{A})-k_k(\mathbf{A})\big] - 2 \qquad (6.34)$$
$$A_{ij} = 0: \quad \Delta_{ijk}|\Phi_{\mathbf{A}}| = 2\big[k_k(\mathbf{A})-k_j(\mathbf{A})\big] - 2 \qquad (6.35)$$

Formula for graph mobilities $|\Phi_{\mathbf{A}}|$ based on edge swaps. Now we turn to the family of so-called edge swaps $F_{ijk\ell}$ (or Seidel switches), which are the most complicated of the move sets in Table 6.1, designed to leave invariant the values of all N degrees in any simple nondirected graph. Again we need to be careful in specifying exactly each distinct edge swap, without overcounting, and this requires a bit more bookkeeping. We can now enumerate the members of the set $\Phi_{\mathbf{A}}$ of edge swaps as follows. We consider all quadruplets (i, j, k, ℓ) of distinct indices, ranked according to $i < j < k < \ell$, and then enumerate the different ways to carry out auto-invertible edge swaps involving each such quadruplet. It turns out that for each ranked quadruplet there are three distinct potential edge swaps, which we will label by $F_{ijk\ell;\alpha}$, where $\alpha \in \{1, 2, 3\}$:

edge swap $F_{ijk\ell;1}$:

edge swap $F_{ijk\ell;2}$:

edge swap $F_{ijk\ell;3}$:

The conditions for each $F_{ijk\ell;\alpha}$ to be applicable on a state $\mathbf{A}$ are that the links shown in each left graphlet are present in $\mathbf{A}$ and the links in each corresponding right graphlet are absent from $\mathbf{A}$, or vice versa. They can be written as follows:

$$F_{ijk\ell;1}: \quad A_{ij}A_{k\ell}(1-A_{jk})(1-A_{i\ell}) + (1-A_{ij})(1-A_{k\ell})A_{jk}A_{i\ell} = 1 \qquad (6.36)$$

$$F_{ijk\ell;2}: \quad A_{ij}A_{k\ell}(1-A_{ik})(1-A_{j\ell}) + (1-A_{ij})(1-A_{k\ell})A_{ik}A_{j\ell} = 1 \qquad (6.37)$$

$$F_{ijk\ell;3}: \quad A_{ik}A_{j\ell}(1-A_{jk})(1-A_{i\ell}) + (1-A_{ik})(1-A_{j\ell})A_{jk}A_{i\ell} = 1 \qquad (6.38)$$

Note that the expressions on the right always consist of two terms, each taking values from $\{0,1\}$ but such that they can never be simultaneously non-zero. Hence, each expression in itself takes values from $\{0,1\}$, and hence acts as a binary indicator to tell us whether on state $\mathbf{A}$ the operation $F_{ijk\ell;\alpha}$ is applicable (1) or not (0).

Each operation can affect only the link variables associated with a particular subset of node pairs (r,s), which can be read off directly from the diagrams above. If we define the subset of each $F_{ijk\ell;\alpha}$ as $\mathcal{S}_{ijk\ell;\alpha}$, and keep in mind the symmetry of $\mathbf{A}$, we find

$$\mathcal{S}_{ijk\ell;1} = \{(i,j),(j,i),(k,\ell),(\ell,k),(j,k),(k,j),(i,\ell),(\ell,i)\} \qquad (6.39)$$

$$\mathcal{S}_{ijk\ell;2} = \{(i,j),(j,i),(k,\ell),(\ell,k),(i,k),(k,i),(j,\ell),(\ell,j)\} \qquad (6.40)$$

$$\mathcal{S}_{ijk\ell;3} = \{(j,\ell),(\ell,j),(i,k),(k,i),(j,k),(k,j),(i,\ell),(\ell,i)\} \qquad (6.41)$$

If their respective admissibility conditions are satisfied, each operation will then execute the following operation on $\mathbf{A}$:

$$(F_{ijk\ell;\alpha}\mathbf{A})_{rs} = A_{rs} \quad \text{if } (r,s) \notin \mathcal{S}_{ijk\ell;\alpha} \qquad (6.42)$$

$$(F_{ijk\ell;\alpha}\mathbf{A})_{rs} = 1 - A_{rs} \quad \text{if } (r,s) \in \mathcal{S}_{ijk\ell;\alpha} \qquad (6.43)$$

We can now proceed to the calculation of the number $|\Phi_{\mathbf{A}}|$ of edge swaps that can be executed on a state $\mathbf{A}$. Since we can use the binary expressions $(6.36, 6.37, 6.38)$ as admissibility indicators, we may write

$$|\Phi_{\mathbf{A}}| = \sum_{i<j<k<\ell} \Big\{ A_{ij}A_{k\ell}(1-A_{jk})(1-A_{i\ell}) + (1-A_{ij})(1-A_{k\ell})A_{jk}A_{i\ell}$$

$$+ A_{ij}A_{k\ell}(1-A_{ik})(1-A_{j\ell}) + (1-A_{ij})(1-A_{k\ell})A_{ik}A_{j\ell}$$

$$+ A_{ik}A_{j\ell}(1-A_{jk})(1-A_{i\ell}) + (1-A_{ik})(1-A_{j\ell})A_{jk}A_{i\ell} \Big\} \qquad (6.44)$$

There are $4! = 24$ possible orderings of the four indices (i,j,k,ℓ). Since the expression between the curly brackets is invariant under any such permutation, we can write our formula as a sum over all non-ranked quadruples of distinct indices, which can be simplified further using the equivalence of various terms after summation over indices:

$$|\Phi_\mathbf{A}| = \frac{1}{24} \sum_{i \neq j \neq k \neq \ell} \Big\{ A_{ij}A_{k\ell}(1-A_{jk})(1-A_{i\ell}) + (1-A_{ij})(1-A_{k\ell})A_{jk}A_{i\ell}$$

$$+ A_{ij}A_{k\ell}(1-A_{ik})(1-A_{j\ell}) + (1-A_{ij})(1-A_{k\ell})A_{ik}A_{j\ell}$$

$$+ A_{ik}A_{j\ell}(1-A_{jk})(1-A_{i\ell}) + (1-A_{ik})(1-A_{j\ell})A_{jk}A_{i\ell} \Big\}$$

$$= \frac{1}{4} \sum_{i \neq j \neq k \neq \ell} A_{ij}A_{k\ell}(1-A_{jk})(1-A_{i\ell})$$

$$= \frac{1}{4} \sum_{i \neq j \neq k \neq \ell} \Big\{ A_{ij}A_{k\ell} + A_{ij}A_{k\ell}A_{jk}A_{i\ell} - A_{ij}A_{k\ell}A_{jk} - A_{ij}A_{k\ell}A_{i\ell} \Big\} \qquad (6.45)$$

We next incorporate the condition that all indices in (i, j, k, ℓ) must be distinct via Kronecker deltas, and we simplify terms wherever possible using $A_{rr} = 0$ for all r:

$$|\Phi_\mathbf{A}| = \frac{1}{4} \sum_{ijk\ell} A_{ij}A_{k\ell}(1-\delta_{ik})(1-\delta_{i\ell})(1-\delta_{jk})(1-\delta_{j\ell})$$

$$+ \frac{1}{4} \sum_{ijk\ell} A_{ij}A_{k\ell}A_{jk}A_{i\ell}(1-\delta_{ik})(1-\delta_{j\ell}) - \frac{1}{2} \sum_{ijk\ell} A_{ij}A_{k\ell}A_{jk}(1-\delta_{ik})(1-\delta_{i\ell})(1-\delta_{j\ell})$$

$$= \frac{1}{4} \sum_{ijk\ell} A_{ij}A_{k\ell}\big[1-\delta_{ik}-\delta_{i\ell}-\delta_{jk}-\delta_{j\ell}+\delta_{ik}\delta_{j\ell}+\delta_{i\ell}\delta_{jk}\big]$$

$$+ \frac{1}{4} \sum_{ijk\ell} A_{ij}A_{k\ell}A_{jk}A_{i\ell}\big[1-\delta_{ik}-\delta_{j\ell}+\delta_{ik}\delta_{j\ell}\big]$$

$$- \frac{1}{2} \sum_{ijk\ell} A_{ij}A_{k\ell}A_{jk}\big[1-\delta_{ik}-\delta_{i\ell}-\delta_{j\ell}+\delta_{ik}\delta_{j\ell}\big]$$

$$= \frac{1}{4} \Big[\sum_i k_i(\mathbf{A})\Big]^2 + \frac{1}{4} \sum_i k_i(\mathbf{A}) - \frac{1}{2} \sum_i k_i^2(\mathbf{A})$$

$$- \frac{1}{2} \sum_{ij} A_{ij}k_i(\mathbf{A})k_j(\mathbf{A}) + \frac{1}{2}\mathrm{Tr}(\mathbf{A}^3) + \frac{1}{4}\mathrm{Tr}(\mathbf{A}^4) \qquad (6.46)$$

in which $\mathrm{Tr}(\mathbf{B}) = \sum_i B_{ii}$ (this is called the *trace* of the matrix $\mathbf{B}$). Using edge swaps as our move sets implies that all N degrees are always conserved. Hence, the first line of (6.46) is invariant during the process, and needs to be calculated only once. The non-trivial terms are those in the second line, which will generally *not* remain stationary during the execution of our algorithm. In Appendix F, we calculate formulae for the change in $|\Phi_\mathbf{A}|$ as a consequence of an executed edge swap, involving only modest computational effort, which can be used to keep track of the value of $|\Phi_\mathbf{A}|$ dynamically during the algorithm.

6.5   Solutions of exercises

Exercise 6.1

We start with the Metropolis–Hastings recipe (6.8). Insertion into (6.7) leads to the following requirements:

$$\forall \mathbf{A} \in G, \ \forall F \in \Phi: \quad p(F\mathbf{A}) \min\left\{1, \frac{p(\mathbf{A})}{p(F\mathbf{A})}\right\} = p(\mathbf{A}) \min\left\{1, \frac{p(F\mathbf{A})}{p(\mathbf{A})}\right\} \quad (6.47)$$

For any given $\mathbf{A} \in G$ and $F \in \Phi$, we simply need to work out which are the relevant minima in the equation that is required to hold, for three possible scenarios:

$$p(\mathbf{A}) < p(F\mathbf{A}): \quad p(F\mathbf{A})\big[p(\mathbf{A})/p(F\mathbf{A})\big] = p(\mathbf{A}) \quad (6.48)$$
$$p(\mathbf{A}) = p(F\mathbf{A}): \quad p(F\mathbf{A}) = p(\mathbf{A}) \quad (6.49)$$
$$p(\mathbf{A}) > p(F\mathbf{A}): \quad p(F\mathbf{A}) = p(\mathbf{A})\big[p(F\mathbf{A})/p(\mathbf{A})\big] \quad (6.50)$$

In all three cases, we find the relevant identity to hold, so the Markov process with acceptance probabilities (6.8) indeed obeys detailed balance. Next we insert the Glauber recipe (6.9) into the detailed balance condition (6.7), which gives

$$\forall \mathbf{A} \in G, \ \forall F \in \Phi: \quad \frac{p(F\mathbf{A})p(\mathbf{A})}{p(\mathbf{A}) + p(F\mathbf{A})} = \frac{p(\mathbf{A})p(F\mathbf{A})}{p(F\mathbf{A}) + p(\mathbf{A})} \quad (6.51)$$

This is evidently always true, so the Markov process with acceptance probabilities (6.9) also obeys detailed balance.

Trivial alternative recipes that satisfy the detailed balance condition would be obtained by multiplying (6.8, 6.9) by a constant $\gamma \in (0,1)$. Alternatively, one could generalize the Glauber formula (6.9) to a larger class of acceptable recipes, upon replacing the denominator with any expression $f[p(\mathbf{A}), p(F(\mathbf{A})]$, where $f[x,y] = f[y,x]$ (to ensure that the detailed balance condition holds) and $f[x,y] \geq \min\{x,y\}$ for all $x, y \in [0,1]$ (to make sure the probabilities always remain in the interval $[0,1]$). The standard Glauber prescription corresponds to $f[x,y] = x+y$. □

Exercise 6.2

We start with the modified Metropolis–Hastings recipe (6.13). Insertion into (6.12) now leads to the following requirements:

$$\forall \mathbf{A} \in G, \ \forall F \in \Phi: \quad \frac{p(F\mathbf{A})}{|\Phi_{F\mathbf{A}}|} \min\left\{1, \frac{p(\mathbf{A})/|\Phi_{\mathbf{A}}|}{p(F\mathbf{A})/|\Phi_{F\mathbf{A}}|}\right\} = \frac{p(\mathbf{A})}{|\Phi_{\mathbf{A}}|} \min\left\{1, \frac{p(F\mathbf{A})/|\Phi_{F\mathbf{A}}|}{p(\mathbf{A})/|\Phi_{\mathbf{A}}|}\right\}$$
$$(6.52)$$

For any given $\mathbf{A} \in G$ and $F \in \Phi$, we again need to work out which are the relevant minima in the equation that is required to hold, for three possible scenarios:

$$\frac{p(\mathbf{A})}{|\Phi_{\mathbf{A}}|} < \frac{p(F\mathbf{A})}{|\Phi_{F\mathbf{A}}|}: \quad \frac{p(F\mathbf{A})}{|\Phi_{F\mathbf{A}}|} \frac{p(\mathbf{A})/|\Phi_{\mathbf{A}}|}{p(F\mathbf{A})/|\Phi_{F\mathbf{A}}|} = \frac{p(\mathbf{A})}{|\Phi_{\mathbf{A}}|} \quad (6.53)$$

$$\frac{p(\mathbf{A})}{|\Phi_{\mathbf{A}}|} = \frac{p(F\mathbf{A})}{|\Phi_{F\mathbf{A}}|} : \quad \frac{p(F\mathbf{A})}{|\Phi_{F\mathbf{A}}|} = \frac{p(\mathbf{A})}{|\Phi_{\mathbf{A}}|} \tag{6.54}$$

$$\frac{p(\mathbf{A})}{|\Phi_{\mathbf{A}}|} > \frac{p(F\mathbf{A})}{|\Phi_{F\mathbf{A}}|} : \quad \frac{p(F\mathbf{A})}{|\Phi_{F\mathbf{A}}|} = \frac{p(\mathbf{A})}{|\Phi_{\mathbf{A}}|} \frac{p(F\mathbf{A})/|\Phi_{F\mathbf{A}}|}{p(\mathbf{A})/|\Phi_{\mathbf{A}}|} \tag{6.55}$$

The relevant condition again holds in all cases, so the modified Markov process with acceptance probabilities (6.13) obeys detailed balance. Next we insert the modified Glauber recipe (6.14) into the detailed balance condition (6.12), which gives

$$\forall \mathbf{A} \in G, \ \forall F \in \Phi : \quad \frac{\left(p(F\mathbf{A})/|\Phi_{F\mathbf{A}}|\right)\left(p(\mathbf{A})/|\Phi_{\mathbf{A}}|\right)}{p(\mathbf{A})/|\Phi_{\mathbf{A}}|+p(F\mathbf{A})/|\Phi_{F\mathbf{A}}|} = \frac{\left(p(\mathbf{A})/|\Phi_{\mathbf{A}}|\right)\left(p(F\mathbf{A})/|\Phi_{F\mathbf{A}}|\right)}{p(F\mathbf{A})/|\Phi_{F\mathbf{A}}|+p(\mathbf{A})/|\Phi_{\mathbf{A}}|} \tag{6.56}$$

This is evidently always true, so the modified Markov process with acceptance probabilities (6.14) also obeys detailed balance. $\square$

Exercise 6.3

We assume that the chosen sequence $\mathbf{k} = (k_1, \ldots, k_N)$ has been subjected to and passed the graphicality test (4.23). Once more we can work with the set Φ of simple elementary moves $\{F_{ij}\}$, defined by (6.18, 6.19), and (6.20) now becomes

$$\Delta_{ij}(\mathbf{A}) = \sum_{r=1}^{N} \lambda_r \big[k_r(F_{ij}\mathbf{A}) - k_r(\mathbf{A}) \big]$$

$$= (1 - 2A_{ij}) \sum_{r=1}^{N} \lambda_r (\delta_{ir} + \delta_{jr}) = (1 - 2A_{ij})(\lambda_i + \lambda_j) \tag{6.57}$$

Substitution into the move acceptance probabilities (6.16, 6.17) now leads to

$$\text{Metropolis–Hastings}: \quad \mathcal{A}[\mathbf{A} \to F\mathbf{A}] = \min\left\{1, e^{(1-2A_{ij})(\lambda_i+\lambda_j)}\right\} \tag{6.58}$$

$$\text{Glauber}: \quad \mathcal{A}[\mathbf{A} \to F\mathbf{A}] = \left(1 + e^{-(1-2A_{ij})(\lambda_i+\lambda_j)}\right)^{-1} \tag{6.59}$$

We provide the pseudocode to generate simple nondirected N-node graphs with a prescribed average degree sequence for the Metropolis–Hastings choice in Algorithm 2. In the regime where $k_i \ll N$ for all i, the parameters $\{\lambda_i\}$ can be evaluated explicitly, giving $\exp(\lambda_i) = k_i/\sqrt{N\langle k\rangle} + \mathcal{O}(N^{-3/2})$, with $\langle k\rangle = N^{-1}\sum_{i \leq N} k_i$. Here we may replace in the above recipes $e^{\lambda_i + \lambda_j} \to k_i k_j / N\langle k\rangle$. In fact, in the latter regime it would be rather inefficient to use MCMC samplers, since according to (4.22) we could generate all link variables $\{A_{ij}\}$ directly and independently in one go according to $p(A_{ij}) = (k_i k_j / N\langle k\rangle)\delta_{A_{ij},1} + (1 - k_i k_j / N\langle k\rangle)\delta_{A_{ij},0}$. $\square$

Exercise 6.4

We have not (yet) considered ERGM ensembles in which the $p(k|\mathbf{A})$ for $k \in \{1, \ldots, N\}$ are themselves chosen to be the constrained observables, but we can generate the desired graphs in a two-stage process. We first generate N degrees independently at random from $p(k)$, subject to the graphicality condition (4.23). Having found an acceptable degree sequence, we can proceed according to the protocol of the previous exercise. This leads to Algorithm 3. For values $\gamma < 2$, it will be increasingly difficult to find graphical degree sequences, so this algorithm will slow down considerably.   □

Exercise 6.5

Again we construct an MCMC algorithm based on the set Φ of simple elementary moves $\{F_{ij}\}$, defined by (6.18,6.19). Inserting the definitions of the two contained observables $\Omega_1(\mathbf{A}) = \sum_{r<s} A_{rs}$ and $\Omega_2(\mathbf{A}) = \sum_{r<\ell<s} A_{r\ell} A_{\ell s} A_{sr}$ of the ERGM (4.61) into formula (6.20) now gives

$$\Delta_{ij}(\mathbf{A}) = \lambda_1(1-2A_{ij}) + \lambda_2 \sum_{r<\ell<s} \Big[(F_{ij}A_{r\ell})(F_{ij}A_{\ell s})(F_{ij}A_{sr}) - A_{r\ell}A_{\ell s}A_{sr} \Big]$$

$$= \lambda_1(1-2A_{ij}) + \lambda_2(1-2A_{ij}) \sum_r A_{ir}A_{jr} \tag{6.60}$$

The second term in the last line follows from the following argument: the sum $\sum_r A_{ir}A_{jr}$ equals both the number of 'lost' triangles upon removing an existing link $A_{ij} = 1$, and also the number of 'gained' triangles upon creating a link, where we presently have $A_{ij} = 0$. Substitution into the move acceptance probabilities (6.16,6.17) then gives

$$\text{Metropolis–Hastings :} \quad \mathcal{A}[\mathbf{A} \to F\mathbf{A}] = \min\Big\{ 1, e^{(1-2A_{ij})[\lambda_1 + \lambda_2 \sum_r A_{ir}A_{jr}]} \Big\} \tag{6.61}$$

$$\text{Glauber :} \quad \mathcal{A}[\mathbf{A} \to F\mathbf{A}] = \Big(1 + e^{-(1-2A_{ij})[\lambda_1 + \lambda_2 \sum_r A_{ir}A_{jr}]} \Big)^{-1} \tag{6.62}$$

As with the two-star ensemble, also here we have no explicit solutions of (λ_1, λ_2) in terms of the desired average numbers (L, T) of links and triangle, except in special limits and upon making approximations. Hence, we retain the pair (λ_1, λ_2) as control parameters, and give the Metropolis–Hastings MCMC pseudocode for generating simple nondirected N-node graphs $\mathbf{A}$ randomly from (4.61) in Algorithm 5. This MCMC algorithm was used to produce the data in Figure 4.5.   □

7

Graphs with hard constraints: further applications and extensions

Introduction. In this chapter, we work out details of implementation of the Markov Chain Monte Carlo (MCMC) approach to generate hard-constrained random graphs. Specifying hard constraints complicates the definition, selection and counting of appropriate candidate moves. There is a clear need to speed up the generally slow MCMC algorithms, but shortcuts have to be chosen and implemented with great care to avoid introducing unwanted modifications of the sampling probabilities.

7.1 Uniform versus non-uniform sampling of candidate moves

Overview: In this section, we dive into the algorithmic technicalities of sampling candidate moves uniformly. We consider three apparently similar ways to sample hinge flips, and use them to randomize a graph. The target is a random graph with Poissonian degree distribution, but depending on the way we sample hinge flips, we may find something very different. These investigations illustrate the importance of ensuring that candidate moves are sampled uniformly, otherwise the acceptance probabilities need to be appropriately corrected.

MCMC algorithms for non-uniform candidate move sampling. All MCMC algorithms developed so far require that the candidate moves F are sampled uniformly, either from the set Φ of all possible moves (for switch-and-hold MCMC) or from the set $\Phi_\mathbf{A}$ of presently applicable moves (for mobility-based MCMC). Recall that with mobility-based MCMC only legal moves are proposed and it is necessary to adjust the move acceptance probability on the basis of the number of legal moves that can act on the configuration. Switch-and-hold MCMC works by regarding the proposal of any illegal move as an iteration spent in the current configuration. All these MCMC versions evolve towards the correct probabilities $p(\mathbf{A})$, but they tend to become slow in the two extremes: in sparse graphs, the likelihood of drawing a connected node pair is very small; in the opposite regime of nearly fully connected graphs, the likelihood of picking two disconnected nodes is very small. In both cases, the mobility-based MCMC algorithm will slow down because allowed moves will be harder to find, and the switch-and-hold MCMC algorithm will be slow to converge because it will spend most of its iterations in the same configurations.

Before we begin the discussion of how to (safely) optimize move sampling, we will explicitly describe the basic sampling procedures for the set Φ of hinge flips $\{F_{ijk}\}$

Generating Random Networks and Graphs. First Edition. A.C.C. Coolen, A. Annibale, and E.S. Roberts. © A.C.C. Coolen, A. Annibale, and E.S. Roberts 2017. Published in 2017 by Oxford University Press. DOI: 10.1093/acprof:oso/9780198709893.001.0001

(as described in Table 6.1). For the mobility-based MCMC algorithm, we enumerate at each iteration all the triplets (i, j, k) such that $A_{ij} = 1$ and $A_{ik} = 0$, which would constitute the set $\Phi_{\mathbf{A}}$, and then draw one of these triplets uniformly at random. In the case of switch-and-hold MCMC, we would draw at each iteration step a triplet (i, j, k) from the $|\Phi| = N(N-1)(N-2)$ possible triplets. The triplet is rejected if it does not satisfy $A_{ij} = 1$ and $A_{ik} = 0$; the network remains in its current state, and another triplet is drawn.

It is possible to use algorithm shortcuts to make the sampling of allowed candidate moves more efficient, but one should think very carefully about the biases that these can potentially introduce. Shortcuts usually take the form of alternative non-uniform ways to generate candidate moves F, so we have to generalize our derivation of the MCMC equations to allow for non-uniform move proposal probabilities. If we denote with $P(F|\mathbf{A})$ the likelihood that we propose move $F \in \Phi$ when in state $\mathbf{A}$, the transition probabilities (6.3) for the switch-and-hold and (6.11) for the mobility-based based algorithms now take the following forms, respectively

$$W[\mathbf{A}' \to \mathbf{A}] = \sum_{F \in \Phi} P(F|\mathbf{A}')\left\{\mathcal{A}[\mathbf{A}' \to F\mathbf{A}']\delta_{\mathbf{A}, F\mathbf{A}'} + \left(1 - \mathcal{A}[\mathbf{A}' \to F\mathbf{A}']\right)\delta_{\mathbf{A}, \mathbf{A}'}\right\} \tag{7.1}$$

$$W[\mathbf{A}' \to \mathbf{A}] = \sum_{F \in \Phi_{\mathbf{A}'}} P(F|\mathbf{A}')\left\{\mathcal{A}[\mathbf{A}' \to F\mathbf{A}']\delta_{\mathbf{A}, F\mathbf{A}'} + \left(1 - \mathcal{A}[\mathbf{A}' \to F\mathbf{A}']\right)\delta_{\mathbf{A}, \mathbf{A}'}\right\} \tag{7.2}$$

In (7.1) we have $\sum_{F \in \Phi} P(F|\mathbf{A}') = 1$, and we reject via the move acceptance probabilities those moves that violate our constraints. In (7.2) we have $\sum_{F \in \Phi_{\mathbf{A}}} P(F|\mathbf{A}') = 1$, and we need not reject any moves.

If one substitutes (7.1) and (7.2) into the detailed balance condition (6.5), the non-uniform move proposal probabilities $P(F|\mathbf{A}')$ will alter the MCMC move acceptance probabilities. For the switch-and-hold MCMC, the conventional Metropolis–Hastings and Glauber recipes (6.8, 6.9) for the acceptance probabilities of those moves that can be executed on state $\mathbf{A}$, i.e. the moves $F \in \Phi_{\mathbf{A}}$, become

$$\text{Metropolis–Hastings}: \quad \mathcal{A}[\mathbf{A} \to F\mathbf{A}] = \min\left\{1, \frac{p(F\mathbf{A})P(F|F\mathbf{A})}{p(\mathbf{A})P(F|\mathbf{A})}\right\} \tag{7.3}$$

$$\text{Glauber}: \quad \mathcal{A}[\mathbf{A} \to F\mathbf{A}] = \frac{p(F\mathbf{A})P(F|F\mathbf{A})}{p(F\mathbf{A})P(F|F\mathbf{A}) + p(\mathbf{A})P(F|\mathbf{A})} \tag{7.4}$$

The corresponding Metropolis–Hastings and Glauber recipes (6.13, 6.14) for the mobility-based MCMC also become (7.3) and (7.4), but here the move proposal probabilities are calculated very differently, and normalized via summation only over the set of *instantaneously feasible* moves. To use sampling shortcuts, we must be able to precisely quantify $P(F|\mathbf{A})$, the probability of picking each candidate move. Failure to do this will result in our MCMC process generating graphs with probabilities that are different from our original target $p(\mathbf{A})$.

Illustration – nondirected graphs with hard-constrained number of links. Let us consider the generation of random graphs from the ensemble (5.16), which describes nondirected

PROTOCOL 1:

(i) pick a site j with $k_j(\mathbf{A}) > 0$
(ii) pick a site $i \in \partial_j(\mathbf{A})$
(iii) pick a site $k \notin \partial_i(\mathbf{A}) \cup \{i\}$

PROTOCOL 2:

(i) pick two disconnected sites
     (i, k) with $k_i(\mathbf{A}) > 0$
(ii) pick a site $j \in \partial_i(\mathbf{A})$

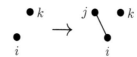

PROTOCOL 3:

(i) pick two connected sites (i, j)
     and a third site k
(ii) while $A_{ik} = 1$ return to (i)

Table 7.1 Three possible ways to sample candidates from the set $\Phi_{\mathbf{A}}$ of allowed hinge flip moves in nondirected simple graphs. Here $\partial_i(\mathbf{A})$ denotes the set of nodes connected to i. For a move F_{ijk} to be allowed, the graph must have $A_{ij} = 1$ and $A_{ik} = 0$. At any stage, the nodes and/or links are picked uniformly from the specified sets. All three protocols lead to different probabilities $P(F_{ijk}|\mathbf{A})$ for finding given candidate moves, so they require different move acceptance probabilities to ensure that the corresponding MCMC processes will generate graphs with the correct probabilities.

simple N-node graphs in which the total number of links is fixed to be L and all allowed graphs are given equal probability:

$$G = \{0, 1\}^{N(N-1)/2}, \qquad p(\mathbf{A}) = \frac{\delta_{L, \sum_{i<j} A_{ij}}}{Z(L)}, \qquad Z(L) = \sum_{\mathbf{A} \in G} \delta_{L, \sum_{i<j} A_{ij}} \qquad (7.5)$$

To generate graphs from this ensemble, we can follow the recipes of Chapter 6 and construct graph-randomizing Markov chains that preserve the total number of links. To achieve this, we can choose for our moves the hinge flips $\{F_{ijk}\}$ described in Table 6.1. The mobility $|\Phi_{\mathbf{A}}|$ for this choice of elementary moves was calculated in (6.31), and can be written in terms of the degrees $k_i(\mathbf{A}) = \sum_j A_{ij}$ as

$$|\Phi_{\mathbf{A}}| = 2L(N-1) - \sum_i k_i^2(\mathbf{A}) = N(N-1)\langle k \rangle - \sum_i k_i^2(\mathbf{A}) \qquad (7.6)$$

We now investigate three different ways of sampling moves F_{ijk}. They are specified and illustrated in Table 7.1. The protocols in this table are just three possible choices;

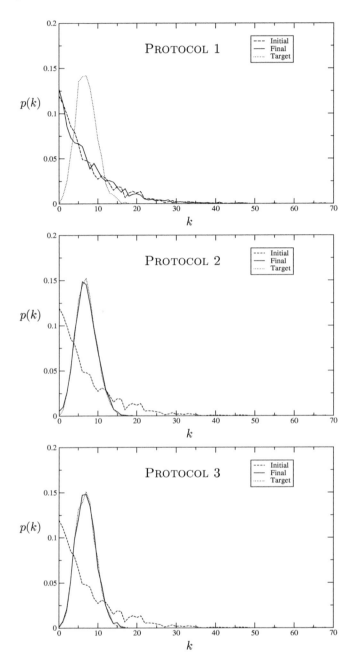

Fig. 7.1 Degree distributions of nondirected simple graphs with $N = 3000$ and $\langle k \rangle = 7$, before and after hinge flip randomization via the MCMC process (6.13), where applicable moves are proposed with uniform probabilities. The initial graph has a power-law degree distribution. The three panels correspond to the three protocols for picking candidate moves described in Table 7.1. Dashed lines: $p(k)$ of the initial (seed) graph. Dotted lines: $p(k)$ of the targeted ensemble distribution $p(\mathbf{A})$. Solid lines: Observed $p(k)$ after the equilibration of the MCMC process.

we could list many more. They may at first glance seem equivalent, but we will see that they lead to drastically different results if one does not take care to account for the differences in the associated move proposal probabilities $P(F_{ijk}|\mathbf{A})$.

In Figure 7.1 we illustrate the consequences of assuming (incorrectly) that all three protocols in Table 7.1 sample the candidate moves from $\Phi_{\mathbf{A}}$ uniformly, and of using the standard mobility-based MCMC acceptance probabilities (6.13,6.14) with formula (7.6) for the graph mobilities instead of (7.3,7.4). In these randomization simulations, the graphs had size $N = 3000$, with average degree $\langle k \rangle = 7$ (these numbers are by construction conserved during the MCMC process). They were initialized with a power-law degree distribution. Strikingly, if we use Protocol 1 to sample the candidate hinge flip moves F_{ijk}, the final degree distribution hardly differs from the initial one (top panel), while both Protocol 2 (middle panel) and Protocol 3 (bottom panel) show a reasonable convergence to the Poisson degree distribution that we would expect to find for graphs of ensemble (7.5). This confirms that what might perhaps seem to be minor implementation details, i.e. precisely how one codes the picking of candidate moves in the MCMC, can have a significant impact on the statistical features of the graphs being generated. The three protocols in Table 7.1 may seem equivalent, but they clearly are not.

Uniform sampling from $\Phi_{\mathbf{A}}$ would require that each allowed hinge flip F_{ijk} is proposed with the same probability

$$\frac{1}{|\Phi_{\mathbf{A}}|} = \frac{1}{N(N-1)\langle k \rangle - \sum_i k_i^2(\mathbf{A})} \tag{7.7}$$

The differences between protocols that are visible in Figure 7.1 are due to the fact that in Protocols 1 and 2 the candidate moves are *not* sampled uniformly. While in Protocol 2 the bias introduced is too small to be noticeable, in Protocol 1 it has a big effect. To confirm this explanation, let us determine the likelihood $P(F_{ijk}|\mathbf{A})$ with which one will draw any given candidate move F_{ijk} for Protocol 1, with $p(k|\mathbf{A})$ denoting the instantaneous degree distribution:

$$\text{Protocol 1:} \quad P(F_{ijk}|\mathbf{A}) = P(j|k_j(\mathbf{A})>0) \; P(i|i\in\partial_j(\mathbf{A})) \; P(k|k\notin\partial_i(\mathbf{A}))$$

$$= \frac{1}{\sum_{\ell=1}^{N}(1-\delta_{k_\ell(\mathbf{A}),0})} \frac{1}{k_j(\mathbf{A})} \frac{1}{N-1-k_i(\mathbf{A})}$$

$$= \frac{1}{N[1-p(0|\mathbf{A})]} \frac{1}{k_j(\mathbf{A})} \frac{1}{N-1-k_i(\mathbf{A})}$$

$$= \frac{1}{N^2[1-p(0|\mathbf{A})]} \frac{1}{k_j(\mathbf{A})} + \mathcal{O}(N^{-3}) \tag{7.8}$$

where we denoted by ∂_j the set of all nodes that are connected to j. Even for large N we will not have $P(F_{ijk}|\mathbf{A}) = |\Phi_{\mathbf{A}}|^{-1}$, and an MCMC using the standard acceptance probabilities (6.13,6.14) with formula (7.6) will consequently *not* evolve towards (7.5).

According to (7.3), to target the probabilities of (7.5), which obey $p(F_{ijk}\mathbf{A}) = p(\mathbf{A})$ for any hinge flip F_{ijk}, we should in the case of Protocol 1 have used the following

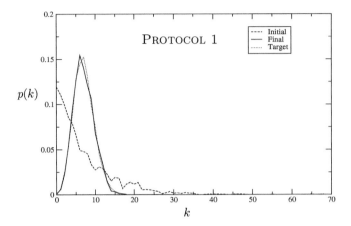

Fig. 7.2 Degree distributions of nondirected simple graphs with $N = 3000$ and $\langle k \rangle = 7$, before and after hinge flip randomization for Protocol 1 in Table 7.1, as in Figure 7.1, but now using the adapted MCMC process (7.3) that incorporates the non-uniform proposal probabilities (7.8) of applicable moves. The target probability distribution is (7.5). Dashed lines: $p(k)$ of the initial (see) graph. Dotted lines: $p(k)$ of the targeted ensemble distribution $p(\mathbf{A})$. Solid lines: Observed $p(k)$ after equilibration of the MCMC process.

modified acceptance probabilities:[1]

$$\mathcal{A}[\mathbf{A} \to F_{ijk}\mathbf{A}] = \min\left\{1, \frac{p(F_{ijk}\mathbf{A})P(F_{ikj}|F_{ijk}\mathbf{A})}{p(\mathbf{A})P(F_{ijk}|\mathbf{A})}\right\}$$

$$= \min\left\{1, \frac{P(F_{ikj}|F_{ijk}\mathbf{A})}{P(F_{ijk}|\mathbf{A})}\right\}$$

$$= \min\left\{1, \frac{k_j(\mathbf{A})}{k_k(F_{ijk}\mathbf{A})} + \mathcal{O}\left(\frac{1}{N}\right)\right\}$$

$$= \min\left\{1, \frac{k_j(\mathbf{A})}{k_k(\mathbf{A})+1} + \mathcal{O}\left(\frac{1}{N}\right)\right\} \tag{7.9}$$

where we used $p(0|F_{ijk}\mathbf{A}) = p(0|\mathbf{A}) + \mathcal{O}(N^{-1})$. Equivalently, we could use the Glauber version (7.3). Upon using these amended acceptance probabilities (7.9), Protocol 1 does indeed converge to a stationary state with the expected Poissonian degree distribution (see Figure 7.2).

In the same manner, we can inspect the likelihood $P(F_{ijk}\mathbf{A})$ of drawing a candidate move when using Protocol 2:

Protocol 2: $P(F_{ijk}|\mathbf{A}) = P(i,k|A_{ik}=0, \; k_i(\mathbf{A})>0) \; P(j|j \in \partial_i(\mathbf{A}))$

$\qquad\qquad\quad = P(i,k|A_{ik}=0) \; P(k_i(\mathbf{A})>0|A_{ik}=0) \; P(j|j \in \partial_i(\mathbf{A}))$

---

[1]We need to keep in mind that our MCMC equations have so far been constructed with the convention that all moves F obey $F^2 = \mathbf{1}$, and hence we must interpret F_{ijk} and F_{ikj} as two possible actions of the *same* hinge flip: the first acting if $A_{ij}(1-A_{ik}) = 1$, and the second if $A_{ik}(1-A_{ij}) = 1$.

$$= \frac{1}{\sum_{r\neq s=1}^{N}(1-A_{rs})}\left[1-p(0|\mathbf{A})+\mathcal{O}\left(\frac{1}{N}\right)\right]\frac{1}{k_i(\mathbf{A})}$$

$$= \frac{1-p(0|\mathbf{A})}{N(N-1)-N\langle k\rangle}\frac{1}{k_i(\mathbf{A})}+\mathcal{O}(N^{-1}) \qquad (7.10)$$

Apparently, in Protocol 2 also we have $P(F_{ijk}|\mathbf{A}) \neq |\Phi_\mathbf{A}|^{-1}$. However, if we now work out the appropriate MCMC move acceptance probabilities (7.3), using once more $p(0|F_{ijk}\mathbf{A}) = p(0|\mathbf{A}) + \mathcal{O}(N^{-1})$, we obtain

$$\mathcal{A}[\mathbf{A}\to F_{ijk}\mathbf{A}] = \min\left\{1, \frac{p(F_{ijk}\mathbf{A})P(F_{ijk}|F_{ijk}\mathbf{A})}{p(F_{ijk}\mathbf{A})P(F_{ijk}|\mathbf{A})}\right\}$$

$$= \min\left\{1, \frac{P(F_{ikj}|F_{ijk}\mathbf{A})}{P(F_{ijk}|\mathbf{A})}\right\}$$

$$= \min\left\{1, \frac{k_i(\mathbf{A})}{k_i(F_{ijk}\mathbf{A})}+\mathcal{O}\left(\frac{1}{N}\right)\right\} = 1-\mathcal{O}\left(\frac{1}{N}\right) \qquad (7.11)$$

The last step follows from the simple fact that the degree $k_i(\mathbf{A})$ of the pivotal node i of the hinge flip F_{ijk} is not affected by the move. Hence, for large N all proposed moves will be accepted in the process. In fact, this is also true when we use the uniform move proposal probabilities (7.7), since $|\Phi_\mathbf{A}|^{-1} = (N^2\langle k\rangle)^{-1}+\mathcal{O}(N^{-1})$. For Protocol 2, the differences between the MCMC of Figure 7.1 and the version that would have led to the correct $p(\mathbf{A})$ are sub-leading in the system size N. This explains why the (in principle incorrect) graph randomization in Figure 7.1 still performed well in practice. For small graphs, in contrast, the differences may well become important. Only Protocol 3 in Table 7.1 represents a fully correct way of sampling uniformly from the set $\Phi_\mathbf{A}$ of applicable hinge flip moves.

Implementation options for correct sampling. The above discussion shows that precision in matching move selection protocols to their corresponding move acceptance probabilities is not a luxury. If we wish to estimate via MCMC sampling the average value of some observable $\Omega(\mathbf{A})$ over all random nondirected graphs with a prescribed number of links and with *uniform probabilities* $p(\mathbf{A}) = 1/|G|$, then we can use one of the following routes. Which of these will be most time efficient will depend on the dimensions (N, L) of the graphs under consideration.

The first route involves sampling moves homogeneously only from the set $\Phi_\mathbf{A}$ of applicable moves, following Protocol 3 of Table 7.1; here we need the non-trivial state-dependent move acceptance probabilities derived in Section 6.3, with the mobilities $|\Phi_\mathbf{A}|$ of the hinge flip move set, as given by (6.31). This algorithm samples correctly for any graph size N, provided it is run for long enough, and its pseudocode is provided in Algorithm 13 of Appendix J.

The second route involves sampling inhomogeneously from the set of applicable moves, as in Protocol 1 of Table 7.1, but now with the correct corresponding move acceptance probabilities as given in (7.9). This algorithm was used to produce Figure 7.2, and will produce correct estimates of $\langle\Omega(\mathbf{A})\rangle$ for sufficiently large N, and provided it is run for long enough. Its pseudocode is provided in Algorithm 14 of Appendix J.

The third route is to run a switch-and-hold MCMC algorithm, based on homogeneous sampling from the set of all moves (whether applicable or non-applicable). Any applicable move is executed, whereas selecting a non-applicable move leads to the next state being identical to the current one. This algorithm will produce correct averages for any value of N, provided it is run for long enough. Its pseudocode is given in Algorithm 15 of Appendix J.

In the regime of sparse graphs, the above three methods all provide much more efficient ways to sample hinge flips than selecting three nodes at random, as they essentially sample *a link and a node* as opposed to sampling *three nodes*. Let us compare the three methods. The third (switch-and-hold) algorithm does not require calculation of acceptance probabilities at each iteration step, and hence its individual iteration steps will take less time than those of the first algorithm, even though the two methods sample moves in the same way. However, one has to bear in mind that the actual number of iterations needed to achieve a certain number of *executed* moves may be much larger in the switch-and-hold algorithm, due to the fact that in switch-and-hold many successive iteration steps may be spent in the same configuration if this configuration allows for only a small number of acceptable moves. This will happen in the regime of very dense graphs for the sampling method used in the above (switch-and-hold) algorithm, where a link and a node are sampled uniformly at random, while it will happen in *both* the regimes of sparse and very dense graphs with the *naive* method, which draws uniformly at random three nodes. Further, in general, any algorithm that is based on homogeneous move sampling will be difficult to equilibrate towards the true target probability measure in either the sparse regime or the dense regime or possibly both. It is therefore always advisable to define and monitor observable quantities that can indicate whether equilibration has been achieved. One could keep track of the total number of executed switches (which must be significantly larger than the number L of links), or count the fraction of links that have not yet switched (which must be close to zero).

When uniform sampling is not effective, non-uniform sampling may be a better option. For example, one may draw an edge (i, j), and then a site k uniformly at random among the $N - k_i - 1$ nodes which are not connected to i and are not i itself. This amounts to sampling move F_{ijk} non-uniformly with probability $1/[N\langle k\rangle(N-1-k_i)]$. Since the degree k_i of the hinge does not change under the flip, this probability will drop out of the detailed balance conditions and will not require computing corrections to restore detailed balance. However, in general, non-uniform samplings do require computing the acceptance probabilities that restore detailed balance at each iteration, with the associated computational cost, which may be a price worth paying if applicable moves are difficult to find.

7.2   Graphs with a prescribed degree distribution and number of links

Overview: In this section, we derive an algorithm, based on hinge flips, to generate graphs with a prescribed degree distribution and a hard-constrained number of links. This algorithm targets the desired degree distribution directly and does not require the prior generation of a degree sequence from that distribution. Algorithmic sampling of hinge flips must be handled with the usual due care.

Alternative route for building degree statistics into graphs. In Exercise 6.3 we constructed an algorithm for producing graphs with a soft-constrained degree sequence. It is easy to implement once the degree sequence is known, but the generation of a valid degree sequence from a given degree distribution can be difficult, since randomly generated degree sequences may be non-graphical. For example, for power-law degree distributions $p(k) \sim k^{-\gamma}$ with $1 < \gamma < 2$, the graphicality condition (4.23) requires that the largest degree obeys $k_{\max} \leq \sqrt{N\langle k \rangle}$. Given that $\langle k \rangle \sim k_{\max}^{2-\gamma}$, one must choose $k_{\max} \sim N^{1/\gamma}$ to have a finite average degree for large $N \to \infty$. If we draw degrees at random from $p(k) \sim k^{-\gamma}$, the likelihood of generating a degree larger than $k_{\max}$ is $\mathcal{O}(N^{(1-\gamma)/\gamma})$ and the likelihood of generating a graphical degree sequence is therefore $(1 - N^{(1-\gamma)/\gamma})^N \simeq \exp(-N^{1/\gamma})$, which vanishes exponentially fast as N becomes large. This algorithm thus becomes unfeasible for $\gamma < 2$. An alternative recipe for generating graphs with a prescribed degree distribution, without the need to generate the degree sequence first, would therefore be extremely useful.

In Section 5.3 we found that one can generate graphs with a prescribed degree distribution $p(k)$ also in the following way[2]. We choose $L = \frac{1}{2} N \sum_k k p(k)$ and define

$$p(\mathbf{A}|L, p) = \sum_{k_1 \ldots k_N} p(\mathbf{A}|\mathbf{k}) p(\mathbf{k}|L, p) \qquad (7.12)$$

with

$$p(\mathbf{k}|L, p) = \frac{\delta_{L, \frac{1}{2} \sum_i k_i}}{\tilde{Z}(L, p)} \left[\prod_{i=1}^N p(k_i) \right], \qquad \tilde{Z}(L, p) = \sum_{k_1 \ldots k_N} \delta_{L, \frac{1}{2} \sum_i k_i} \left[\prod_{i=1}^N p(k_i) \right] \quad (7.13)$$

$$p(\mathbf{A}|\mathbf{k}) = \frac{\prod_i \delta_{k_i, \sum_j A_{ij}}}{Z(\mathbf{k})}, \qquad Z(\mathbf{k}) = \sum_{\mathbf{A} \in G} \prod_i \delta_{k_i, \sum_j A_{ij}} \qquad (7.14)$$

After carrying out the summation over $\{k_1, \ldots, k_N\}$, we can equivalently write

$$p(\mathbf{A}|L, p) = \frac{\left[\prod_i p(k_i(\mathbf{A})) \right] \delta_{L, \frac{1}{2} \sum_i k_i(\mathbf{A})}}{\tilde{Z}(L, p) Z(\mathbf{k}(\mathbf{A}))} \qquad (7.15)$$

We show in Appendix G that

$$Z(\mathbf{k}) = \left(\prod_i \frac{1}{k_i!} \right) e^{\frac{1}{2} N \langle k \rangle [\log(N\langle k \rangle) - 1] + \mathcal{O}(\log N)} \qquad (7.16)$$

With this expression, and upon absorbing all terms that only depend on $\langle k \rangle$ into a new normalization factor $Z(L, p)$, we obtain

$$p(\mathbf{A}|L, p) = \frac{1}{Z(L, p)} \left[\prod_i p(k_i(\mathbf{A})) k_i(\mathbf{A})! \right] \delta_{L, \frac{1}{2} \sum_i k_i(\mathbf{A})} \qquad (7.17)$$

---

[2]In the language of Section 5.3 we choose $g(k) = p(k)$, so the soft-constrained degree distribution is compatible with the imposed number of links and therefore $\alpha = 0$.

MCMC algorithm implementation. We can now build Markov chains that will equilibrate to the above measure $p(\mathbf{A}|L,p)$. Given that we have a hard constraint in (7.17) on the total number L of links, we choose again as elementary moves the hinge flips F_{ijk} (see Table 6.1), and we initialize the algorithm to any graph $\mathbf{A}_0$ with L links. If the moves are drawn uniformly at random from the set $\Phi_\mathbf{A}$ of allowed moves, i.e. if we follow Protocol 3 in Table 7.1, we can use in the Metropolis–Hastings recipe (7.3) the move proposal probabilities $P(F_{ijk}|\mathbf{A}) = 1/|\Phi_\mathbf{A}|$, which follow directly from (7.6). Hence, our MCMC will be defined by

$$A[\mathbf{A} \to F_{ijk}\mathbf{A}] = \min\left\{1, \frac{p(F_{ijk}\mathbf{A})/|\Phi_{F_{ijk}\mathbf{A}}|}{p(\mathbf{A})/|\Phi_\mathbf{A}|}\right\}$$

$$= \min\left\{1, \frac{|\Phi_\mathbf{A}|}{|\Phi_{F_{ijk}\mathbf{A}}|} \prod_{\ell=1}^{N} \frac{p(k_\ell(F_{ijk}\mathbf{A}))k_\ell(F_{ijk}\mathbf{A})!}{p(k_\ell(\mathbf{A}))k_\ell(\mathbf{A})!}\right\} \quad (7.18)$$

The hinge flip F_{ijk} alters only the degrees of nodes j and k, via $k_j(F_{ijk}\mathbf{A}) = k_j(\mathbf{A})-1$ and $k_k(F_{ijk}\mathbf{A}) = k_k(\mathbf{A})+1$. Our move acceptance probabilities thus simplify to

$$A[\mathbf{A} \to F_{ijk}\mathbf{A}] = \min\left\{1, \frac{|\Phi_\mathbf{A}|}{|\Phi_{F_{ijk}\mathbf{A}}|} \frac{p(k_j(\mathbf{A})-1)}{p(k_j(\mathbf{A}))k_j(\mathbf{A})} \frac{p(k_k(\mathbf{A})+1)[k_k(\mathbf{A})+1]}{p(k_k(\mathbf{A}))}\right\} \quad (7.19)$$

This can also be written as

$$A[\mathbf{A} \to F_{ijk}\mathbf{A}] = \min\left\{1, \frac{f(k_j(\mathbf{A})-1)f(k_k(\mathbf{A})+1)}{f(k_j(\mathbf{A}))f(k_k(\mathbf{A}))} \frac{|\Phi_\mathbf{A}|}{|\Phi_{F_{ijk}\mathbf{A}}|}\right\} \quad (7.20)$$

with the shorthand $f(q) = q!p(q)$. We note that for Poissonian degree distributions, i.e. for $p(q) = e^{-\bar{k}}\bar{k}^q/q!$, one would find

$$\frac{f(k_j(\mathbf{A})-1)f(k_k(\mathbf{A})+1)}{f(k_j(\mathbf{A}))f(k_k(\mathbf{A}))} = \frac{\bar{k}^{k_j(\mathbf{A})-1}\bar{k}^{k_k(\mathbf{A})+1}}{\bar{k}^{k_j(\mathbf{A})}\bar{k}^{k_k(\mathbf{A})}} = 1 \quad (7.21)$$

and recipe (7.19) reduces to the one corresponding to the generation via hinge flips of random graphs with a constrained number of links and uniform probabilities (since those will indeed typically exhibit Poissonian degree statistics).

Recipe (7.19), whose switch-and-hold version[3] was first introduced in [25], based on previous work done in [64] and [34], can be used to generate graphs with arbitrary degree statistics, including power-law distributions $p(k) \sim k^{-\gamma}$. In Figure 7.3 we use this algorithm to generate graphs with power-law degree distributions, by randomizing an initial Poissonian graph with $N = 3000$ nodes and average connectivity $\langle k \rangle = 7$. In the top left panel, we use Protocol 3 of Table 6.1 to sample moves uniformly from

---

[3] As usual, the switch-and-hold version is obtained by removing $|\Phi_\mathbf{A}|$ and $|\Phi_{F_{ijk}\mathbf{A}}|$ from (7.19), and sampling at each step a node and a link uniformly from the sets of all nodes and all links, respectively. If F_{ijk} is feasible, it is executed with the stated acceptance probabilities. All non-feasible ones imply that the current state $\mathbf{A}$ is maintained. Ensemble averages are calculated as averages over all iteration steps.

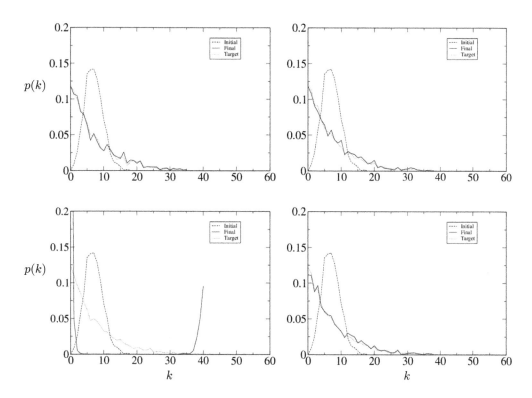

Fig. 7.3 Degree distributions of nondirected simple graphs with $N = 3000$ and $\langle k \rangle = 7$, before and after hinge flip MCMC randomization, targeting the measure (7.17) with a power-law degree distribution, following initialization to an Erdös–Rènyi graph. The four panels correspond to different move selection protocols and move acceptance probabilities. Dashed lines: Initial $p(k)$. Dotted lines: $p(k)$ of the targeted ensemble distribution (7.17). Solid lines: Observed $p(k)$ after equilibration of the MCMC process. Top left: Candidate moves are sampled according to Protocol 3 in Table 6.1, and the acceptance probabilities used are (7.19). Top right: Moves are sampled according to Protocol 1, and accepted with probabilities (7.22). Both MCMC versions should produce correct sampling, and are indeed seen to do so. In the bottom left panel, moves are sampled according to Protocol 1 but accepted with the probabilities (7.19); here we observe that the non-uniform sampling of candidate moves, now uncorrected in the acceptance probabilities, generates a strong bias. In the bottom right panel moves are sampled according to Protocol 2 and accepted with the probabilities (7.20).

the set of allowed moves, and this is indeed seen to produce random graphs with the targeted power-law degree distribution.

Using non-uniform sampling of moves F_{ijk} will require modifying the previous probabilities (7.19). Here we need to be careful, as before, with the requirement that $F^2 = \mathbf{I}$. For hinge flips this requirement implies that F_{ijk} and F_{ikj} are to be interpreted as two versions of the same operation, with the proviso that after executing F_{ijk} we go back to the previous state via F_{ikj}, and vice versa. A general expression of the form

$p(F|F\mathbf{A})$ would therefore become either $p(F_{ikj}|F_{ijk}\mathbf{A})$ or $p(F_{ijk}|F_{ikj}\mathbf{A})$. If we use Protocol 1 of Table 6.1, each move is drawn with likelihood (7.8), and upon inserting these sampling probabilities into formula (7.3) for the correct Metropolis–Hastings acceptance probabilities, we obtain for sufficiently large graphs

$$\mathcal{A}[\mathbf{A}\to F_{ijk}\mathbf{A}] = \min\left\{1, \frac{p(F_{ijk}\mathbf{A})P(F_{ikj}|F_{ijk}\mathbf{A})}{p(\mathbf{A})P(F_{ijk}|\mathbf{A})}\right\}$$

$$= \min\left\{1, \frac{p(F_{ijk}\mathbf{A})k_j(\mathbf{A})}{p(\mathbf{A})k_k(F_{ijk}\mathbf{A})}\right\}$$

$$= \min\left\{1, \frac{p(F_{ijk}\mathbf{A})k_j(\mathbf{A})}{p(\mathbf{A})[k_k(\mathbf{A})+1]}\right\}$$

$$= \min\left\{1, \frac{p(k_j(\mathbf{A})-1)p(k_k(\mathbf{A})+1)}{p(k_j(\mathbf{A}))p(k_k(\mathbf{A}))}\right\} \tag{7.22}$$

Here we used $p(0|F_{ijk}\mathbf{A}) = p(0|\mathbf{A}) + \mathcal{O}(N^{-1})$. In the top right panel of Fig. 7.3, we show the result of carrying out MCMC simulations in which candidate moves are sampled according to Protocol 1 and accepted with probability (7.22), and we see that the process indeed converges to the desired degree distribution.

In contrast, if we fail to take into account the non-trivial effects of non-uniform move sampling, and use Protocol 1 of Table 6.1 with the move acceptance probabilities (7.19) (which were derived for MCMC randomization with uniformly sampled moves), we will observe a strong bias in the resulting graph statistics. See the bottom left panel of Figure 7.3. Finally, when using Protocol 2 of Table 6.1, which draws each move with likelihood (7.10), one should according to (7.3) use for large graphs the following acceptance probabilities:

$$\mathcal{A}[\mathbf{A}\to F\mathbf{A}] = \min\left\{1, \frac{p(F_{ijk}\mathbf{A})P(F_{ikj}|F_{ijk}\mathbf{A})}{p(\mathbf{A})P(F_{ijk}|\mathbf{A})}\right\}$$

$$= \min\left\{1, \frac{p(F_{ijk}\mathbf{A})k_i(\mathbf{A})}{p(\mathbf{A})P(F_{ijk}|\mathbf{A})k_i(F_{ijk}\mathbf{A})}\right\}$$

$$= \min\left\{1, \frac{p(k_j(\mathbf{A})-1)}{p(k_j(\mathbf{A}))k_j(\mathbf{A})}\frac{p(k_k(\mathbf{A})+1)[k_k(\mathbf{A})+1]}{p(k_k(\mathbf{A}))}\right\} \tag{7.23}$$

The MCMC simulations again confirm that this process samples our graphs correctly, as shown in the bottom right panel of Figure 7.3.

Exercise 7.1 Write a pseudocode to generate via hinge flip randomization nondirected simple random graphs with a given degree distribution $p(k)$ and a hard-constrained number L of links, with uniform probabilities, and with candidate moves F_{ijk} being sampled according to Protocols 1, 2 or 3, respectively.

Exercise 7.2 Define a recipe to generate large nondirected finitely connected random graphs with *expected* degree sequence $\mathbf{k}$ and a hard-constrained number L of links.

7.3 Ensembles with controlled degrees and degree correlations

Overview: In this section, we show that ensemble (7.28) is tailored, for large N, to the production of graphs with prescribed degree sequence $\mathbf{k}$ and relative degree correlations $\Pi(k, k') = Q(k, k')/[F(k|Q)F(k'|Q)]$, where $F(k|Q)$ solves the self-consistency equation (7.34). Conversely, to generate large random graphs with prescribed degree sequence $\mathbf{k}$ and degree correlations $W(k, k')$, one should use ensemble (7.28) with $Q(k, k') = W(k, k')/[p(k)p(k')]$ (the canonical kernel) or any other kernel related to the canonical kernel via a separable transformation.

Defining ensembles of graphs with hard-constrained degrees. We have illustrated in Section 7.2 how to generate simple nondirected graphs with soft-constrained degrees. In this section, we turn to the generation of simple nondirected graphs with a hard-constrained degree sequence $\mathbf{k} = (k_1, \ldots, k_N)$. The standard maximum entropy ensemble to describe such graphs is

$$p(\mathbf{A}|\mathbf{k}) = \frac{1}{Z(\mathbf{k})} \delta_{\mathbf{k}, \mathbf{k}(\mathbf{A})}, \qquad Z(\mathbf{k}) = \sum_{\mathbf{A} \in G} \delta_{\mathbf{k}, \mathbf{k}(\mathbf{A})} \tag{7.24}$$

where $\delta_{\mathbf{k}, \mathbf{k}(\mathbf{A})} = \prod_i \delta_{k_i, k_i(\mathbf{A})}$, and $k_i(\mathbf{A}) = \sum_j A_{ij}$. This ensemble (7.24) can be written in the superficially different but mathematically identical alternative form

$$p(\mathbf{A}|\mathbf{k}) = \frac{1}{\tilde{Z}(\mathbf{k})} \prod_{i<j} \left[\frac{\langle k \rangle}{N} \delta_{A_{ij}, 1} + \left(1 - \frac{\langle k \rangle}{N}\right) \delta_{A_{ij}, 0} \right] \delta_{\mathbf{k}, \mathbf{k}(\mathbf{A})}, \tag{7.25}$$

where $\langle k \rangle = N^{-1} \sum_i k_i$, and $\tilde{Z}(\mathbf{k})$ is the appropriate normalization constant. The claim that (7.24) is identical to (7.25) is confirmed in a few steps, starting from (7.25):

$$p(\mathbf{A}|\mathbf{k}) = \frac{1}{\tilde{Z}(\mathbf{k})} \left(\frac{\langle k \rangle}{N}\right)^{\sum_{i<j} A_{ij}} \left(1 - \frac{\langle k \rangle}{N}\right)^{\frac{1}{2}N(N-1) - \sum_{i<j} A_{ij}} \delta_{\mathbf{k}, \mathbf{k}(\mathbf{A})}$$

$$= \frac{1}{\tilde{Z}(\mathbf{k})} \left(\frac{\langle k \rangle}{N}\right)^{\frac{1}{2} \sum_i k_i} \left(1 - \frac{\langle k \rangle}{N}\right)^{\frac{1}{2}N(N-1) - \frac{1}{2} \sum_i k_i} \delta_{\mathbf{k}, \mathbf{k}(\mathbf{A})}$$

$$= \frac{\delta_{\mathbf{k}, \mathbf{k}(\mathbf{A})}}{Z(\mathbf{k})}, \qquad Z(\mathbf{k}) = \frac{\tilde{Z}(\mathbf{k})}{\left(\frac{\langle k \rangle}{N}\right)^{\frac{1}{2} \sum_i k_i} \left(1 - \frac{\langle k \rangle}{N}\right)^{\frac{1}{2}N(N-1) - \frac{1}{2} \sum_i k_i}} \tag{7.26}$$

One could alternatively be tempted to define a hard-constrained ensemble by simply complementing the definition (4.22) for ensembles with soft-constrained degrees with an additional hard constraint on the degree sequence. This would give

$$p(\mathbf{A}|\mathbf{k}) = \frac{1}{Z'(\mathbf{k})} \prod_{i<j} \left[\frac{k_i k_j}{N \langle k \rangle} \delta_{A_{ij}, 1} + \left(1 - \frac{k_i k_j}{N \langle k \rangle}\right) \delta_{A_{ij}, 0} \right] \delta_{\mathbf{k}, \mathbf{k}(\mathbf{A})} \tag{7.27}$$

with $Z'(\mathbf{k})$ again denoting a normalization constant. In view of the above equivalence of (7.24) and (7.25), one may wonder whether by some suitable transformation the ensemble (7.27) can perhaps also be shown to be identical to (7.24).

Ensembles with controlled degree correlations. Such questions lead us to inspect in more detail the class of nondirected simple graph ensembles that are defined by probabilities of the following generic form [146]:

$$p(\mathbf{A}|\mathbf{k}, Q) = \frac{1}{Z(\mathbf{k}, Q)} \prod_{i<j} \left[\frac{\langle k \rangle}{N} Q(k_i, k_j) \delta_{A_{ij},1} + \left(1 - \frac{\langle k \rangle}{N} Q(k_i, k_j) \right) \delta_{A_{ij},0} \right] \delta_{\mathbf{k},\mathbf{k}(\mathbf{A})}$$

(7.28)

Here the degrees are, as before, prescribed via a hard constraint, and in addition the link probabilities are modified by some symmetric function $Q(k, k')$ of the degrees of the two vertices involved. Clearly, to have non-negative probabilities we must require that $Q(k, k') \geq 0$ for all (k, k'). As always, $Z(\mathbf{k}, Q)$ denotes the relevant normalization constant. In order to ensure that for large N such graphs can actually be found, we need to choose the function $Q(k, k')$ such that in formula (7.28) the partial measure

$$W(\mathbf{A}|Q, \mathbf{k}) = \prod_{i<j} \left[\frac{\langle k \rangle}{N} Q(k_i, k_j) \delta_{A_{ij},1} + \left(1 - \frac{\langle k \rangle}{N} Q(k_i, k_j) \right) \delta_{A_{ij},0} \right] \qquad (7.29)$$

(which is itself properly normalized) is consistent with the average connectivity $\langle k \rangle = N^{-1} \sum_{\ell} k_\ell$ imposed by the degree constraint $\prod_i \delta_{k_i, k_i(\mathbf{A})}$. This is achieved when

$$\frac{1}{N} \sum_\ell k_\ell = \sum_{\mathbf{A} \in G} W(\mathbf{A}|Q, \mathbf{k}) \frac{1}{N} \sum_i k_i(\mathbf{A}) = \frac{\langle k \rangle}{N^2} \sum_{i \neq j} Q(k_i, k_j)$$

$$= \langle k \rangle \sum_{k,k'} p(k) p(k') Q(k, k') + \mathcal{O}\left(\frac{1}{N} \right) \qquad (7.30)$$

with $p(k) = N^{-1} \sum_i \delta_{k,k_i}$. Hence, for large N the function $Q(k, k')$ must satisfy[4]

$$Q(k, k') \geq 0 \quad \forall (k, k') \qquad \text{and} \qquad \sum_{k,k'} p(k) p(k') Q(k, k') = 1 \qquad (7.31)$$

It is easy to see that both ensembles (7.27) and (7.25) are special cases of (7.28), obtained upon making the following choices for $Q(k, k')$, respectively:

$$Q(k, k') = \frac{kk'}{\langle k \rangle^2} \qquad \text{and} \qquad Q(k, k') = 1 \qquad (7.32)$$

Both indeed satisfy the required conditions (7.31).

Intuitively one expects that in ensembles of the form (7.28) one can use the freedom in choosing a function $Q(k, k')$ to manipulate degree correlations. Indeed, in Appendix H we show that (7.28) produces for large N graphs in which the average degree correlations $\Pi(k, k') = \sum_{\mathbf{A} \in G} \Pi(k, k'|\mathbf{A}) p(\mathbf{A}|\mathbf{k}, Q)$, with $\Pi(k, k'|\mathbf{A})$ as defined in (2.12), take the values

$$\Pi(k, k') = \frac{Q(k, k')}{F(k|Q) F(k'|Q)} \qquad (7.33)$$

---

[4]One could, of course, also define the ensembles (7.28) with $N-1$ instead of N in the prefactors of $Q(k, k')$, in which case the conditions (7.31) would be met even for small N. Here we retained the definition that was used in the relevant literature, which was mostly concerned with large graphs.

Here the quantities $F(k|Q)$ solve the self-consistency equation

$$F(k|Q) = \frac{1}{\langle k \rangle} \sum_{k'} Q(k, k') \frac{k' p(k')}{F(k'|Q)} \tag{7.34}$$

The canonical kernel $Q(k, k')$. Having to solve equations such as (7.34) in order to find a kernel $Q(k, k')$ that will implement a given pattern $\Pi(k, k')$ of average degree correlations is an unwelcome complication[5], which fortunately can be avoided by using the *canonical kernel* [9]. To show this, we first observe upon eliminating $Q(k, k')$ from (7.34) via (7.33) that (7.34) becomes identical to

$$1 = \frac{1}{\langle k \rangle} \sum_{k'} \Pi(k, k') k' p(k') \tag{7.35}$$

This identity can in fact be derived directly from the microscopic definition (2.12) of $\Pi(k, k'|\mathbf{A})$, and must therefore hold for every single graph of the ensemble (7.28), provided N is sufficiently large[6]. Since $Q(k, k') = \Pi(k, k') F(k|Q) F(k'|Q)$ is always a solution of (7.34), for each degree distribution $p(k)$ and each relative degree correlation function $\Pi(k, k')$, we can always find kernels $Q(k, k')$ such that their associated ensembles (7.28) will for large N be tailored to the production of random graphs with precisely these statistical features. The only remaining condition is that we can find a non-negative function $F(k|Q)$ that satisfies

$$\sum_{k, k'} p(k) p(k') \Pi(k, k') F(k|Q) F(k'|Q) = 1 \tag{7.36}$$

Clearly, such a function always exists. The simplest choice is $F(k|Q) = k/\langle k \rangle$, leading to the kernel $Q^\star(k, k') = \Pi(k, k') kk'/\langle k \rangle^2$. We refer to this choice as the *canonical kernel*. We note, however, that any kernel related to the canonical kernel via the separable transformation, i.e. $Q(k, k') = G(k) Q^\star(k, k') G(k')$ with suitable normalized non-negative functions $G(k)$, will lead to an ensemble with the same average degree correlations $\Pi(k, k')$, albeit with different functions $F(k|Q) = G(k) k/\langle k \rangle$.

We can now understand how the various ensembles discussed at the beginning of this section are related. When $Q(k, k') = 1$ for all (k, k'), we obtain ensemble (7.25). Here equation (7.34) is seen to be solved by $F(k|Q) = 1$ for all k, leading to $\Pi(k, k') = 1$, via (7.33). This confirms that (7.25) produces graphs with strictly uncorrelated degrees. For $Q(k, k') = kk'/\langle k \rangle^2$, we obtain ensemble (7.27). Now (7.34) is solved by $F(k|Q) = k/\langle k \rangle$, which leads again to $\Pi(k, k') = 1$, confirming that (7.27) also produces graphs with uncorrelated degrees. The two ensembles (7.25) and (7.27) generate graphs with identical degrees and identical average degree correlations because they are related via a separable transformation. However, they may still differ in higher order degree correlations.

---

[5] A routine to numerically solve (7.34) is provided in Algorithm 34.

[6] For finite N a typical graph of the ensemble (7.28) will display deviations from (7.33) that are at least of order $\mathcal{O}(N^{-1})$, but possibly of order $\mathcal{O}(N^{-1/2})$ (the typical finite size corrections in empirical averages over $\mathcal{O}(N)$ independent samples).

We also conclude that we can always tailor ensembles of the form (7.28) to produce, for large N, graphs with any desired graphical degree distribution $p(k)$ and any desired graphical degree correlation function $\Pi(k, k')$, without the need to solve further equations, upon choosing the ensemble corresponding to the canonical kernel:

$$p(\mathbf{A}|\mathbf{k}, \Pi) = \frac{1}{Z(\mathbf{k}, \Pi)} \prod_{i<j} \left[\frac{k_i k_j}{N\langle k \rangle} \Pi(k_i, k_j) \delta_{A_{ij},1} + \left(1 - \frac{k_i k_j}{N\langle k \rangle} \Pi(k_i, k_j)\right) \delta_{A_{ij},0} \right] \delta_{\mathbf{k},\mathbf{k}(\mathbf{A})}$$

(7.37)

Note that if $N^{-1} \sum_i \delta_{k,k_i(\mathbf{a})} = p(k)$ and $\langle k \rangle = \sum_k kp(k)$, using

$$\Pi(k, k') = \frac{W(k, k')\langle k \rangle^2}{p(k)p(k')kk'}$$

(7.38)

we can rewrite (7.37) as

$$p(\mathbf{A}|\mathbf{k}, W) = \frac{1}{Z(\mathbf{k}, W)} \prod_{i<j} \left[\frac{\langle k \rangle}{N} \frac{W(k_i, k_j)}{p(k_i)p(k_j)} \delta_{A_{ij},1} + \left(1 - \frac{\langle k \rangle}{N} \frac{W(k_i, k_j)}{p(k_i)p(k_j)}\right) \delta_{A_{ij},0} \right] \delta_{\mathbf{k},\mathbf{k}(\mathbf{A})}$$

(7.39)

Exercise 7.3 Show that the canonical ensemble (7.39) gives the maximum entropy within the subspace of simple nondirected graphs with prescribed degrees and upon imposing as a constraint the average values $W(k, k') = \langle W(k, k'|\mathbf{A}) \rangle$ of the relative degree correlations.

Exercise 7.4 Suppose you have different simple nondirected graphs, not necessarily of the same size. Could you devise a method to measure distances between them, based on their measured degree distributions and degree correlations?

Hint: You may find it useful to be reminded that an information-theoretic distance between two distributions $p(x)$ and $q(x)$ is given by the Kullback–Leibler distance

$$D(p||q) = \sum_x p(x) \log \left[p(x)/q(x) \right]$$

(7.40)

or, in symmetrized form, the Jeffreys divergence

$$D(p||q) = \frac{1}{2} \sum_x p(x) \log \left[p(x)/q(x) \right] + \frac{1}{2} \sum_x q(x) \log \left[q(x)/p(x) \right]$$

(7.41)

Exercise 7.5 Suppose you have a protein interaction network, and you know a number q of genomic and evolutionary properties of its constituent nodes, for example, whether the node was acquired early on in evolution, whether it is highly conserved across species and whether it has a single or multiple copies in the genome. How would you build a null model of the network you observe, if you suspect that the genomic and evolutionary properties of the constituent nodes are important in determining the connectivity of the network?

7.4 Generating graphs with prescribed degrees and correlations

Overview: In this section, we show how to generate graphs from the ensemble (7.28) via repeated application of edge swap moves. We obtain different degree correlations upon inserting different kernels $Q(k, k')$ and show the equivalence of degree correlations obtained for kernels related via separable transformations. We discuss the consequences of sampling edge swap moves uniformly at random from the set of applicable moves as opposed to that of all moves.

MCMC algorithms for generating graphs with prescribed degrees. Constructing MCMC versions that target the generic measure (7.28) via edge-swap randomization is in principle straightforward, but it will often be efficient to first rewrite (7.29) slightly to

$$W(\mathbf{A}|Q, \mathbf{k}) = e^{\sum_{i<j} A_{ij} \log\left[\langle k\rangle Q(k_i,k_j)/N\right] + \sum_{i<j}(1-A_{ij}) \log\left[1-\langle k\rangle Q(k_i,k_j)/N\right]}$$

$$= \left(\prod_{i<j}\left[1 - \frac{\langle k\rangle}{N}Q(k_i, k_j)\right]\right) e^{\sum_{i<j} L_{ij} A_{ij}} \tag{7.42}$$

in which

$$L_{ij} = \log\left(\frac{\langle k\rangle Q(k_i, k_j)/N}{1-\langle k\rangle Q(k_i, k_j)/N}\right) \tag{7.43}$$

Apparently we can now write (7.28) also as

$$p(\mathbf{A}|\mathbf{k}, Q) = \frac{\delta_{\mathbf{k}, \mathbf{k}(\mathbf{A})}}{Z'(\mathbf{k}, Q)} e^{\sum_{i<j} L_{ij} A_{ij}} \tag{7.44}$$

Since the normalization constant $Z'(\mathbf{k}, Q)$ will drop out of our acceptance probabilities, and the $\frac{1}{2}N(N-1)$ quantities $\{L_{ij}\}$ need to be calculated only once prior to execution of the algorithm, this alternative form (7.44) of our target measure simplifies notation and computer code, and speeds up our algorithms.

The simplest randomization algorithms that preserve all degrees $\{k_1(\mathbf{A}), \ldots, k_N(\mathbf{A})\}$ of $\mathbf{A}$ are based on the use of 'edge swaps' $\{F_{ijk\ell;\alpha}\}$; see Table 6.1. We have already studied the properties of these elementary moves in Section 6.4. Our standard MCMC move acceptance probabilities of the Metropolis–Hastings or Glauber form will be (7.3, 7.4) for both switch-and-hold execution and mobility-based execution, although move proposal probabilities are normalized differently.

If all candidate edge swaps are drawn with uniform probabilities from the set of all moves, one will have move proposal probabilities $P(F_{ijk\ell;\alpha}|\mathbf{A})$ in (7.3,7.4) that are independent of $\mathbf{A}$, and they will simply drop out. If the swap $F_{ijk\ell;\alpha}$ is applicable on the current state $\mathbf{A}$, it will be accepted in the Metropolis–Hastings version of the switch-and-hold MCMC with the following probability:

$$\mathcal{A}[\mathbf{A} \to F_{ijk\ell;\alpha}\mathbf{A}] = \min\left\{1, \frac{p(F_{ijk\ell;\alpha}\mathbf{A})}{p(\mathbf{A})}\right\}$$

$$= \min\left\{1, e^{\sum_{r<s} L_{rs}\left[(F_{ijk\ell;\alpha}\mathbf{A})_{rs} - A_{rs}\right]}\right\}$$

$$= \min\left\{1, e^{\sum_{r<s,(r,s)\in\mathcal{S}_{ijk\ell;\alpha}} L_{rs}(1-2A_{rs})}\right\} \qquad (7.45)$$

The sets $\mathcal{S}_{ijk\ell;\alpha}$ define those index pairs whose connectivities are affected by the move $F_{ijk\ell;\alpha}$, and are given in $(6.39, 6.40, 6.41)$. If the edge swap is not applicable, the current state $\mathbf{A}$ is maintained. The resulting pseudocode is given in Algorithm 19 of Appendix J. If our candidate edge swaps are drawn with uniform probabilities only from the set $\Phi_{\mathbf{A}}$ of currently applicable edge swaps, one will have to use formulae $(7.3, 7.4)$, with the move proposal probabilities $P(F_{ijk\ell}|\mathbf{A}) = 1/|\Phi_{\mathbf{A}}|$, in which $|\Phi_{\mathbf{A}}|$ (the total number of edge swaps that can be carried out on the state $\mathbf{A}$) is given by formula (6.46). The Metropolis–Hastings version of this mobility-based MCMC will then have the following move acceptance probabilities:

$$\mathcal{A}[\mathbf{A}\to F_{ijk\ell;\alpha}\mathbf{A}] = \min\left\{1, \frac{|\Phi_{\mathbf{A}}|}{|\Phi_{F_{ijk\ell;\alpha}\mathbf{A}}|} e^{\sum_{r<s,(r,s)\in\mathcal{S}_{ijk\ell;\alpha}} L_{rs}(1-2A_{rs})}\right\} \qquad (7.46)$$

This leads to the pseudocode given in Algorithm 21 of Appendix J.

However, similar to the situation with hinge flips as studied in the beginning of this chapter (see, e.g. Table 7.1), here also the question of how to sample the candidate moves $F_{ijk\ell;\alpha}$ strictly *uniformly* is not trivial, while failure to sample uniformly can have disastrous consequences. We will return to this question later, as well as possible options for computationally more efficient inhomogeneous sampling of moves, where we will no longer have $P(F_{ijk\ell;\alpha}|\mathbf{A}) = 1/|\Phi_{\mathbf{A}}|$.

Generation of large graphs with uncorrelated degrees. The question of how to generate graphs with a given degree sequence and a guaranteed lack of degree correlations is quite relevant to the theorist, since many problems in complex network theory can be solved exactly only for uncorrelated graphs. The ability to generate uncorrelated graphs is therefore crucial to benchmark theoretical results. Graphs with uncorrelated degrees have $\Pi(k,k'|\mathbf{A}) = 1$ for all (k,k'), in terms of the definition (2.12). We saw in the previous section that ensembles of the generic form (7.28) will produce graphs that on average[7] exhibit uncorrelated degrees for all expressions $Q(k,k')$ of the form $Q(k,k') = F(k)F(k')$, with $\sum_k p(k)F(k) = 1$. The ensembles corresponding to different choices of $F(k)$ are not identical, and only for $F(k) = 1$ will we obtain the maximum entropy ensemble, given the imposed degree sequence. However, the differences in the probabilities that they assign to graphs, and hence also the differences in their Shannon entropies, are sub-leading in N.

The two examples (7.25) and (7.27) are both of the form $Q(k,k') = F(k)F(k')$ that guarantees uncorrelated degrees, and correspond to, respectively, $F(k) = 1$ and $F(k) = k/\langle k\rangle$. In the former case (7.25), where all states with the correct degree sequence are equally probable, we should recover move acceptance probabilities that are equal to one for switch-and-hold MCMC, and dependent only on the mobilities $|\Phi_{\mathbf{A}}|$ for MCMC with candidate moves drawn strictly from $\Phi_{\mathbf{A}}$. This is indeed the

---

[7]Although in these ensembles we will find absence of degree correlations on average, individual graphs $\mathbf{A}$ generated by them will exhibit small deviations from $\Pi(k,k'|\mathbf{A}) = 1 \ \forall \ k,k'$ that vanish for $N\to\infty$.

case. Upon calculating the quantities (7.43) for $Q(k,k') = 1$, we obtain the trivial result $L_{ij} = \log[\langle k \rangle / (N - \langle k \rangle)]$, and this gives

$$\sum_{r<s,(r,s)\in\mathcal{S}_{ijk\ell;\alpha}} L_{rs}(1-2A_{rs}) = \log\left(\frac{\langle k \rangle}{N-\langle k \rangle}\right) \sum_{r<s,(r,s)\in\mathcal{S}_{ijk\ell;\alpha}} (1-2A_{rs}) = 0 \qquad (7.47)$$

(since in each edge swap two affected node pairs will have $A_{rs} = 0$, and two will have $A_{rs} = 1$). This indeed leads to the aforementioned trivial acceptance probabilities. If we target instead the measure (7.27), we find that $L_{ij} = \log[k_i k_j / (N\langle k \rangle - k_i k_j)]$. Switch-and-hold MCMC sampling will now target (7.27) via the acceptance probabilities

$$\mathcal{A}[\mathbf{A} \to F_{ijk\ell;\alpha}\mathbf{A}] = \min\left\{1, \prod_{r<s,(r,s)\in\mathcal{S}_{ijk\ell;\alpha}} \left(\frac{k_r k_s}{N\langle k \rangle - k_r k_s}\right)^{1-2A_{rs}}\right\} \qquad (7.48)$$

which translates for the three flavours $\alpha \in \{1,2,3\}$ into

$$\mathcal{A}[\mathbf{A} \to F_{ijk\ell;1}\mathbf{A}] = \min\left\{1, \frac{\left(1-\frac{k_i k_j}{N\langle k \rangle}\right)\left(1-\frac{k_k k_\ell}{N\langle k \rangle}\right)}{\left(1-\frac{k_j k_k}{N\langle k \rangle}\right)\left(1-\frac{k_i k_\ell}{N\langle k \rangle}\right)}\right\} \qquad (7.49)$$

$$\mathcal{A}[\mathbf{A} \to F_{ijk\ell;2}\mathbf{A}] = \min\left\{1, \frac{\left(1-\frac{k_i k_j}{N\langle k \rangle}\right)\left(1-\frac{k_k k_\ell}{N\langle k \rangle}\right)}{\left(1-\frac{k_j k_\ell}{N\langle k \rangle}\right)\left(1-\frac{k_i k_k}{N\langle k \rangle}\right)}\right\} \qquad (7.50)$$

$$\mathcal{A}[\mathbf{A} \to F_{ijk\ell;3}\mathbf{A}] = \min\left\{1, \frac{\left(1-\frac{k_j k_\ell}{N\langle k \rangle}\right)\left(1-\frac{k_i k_k}{N\langle k \rangle}\right)}{\left(1-\frac{k_j k_k}{N\langle k \rangle}\right)\left(1-\frac{k_i k_\ell}{N\langle k \rangle}\right)}\right\} \qquad (7.51)$$

As always, in MCMC versions where moves are drawn uniformly from the set $\Phi_{\mathbf{A}}$ of those that can be executed on the instantaneous state $\mathbf{A}$, the relevant probabilities are modified by the ratio of the mobilities of the states before and after the proposed move, giving

$$\mathcal{A}[\mathbf{A} \to F_{ijk\ell;1}\mathbf{A}] = \min\left\{1, \frac{|\Phi_{\mathbf{A}}|}{|\Phi_{F_{ijk\ell;1}\mathbf{A}}|} \frac{\left(1-\frac{k_i k_j}{N\langle k \rangle}\right)\left(1-\frac{k_k k_\ell}{N\langle k \rangle}\right)}{\left(1-\frac{k_j k_k}{N\langle k \rangle}\right)\left(1-\frac{k_i k_\ell}{N\langle k \rangle}\right)}\right\} \qquad (7.52)$$

$$\mathcal{A}[\mathbf{A} \to F_{ijk\ell;2}\mathbf{A}] = \min\left\{1, \frac{|\Phi_{\mathbf{A}}|}{|\Phi_{F_{ijk\ell;2}\mathbf{A}}|} \frac{\left(1-\frac{k_i k_j}{N\langle k \rangle}\right)\left(1-\frac{k_k k_\ell}{N\langle k \rangle}\right)}{\left(1-\frac{k_j k_\ell}{N\langle k \rangle}\right)\left(1-\frac{k_i k_k}{N\langle k \rangle}\right)}\right\} \qquad (7.53)$$

$$\mathcal{A}[\mathbf{A} \to F_{ijk\ell;3}\mathbf{A}] = \min\left\{1, \frac{|\Phi_{\mathbf{A}}|}{|\Phi_{F_{ijk\ell;3}\mathbf{A}}|} \frac{\left(1-\frac{k_j k_\ell}{N\langle k \rangle}\right)\left(1-\frac{k_i k_k}{N\langle k \rangle}\right)}{\left(1-\frac{k_j k_k}{N\langle k \rangle}\right)\left(1-\frac{k_i k_\ell}{N\langle k \rangle}\right)}\right\} \qquad (7.54)$$

In Figure 7.4 we show the results of Markovian dynamics tailored to target the network distributions (7.27) and (7.24). An initial graph $\mathbf{A}_0$ was constructed with a non-Poissonian degree distribution and non-trivial degree correlations, i.e. $W(k,k') \neq W(k)W(k')$, as shown in Figure 7.4 (panels a and b). After iterating the Markov chain until equilibrium, degree correlations are seen to smoothen in the same fashion for both the MCMC versions that target (7.27) (left panels) and (7.24) (right panels). We sampled edge swap moves according to three different protocols:

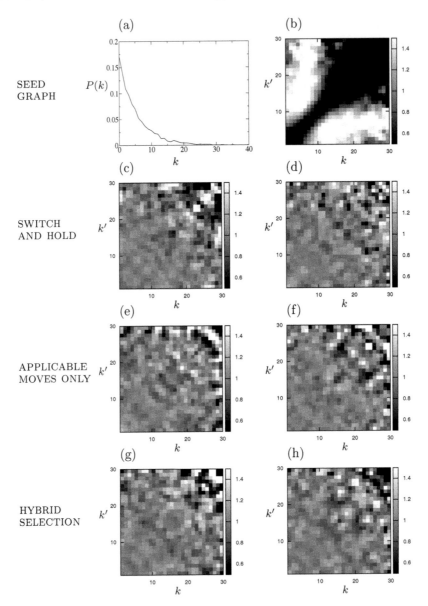

Fig. 7.4 Results of edge-swap based MCMC randomization dynamics, tailored to generating uncorrelated graphs. Panel (a): Degree distribution of the (randomly generated) initial graph $\mathbf{A}_0$, with $N = 4000$ and $\langle k \rangle = 5$. Panel (b): Relative degree correlations $\Pi(k, k'|\mathbf{A}_0)$ of the initial graph. Panels (c,e,g): Relative degree correlations after termination of MCMC, targeting the distribution (7.27). Panels (d,f,h): Relative degree correlations after termination of MCMC, targeting the distribution (7.24). Switch-and-hold MCMC was used in panels (c,d), MCMC with moves drawn strictly from the set of applicable moves was used in panels (c,f), and a hybrid algorithm, to be described in Section 7.7, was used in panels (g,h). In each simulation, the number of proposed moves per bond was 1000.

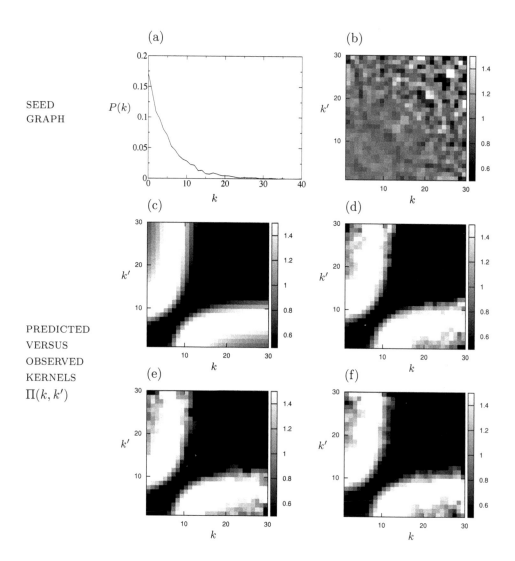

SEED
GRAPH

PREDICTED
VERSUS
OBSERVED
KERNELS
$\Pi(k, k')$

Fig. 7.5 Results of edge-swap based MCMC randomization dynamics, tailored to generating degree-correlated graphs from the ensemble measure (7.28), with the disassortative kernel $Q(k, k')$ given in (7.61). Panel (a): Degree distribution of the (randomly generated) initial graph $\mathbf{A}_0$, with $N = 4000$ and $\langle k \rangle = 5$. Panel (b): Relative degree correlations $\Pi(k, k'|\mathbf{A}_0)$ of the initial graph. Panel (c): Predicted average relative degree correlations $\Pi(k, k')$ computed via formulae (7.33, 7.34). Panel (d,e,f): Relative degree correlations $\Pi(k, k'|\mathbf{A})$ after termination of three MCMC algorithm versions, after 1000 proposed moves per link (uniform selection from the set of all allowed moves (d); uniform selection from the set of all moves (e); and the hybrid algorithm of Section 7.7 (f)).

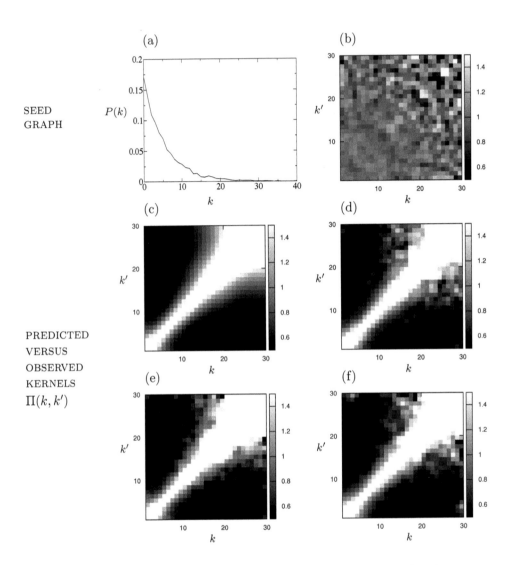

Fig. 7.6 Results of edge-swap based MCMC randomization dynamics, tailored to generating degree-correlated graphs from the ensemble measure (7.28), with the assortative kernel $Q(k, k')$ given in (7.62). Panel (a): Degree distribution of the (randomly generated) initial graph $\mathbf{A_0}$, with $N = 4000$ and $\langle k \rangle = 5$. Panel (b): Relative degree correlations $\Pi(k, k'|\mathbf{A_0})$ of the initial graph. Panel (c): Predicted relative degree correlations $\Pi(k, k')$ computed via formulae (7.33, 7.34). Panel (d,e,f): Relative degree correlations $\Pi(k, k'|\mathbf{A})$ measured after termination of three distinct MCMC algorithm versions, after 1000 proposed moves per link (uniform selection from the set of all allowed moves (d); uniform selection from the set of all moves (e); and the hybrid algorithm of Section 7.7 (f)).

uniform sampling from the set of all moves (the switch-and-hold algorithm, panels c and d), uniform sampling from the set of allowed moves (mobility-based algorithm, panels e and f) and a specific type of biased sampling from the set of all moves, leading to the hybrid algorithm to be discussed in Section 7.7 (panels g and h). All target measures and all MCMC versions, provided they are executed with their correct move acceptance probabilities, are seen to produce the desired uncorrelated graphs.

Generation of graphs with correlated degrees. To produce large graphs with prescribed degrees and non-trivial degree correlations, encoded in the ratio $\Pi(k, k')$, we can use the canonical kernel $Q(k, k') = \Pi(k_i, k_j)k_i k_j/\langle k \rangle^2$ and sample from the ensemble (7.37), which is guaranteed to produce the requested degree correlations on average. Thus, we can simply run edge-swap based MCMC algorithms, with either the acceptance probabilities (7.45) (for switch-and-hold dynamics), or (7.46) (if we only select applicable edge swaps). Working out these formulae for the three edge swap flavours $\alpha = 1, 2, 3$ gives for switch-and-hold MCMC:

$$\mathcal{A}[\mathbf{A} \to F_{ijk\ell;1}\mathbf{A}] = \min\left\{1, e^{L_{jk}+L_{i\ell}-L_{ij}-L_{k\ell}}\right\} \tag{7.55}$$

$$\mathcal{A}[\mathbf{A} \to F_{ijk\ell;2}\mathbf{A}] = \min\left\{1, e^{L_{j\ell}+L_{ik}-L_{ij}-L_{k\ell}}\right\} \tag{7.56}$$

$$\mathcal{A}[\mathbf{A} \to F_{ijk\ell;3}\mathbf{A}] = \min\left\{1, e^{L_{jk}+L_{i\ell}-L_{j\ell}-L_{ik}}]\right\} \tag{7.57}$$

with the shorthands (7.43). When moves are drawn strictly from the set of allowed move, the correct acceptance probabilities are

$$\mathcal{A}[\mathbf{A} \to F_{ijk\ell;1}\mathbf{A}] = \min\left\{1, \frac{|\Phi_{\mathbf{A}}|}{|\Phi_{F_{ijk\ell;1}\mathbf{A}}|}e^{L_{jk}+L_{i\ell}-L_{ij}-L_{k\ell}}\right\} \tag{7.58}$$

$$\mathcal{A}[\mathbf{A} \to F_{ijk\ell;2}\mathbf{A}] = \min\left\{1, \frac{|\Phi_{\mathbf{A}}|}{|\Phi_{F_{ijk\ell;2}\mathbf{A}}|}e^{L_{j\ell}+L_{ik}-L_{ij}-L_{k\ell}}\right\} \tag{7.59}$$

$$\mathcal{A}[\mathbf{A} \to F_{ijk\ell;3}\mathbf{A}] = \min\left\{1, \frac{|\Phi_{\mathbf{A}}|}{|\Phi_{F_{ijk\ell;3}\mathbf{A}}|}e^{L_{jk}+L_{i\ell}-L_{j\ell}-L_{ik}}]\right\} \tag{7.60}$$

We illustrate the above generation algorithms via application to two examples of non-trivial kernels $Q(k, k')$, both satisfying the conditions (7.31). The first is

$$Q(k, k') = \frac{(k - k')^2}{\sum_{qq' \geq 0} p(q)p(q')(q-q')^2} \tag{7.61}$$

The probability of a link between two nodes (subject to the degree sequence constraint) increases monotonically with the absolute difference between their respective degrees. This choice therefore corresponds to an ensemble that will typically generate *disassortative* graphs. The second is

$$Q(k, k') = \frac{(1 + |k-k'|)^{-1}}{\sum_{qq' \geq 0} p(q)p(q')(1 + |q-q'|)^{-1}} \tag{7.62}$$

Here the probability of a link between two nodes (subject to the degree sequence constraint) decreases monotonically with the absolute difference between their respective degrees. This choice therefore corresponds to an ensemble that will typically generate *assortative* graphs. We note that the normalization denominators in (7.61) and (7.62) need not be calculated, since they drop out of the move acceptance formulae (7.58, 7.59, 7.60) and (7.55, 7.56, 7.57).

The predicted values of $\Pi(k, k')$ are in both cases calculated from expression (7.33), in which the values of $F(k|Q)$ were calculated by solving the self-consistency equation (7.34) numerically (via Algorithm 34). In all simulations, an initial graph $\mathbf{A}_0$ of size $N = 4000$ and with average degree $\langle k \rangle = 5$ was constructed with a non-Poissonian degree distribution and absent degree correlations, i.e. with $\Pi(k, k'|\mathbf{A}_0) \approx 1$. After iterating the MCMC algorithms until equilibrium is expected to be reached, for which we used the criterion of having generated 1000 proposed moves per link, the observed degree correlations are seen to be in excellent agreement with their target values, given by (7.33). Results for the disassortative kernel (7.61) are shown in Figure 7.5, for three types of move sampling: uniform random selection from the set of allowed moves (with acceptance probabilities (7.58, 7.59, 7.60), uniform random selection from the set of all moves (switch-and-hold MCMC, with acceptance probabilities (7.55, 7.56, 7.57), and the biased protocol to be discussed in Section 7.7. Similar results for the assortative kernel (7.62) are shown in Figure 7.6. Clearly, for both kernels, the various algorithms succeed in generating graphs with the desired degree statistics, including correlations, up to finite size fluctuations.

7.5   Edge swaps revisited

Overview: In this section, we turn to an alternative sampling of candidate edge-swap moves which is more efficient for sparse graphs. This entails drawing uniformly at random a pair of ordered links as opposed to a quadruplet of nodes.

Alternative formulation. Edge swaps are the simplest moves in nondirected simple graphs that preserve the values of all degrees. At each step in an MCMC algorithm based on edge swaps, one has to draw a quadruplet (i, j, k, ℓ) with $i < j < k < \ell$ uniformly at random, followed by selecting randomly one of the three possible candidate edge swaps $F_{ijk\ell;\alpha}$. Proposed moves are subsequently accepted or rejected with appropriate probabilities. However, for finitely connected graphs, where the likelihood of finding a link between two randomly drawn sites is $\mathcal{O}(N^{-1})$, it will be much more efficient to draw two links (i, j) and (k, ℓ) at random, as opposed to drawing an ordered quadruplet (i, j, k, ℓ) of sites that happen to harbour two pairwise links (which is required for an edge swap to be applicable). This leads us to the introduction of the following auto-invertible move acting on ordered links, or *ordered* edge swap $F_{ijk\ell}$, which interchanges the (ordered) links $\{(i, j), (k, \ell)\}$ with $\{(i, \ell), (k, j)\}$. We no longer order the four node labels (i, j, k, ℓ), but only require that they must be distinct. This alternative definition does not affect the set of moves being considered – we continue to work with edge swaps – but only how these edge swaps will be labelled, selected and counted.

We list below the possible ordered edge-swap moves that can act on the quadruplet (i, j, k, ℓ). Because of the ordering given to the links, it is useful to draw an arrow on them[8] in such a way that this arrow goes from a to b in the link (a, b). This gives

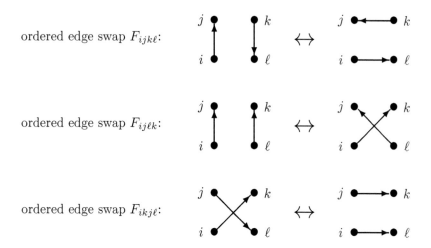

ordered edge swap $F_{ijk\ell}$:

ordered edge swap $F_{ij\ell k}$:

ordered edge swap $F_{ikj\ell}$:

One can verify easily in these diagrams that moves $F_{ijk\ell}$, $F_{ji\ell k}$ act in an identical way on undirected graphs. However, we will regard these moves as two distinct ordered edge swaps. In contrast, move $F_{k\ell ij}$ gives the same ordered edge swap as $F_{ijk\ell}$, as the order in which the two ordered edges are picked does not matter. The conditions for an ordered edge swap $F_{ijk\ell}$ to be applicable on a state $\mathbf{A}$ are that $A_{ij} = A_{k\ell} = 1$ and $A_{jk} = A_{i\ell} = 0$, and each such swap can affect only the link variables associated with node pairs (r, s) in the set

$$S_{ijk\ell} = \{(i, j), (j, i), (k, \ell), (\ell, k), (j, k), (k, j), (i, \ell), (\ell, i)\} \tag{7.63}$$

If their respective admissibility conditions are satisfied, each operation will then execute the following operation on $\mathbf{A}$:

$$(F_{ijk\ell}\mathbf{A})_{rs} = A_{rs} \quad \text{if } (r, s) \notin S_{ijk\ell} \tag{7.64}$$

$$(F_{ijk\ell}\mathbf{A})_{rs} = 1 - A_{rs} \quad \text{if } (r, s) \in S_{ijk\ell} \tag{7.65}$$

Mobility with respect to ordered edge swaps. We can now calculate the number $|\Phi_{\mathbf{A}}|$ of ordered edge swaps that can be executed on a state $\mathbf{A}$. According to the present definitions, as soon as we have picked the ordered pairs (i, j) and (k, ℓ) there is only one possible edge swap, i.e. $F_{ijk\ell}$. However, there are $4! = 24$ permutations of the four indices (i, j, k, ℓ), and each permutation leads to a different ordered edge swap, except for the invariance under the permutation of pairs $(i, j), (k, \ell) \leftrightarrow (k, \ell), (i, j)$. Hence, we may write

---

[8]Note that these arrows do not imply that we work with directed graphs. They only play a bookkeeping role by allowing us to keep track of whether a node pair was selected as (i, j) or (j, i) (although this representation will form the basis later for defining edge swaps in directed graphs).

$$|\Phi_{\mathbf{A}}| = \frac{1}{2} \sum_{i \neq j \neq k \neq \ell} A_{ij} A_{k\ell} (1 - A_{jk})(1 - A_{i\ell}) \qquad (7.66)$$

where the factor $1/2$ prevents the overcounting caused by the invariance under the order in which the two links are selected. Comparison with intermediate stages of the derivation of (6.46) shows that this is twice the mobility of our earlier (undirected) definition of the edge-swap set. Indeed, for each undirected edge swap $\{[i,j],[k,\ell]\} \rightarrow \{[i,\ell],[j,k]\}$ (with $[i,j] = (i,j)$ for $i < j$ and $[i,j] = (j,i)$ for $j < i$), we have now two directed edge swaps $\{(i,j),(k,\ell)\} \rightarrow \{(i,\ell),(k,j)\}$ and $\{(j,i),(\ell,k)\} \rightarrow \{(\ell,i),(j,k)\}$. Therefore, we have

$$2|\Phi_{\mathbf{A}}| = \left[\sum_i k_i(\mathbf{A}) \right]^2 + \sum_i k_i(\mathbf{A}) - 2\sum_i k_i^2(\mathbf{A})$$
$$- 2\sum_{ij} A_{ij} k_i(\mathbf{A}) k_j(\mathbf{A}) + 2\text{Tr}(\mathbf{A}^3) + \text{Tr}(\mathbf{A}^4) \qquad (7.67)$$

in which only the second line can vary during the MCMC process, since all degrees are conserved. Since our move acceptance probabilities will require the instantaneous values of the mobility (unless we run a switch-and-hold version), a quantity that we would not want to calculate from scratch after each step of the dynamics, we keep track of the value of $|\Phi_{\mathbf{A}}|$ by evaluating its change after an executed move, via

$$2\Delta_{ijk\ell}|\Phi_{\mathbf{A}}| = 2|\Phi_{F_{ijk\ell}\mathbf{A}}| - 2|\Phi_{\mathbf{A}}|$$
$$= -2\sum_{rs} k_r k_s [(F_{ijk\ell}\mathbf{A})_{rs} - A_{rs}] + 2\Delta_{ijk\ell}\text{Tr}(\mathbf{A}^3) + \Delta_{ijk\ell}\text{Tr}(\mathbf{A}^4)$$
$$= -4\Big[k_i k_j (1 - 2A_{ij}) + k_k k_\ell (1 - 2A_{k\ell}) + k_j k_k (1 - 2A_{jk}) + k_i k_\ell (1 - 2A_{i\ell}) \Big]$$
$$+ 2\Delta_{ijk\ell}\text{Tr}(\mathbf{A}^3) + \Delta_{ijk\ell}\text{Tr}(\mathbf{A}^4) \qquad (7.68)$$

At this point we benefit from the alternative definition of our edge swaps: within the present formulation we know exactly the values of all A_{rs} with $(r,s) \in \mathcal{S}_{ijk\ell}$ (see diagrams). Specifically, we know that if we execute the move $F_{ijk\ell}$ then $A_{ij} = A_{k\ell} = 1$ and $A_{kj} = A_{i\ell} = 0$. Hence

$$2\Delta_{ijk\ell}|\Phi_{\mathbf{A}}| = 4(k_i k_j + k_k k_\ell - k_j k_k - k_i k_\ell) + 2\Delta_{ijk\ell}\text{Tr}(\mathbf{A}^3) + \Delta_{ijk\ell}\text{Tr}(\mathbf{A}^4) \qquad (7.69)$$

The effects of a move $F_{ijk\ell}$ on $\text{Tr}(\mathbf{A}^3)$ and $\text{Tr}(\mathbf{A}^4)$ are calculated in exactly the same way as was done in Appendix F. Again, the present edge-swap definition allows us to work out these expressions further, by writing explicitly the sum over all eight pairs $(r,s) \in \mathcal{S}_{ijk\ell}$ and substituting the known values of each corresponding A_{rs}. Alternatively, one can exploit the present definitions more explicitly and work out $\Delta_{ijk\ell}\text{Tr}(\mathbf{A}^3)$ and $\Delta_{ijk\ell}\text{Tr}(\mathbf{A}^4)$ from scratch. This is done in Appendix I, and results in the appealing simple formulae

$$\Delta_{ijk\ell}\text{Tr}(\mathbf{A}^3) = 6 \sum_{p \notin \{i,j,k,\ell\}} (A_{kp} - A_{ip})(A_{jp} - A_{\ell p}) \qquad (7.70)$$

$$\Delta_{ijk\ell} \mathrm{Tr}(\mathbf{A}^4) = 8\Bigg[\sum_{p\notin\{j,k,\ell\}} \sum_{r\notin\{i,j,k\}} \bar{\delta}_{(p,r),(i,\ell)} A_{kp} A_{pr} A_{rj}$$

$$- \sum_{p\notin\{i,j,k\}} \sum_{r\notin\{\ell,i,j\}} \bar{\delta}_{(p,r),(\ell,k)} A_{jp} A_{pr} A_{ri}$$

$$+ \sum_{p\notin\{i,j,\ell\}} \sum_{r\notin\{i,k,\ell\}} \bar{\delta}_{(p,r),(k,j)} A_{ip} A_{pr} A_{r\ell}$$

$$- \sum_{p\notin\{i,k,\ell\}} \sum_{r\notin\{j,k,\ell\}} \bar{\delta}_{(p,r),(j,i)} A_{\ell p} A_{pr} A_{rk} \Bigg] \qquad (7.71)$$

with $\bar{\delta}_{(a,b),(c,d)} = 1$ if $(a,b) \neq (c,d)$ and $\bar{\delta}_{(a,b),(c,d)} = 0$ if $(a,b) = (c,d)$. The above formulae can then be used in the by now familiar way to build detailed-balance MCMC algorithms that will equilibrate to any desired graph distribution $p(\mathbf{A})$ with hard-constrained degrees. The Metropolis–Hastings move acceptance probabilities for switch-and-hold MCMC algorithms take the form

$$\mathcal{A}[\mathbf{A} \to F_{ijk\ell}\mathbf{A}] = \min\left\{1, \frac{p(F_{ijk\ell}\mathbf{A})}{p(\mathbf{A})}\right\} \qquad (7.72)$$

For the mobility-based Metropolis–Hastings MCMC, we have

$$\mathcal{A}[\mathbf{A} \to F_{ijk\ell}\mathbf{A}] = \min\left\{1, \frac{p(F_{ijk\ell}\mathbf{A})|\Phi_{\mathbf{A}}|}{p(\mathbf{A})|\Phi_{F_{ijk\ell}\mathbf{A}}|}\right\} \qquad (7.73)$$

The pseudocodes for both switch-and-hold and mobility-based Metropolis–Hastings MCMC are provided in Algorithms 20 and 22, respectively.

If we allow for non-uniform drawing of candidate ordered edge swaps, according to some specific recipe $P(F_{ijk\ell}|\mathbf{A})$ that is neither switch-and-hold nor reduces to uniform selection from the set of presently applicable moves, then we simply use (7.3) in its original form. As we have seen with the hinge flip move set, capturing accurately the move proposal protocols in formulae for move proposal probabilities, and incorporating these accurately in the canonical move acceptance probabilities is vital to ensure that the correct measure $p(\mathbf{A})$ will be achieved. Apparently small shortcuts in computer code that are not accounted for can have disastrous consequences; see, e.g. Figure 7.1.

7.6 Non-uniform sampling of allowed edge swaps

Overview: In this section, we introduce a non-uniform sampling of pairs of ordered links which is not only efficient for sparse graphs but also for dense graphs. This entails drawing uniformly at random a first ordered link and drawing a second ordered link uniformly at random among those that are not adjacent to the first ordered link.

Nearly hardcore graphs. As we noted earlier, uniform sampling of pairs of links, as opposed to quadruplets of nodes, make the sampling of candidate moves much more efficient for sparse graphs. However, they can become computationally inefficient in the regime of very dense graphs. To illustrate this, we consider as an example a choice

TYPE A   TYPE B

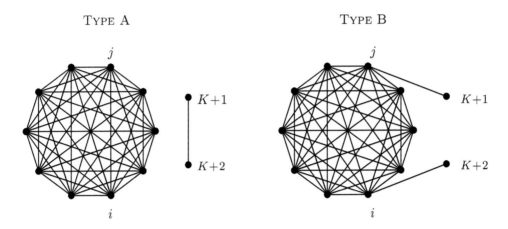

Fig. 7.7 Connected parts of 'nearly hardcore' graphs, as defined by a degree sequence (modulo node permutations) of the form: $k_i = K - 1$ for $i \leq K$, $k_i = 1$ for $i \in \{K+1, K+2\}$, and $k_i = 0$ for $i > K + 2$. In the present figure, $K = 10$ and the nodes $i > K + 2$ with zero degree are not shown. Left: The graph of type A, with $A_{K+1,K+2} = 1$, of which there is only one; here $|\Phi_\mathbf{A}| = K(K-1)$. Right: The graphs of type B, with $A_{K+1,K+2} = 0$, of which there are $K(K-1)$ (one for each choice of k and ℓ); here $|\Phi_\mathbf{A}| = 2(K-1)$.

of degrees $\mathbf{k} = (k_1, \ldots, k_N)$ that corresponds to a 'nearly hardcore' graph (see Figure 7.7), involving a fully connected core of size K and two degree-1 nodes (note that for a fully 'hardcore' graph with $K = N$, there would not be any allowed edge swap):

$$i \leq K: \quad k_i = K-1, \quad i = K+1, K+2: \quad k_i = 1 \tag{7.74}$$

There exist $K(K-1) + 1$ such graphs. Close inspection of their possible realizations reveals only two types, A and B, which are characterized by whether or not the two degree-1 nodes are connected to each other; see Figure 7.7 (where mobilities $|\Phi_\mathbf{A}|$ are defined using the standard edge-swaps definitions of Section 6.4).

- Type A:

$$A_{K+1,K+2} = 1, \quad A_{rs} = 1 \ \forall (r,s), \ r < s \leq K, \quad A_{rs} = 0 \ \text{elsewhere} \tag{7.75}$$

  Here we have the two degree-1 nodes connected to each other, plus a fully connected core of K nodes. There is just one such type-A graph $\mathbf{A}$, which we will call $\mathbf{A_A}$. It allows only for edge swaps involving the two degree-1 nodes and any two nodes taken from the core, so $|\Phi_\mathbf{A}| = K(K-1)$.

- Type B:

$$A_{K+1,K+2} = 0, \quad \exists (i, j \leq K) \text{ such that} \quad A_{ij} = 0, \quad A_{j,K+1} = A_{i,K+2} = 1,$$
$$A_{rs} = 1 \ \forall \ r < s \leq K, \ \{r,s\} \neq \{i,j\}$$
$$A_{rs} = 0 \ \text{elsewhere} \tag{7.76}$$

Here we have de facto carried out an edge swap relative to $\mathbf{A}_A$, and replaced both the link between the two degree-1 nodes and the link between two nodes (i,j) of the core by two new links: one from node j to degree-1 node $K+1$, and one from node i to the other degree-1 node $K+2$. We will call the resulting graph $\mathbf{A}_{B;j,i}$. There are $K(K-1)$ type-B graphs, one for each choice of (i,j), but each of these allows only for $2(K-1)$ edge swaps. In fact, the two outlying edges can be switched among each other in two ways: the one that brings us back to $\mathbf{A}_A$ and the swap that is equivalent to replacing the links $A_{k,K+1}$ and $A_{i,K+2}$ by the links $A_{k,K+2}$ and $A_{i,K+1}$. In addition, each of the outlying edges can be swapped separately in one way with every edge that has a node in common with the other outlying edge (this is easily checked directly). There are $K-2$ such edges. Hence, $|\Phi_\mathbf{A}| = 2+2(K-2) = 2(K-1)$ for each state of type B.

It is clear that sampling links uniformly at random will result most of the times in a pair of edges that cannot be switched. More efficient samplings of randomizing moves can be achieved by drawing uniformly at random a first link (i,j) among those available in the network and a second link (k,ℓ) uniformly among those that are not adjacent to link (i,j). The likelihood of drawing the link (k,ℓ) will then depend on (i,j), leading to a non-uniform sampling of moves, which requires calculations of acceptance probabilities that restore detailed balance at each iteration. These can be tricky to calculate; hence, the challenge is to choose a way of sampling moves which is efficient but at the same time keeps the calculation of acceptance probabilities relatively transparent and easy.

Non-uniform sampling of legal candidate moves. In this section, we propose a non-uniform sampling of legal moves $F_{ijk\ell}$ which is efficient in the sparse as well as in the dense regimes and at the same time allows for a straightforward calculation of the acceptance probabilities.

In the previous section, we have defined the move $F_{ijk\ell}$ as the map between the pair of ordered links $\{(i,j),(k,\ell)\} \to \{(i,\ell),(k,j)\}$. Let us now define as $f_{ijk\ell}$ the map between the ordered pair of ordered links $((i,j),(k,\ell)) \to ((i,\ell),(k,j))$. The likelihood of sampling a legal move $F_{ijk\ell}$ in configuration $\mathbf{A}$ is then

$$P(F_{ijk\ell}|\mathbf{A}) = P(f_{ijk\ell}|\mathbf{A}) + P(f_{k\ell ij}|\mathbf{A}) \tag{7.77}$$

where $P(f_{ijk\ell}|\mathbf{A})$ is the likelihood of sampling a legal move $f_{ijk\ell}$ in graph configuration $\mathbf{A}$. Similarly, we have $P(F_{ji\ell k}|\mathbf{A}) = P(f_{ji\ell k}|\mathbf{A}) + P(f_{\ell k ji}|\mathbf{A})$. Next, we consider the following procedure to sample a pair of ordered links $(i,j),(k,\ell)$: we first draw an ordered link (i,j) uniformly at random among the $N\langle k\rangle$ in the graph, then we draw a second ordered link (k,ℓ) among the $N\langle k\rangle - k_i - \sum_r A_{ir}k_r$ that do not end in i or in one of its neighbours, i.e. such that $\ell \neq i,j$ and $A_{i\ell} = 0$. This pair of links will only be switchable under move $f_{ijk\ell}$ if $A_{kj} = 0$ and k is different from i,j. If this requirement is not fulfilled, we will reject the pair out of hand and pick another pair. The likelihood that the drawn links are switchable is:

$$P(k \neq i,j; A_{jk} = 0 | A_{ij} = A_{\ell k} = 1, A_{i\ell} = 0, \ell \neq i,j) \tag{7.78}$$

$$= \frac{\sum_{ijk\ell} A_{ij} A_{k\ell} (1 - A_{i\ell})(1 - A_{jk}) \bar{\delta}_{ik} \bar{\delta}_{jk} \bar{\delta}_{i\ell} \bar{\delta}_{j\ell}}{\sum_{ijk\ell} A_{ij} A_{k\ell} (1 - A_{i\ell}) \bar{\delta}_{i\ell} \bar{\delta}_{j\ell}}$$

where the numerator is exactly $2|\Phi_{\mathbf{A}}|$, with $|\Phi_{\mathbf{A}}|$ given in (7.66), while the denominator gives

$$\sum_{ijk\ell} A_{ij} A_{k\ell} (1 - A_{i\ell}) \bar{\delta}_{i\ell} \bar{\delta}_{j\ell} = \sum_{ijk\ell} A_{ij} A_{k\ell} (1 - A_{i\ell})(1 - \delta_{i\ell} - \delta_{j\ell} + \delta_{i\ell}\delta_{j\ell})$$

$$= \sum_{ijk\ell} A_{ij} A_{k\ell} (1 - A_{i\ell})(1 - \delta_{i\ell})$$

$$= \sum_{ijk\ell} [A_{ij} A_{k\ell} - A_{ij} A_{k\ell} A_{i\ell})] - \sum_{ijk} A_{ij} A_{ik}$$

$$= \left(\sum_i k_i \right)^2 - \sum_{i\ell} A_{i\ell} k_i k_\ell - \sum_i k_i^2 \tag{7.79}$$

Hence, the likelihood of sampling a legal move $f_{ijk\ell}$ according to the above protocol is

$$P(f_{ijk\ell}|\mathbf{A}) = \frac{1}{N\langle k \rangle} \frac{1}{N\langle k \rangle - k_i - \sum_r A_{ir} k_r} \frac{2|\Phi_{\mathbf{A}}|}{(\sum_i k_i)^2 - \sum_{i\ell} A_{i\ell} k_i k_\ell - \sum_i k_i^2} \tag{7.80}$$

which gives, for the likelihood of sampling a legal move $F_{ijk\ell}$ in configuration $\mathbf{A}$

$$P(F_{ijk\ell}|\mathbf{A}) = \frac{1}{N\langle k \rangle} \left[\frac{1}{N\langle k \rangle - k_i - \sum_r A_{ir} k_r} + \frac{1}{N\langle k \rangle - k_k - \sum_r A_{kr} k_r} \right]$$

$$\times \frac{2|\Phi_{\mathbf{A}}|}{(\sum_i k_i)^2 - \sum_{ij} A_{ij} k_i k_j - \sum_i k_i^2} \tag{7.81}$$

and for move $F_{ji\ell k}$

$$P(F_{ji\ell k}|\mathbf{A}) = \frac{1}{N\langle k \rangle} \left[\frac{1}{N\langle k \rangle - k_j - \sum_r A_{jr} k_r} + \frac{1}{N\langle k \rangle - k_\ell - \sum_r A_{\ell r} k_r} \right]$$

$$\times \frac{2|\Phi_{\mathbf{A}}|}{(\sum_i k_i)^2 - \sum_{ij} A_{ij} k_i k_j - \sum_i k_i^2} \tag{7.82}$$

yielding, for the likelihood of picking a move going from configuration $\mathbf{A} \to \mathbf{A}' = F_{ijk\ell}\mathbf{A}$, in undirected graphs,

$$P(F_{ijk\ell}|\mathbf{A}) + P(F_{ji\ell k}|\mathbf{A}) = \frac{1}{N\langle k \rangle} [R_i^{-1}(\mathbf{A}) + R_j^{-1}(\mathbf{A}) + R_k^{-1}(\mathbf{A}) + R_\ell^{-1}(\mathbf{A})] \frac{2|\Phi_{\mathbf{A}}|}{\sum_s k_s R_s(\mathbf{A})} \tag{7.83}$$

with

$$R_s(\mathbf{A}) = N\langle k \rangle - \sum_r A_{sr} k_r - k_s \tag{7.84}$$

A uniform sampling of graphs is then ensured by choosing the acceptance probabilities

$$\mathcal{A}[\mathbf{A} \to F_{ijk\ell}\mathbf{A}] = \left\{ 1, \frac{\sum_s k_s R_s(\mathbf{A})}{\sum_s k_s R_s(\mathbf{A}')} \frac{|\Phi(\mathbf{A}')|}{|\Phi_{\mathbf{A}}|} \right.$$
$$\left. \times \frac{R_i^{-1}(\mathbf{A}') + R_j^{-1}(\mathbf{A}') + R_k^{-1}(\mathbf{A}') + R_\ell^{-1}(\mathbf{A}')}{R_i^{-1}(\mathbf{A}) + R_j^{-1}(\mathbf{A}) + R_k^{-1}(\mathbf{A}) + R_\ell^{-1}(\mathbf{A})} \right\} \qquad (7.85)$$

We can relate the terms in the numerator with those in the denominator via

$$R_i(\mathbf{A}') = N\langle k \rangle - \sum_r A'_{ir} k_r - k_i = N\langle k \rangle - \sum_r A_{ir} k_r - k_i + k_j - k_\ell$$
$$= R_i(\mathbf{A}) + k_j - k_\ell \qquad (7.86)$$

and, similarly,

$$R_j(\mathbf{A}') = R_j(\mathbf{A}) + k_i - k_k$$
$$R_k(\mathbf{A}') = R_k(\mathbf{A}) + k_\ell - k_j$$
$$R_\ell(\mathbf{A}') = R_\ell(\mathbf{A}) + k_k - k_i \qquad (7.87)$$

and by using $|\Phi(\mathbf{A}')| = |\Phi_{\mathbf{A}}| + \Delta_{ijk\ell}|\Phi_{\mathbf{A}}|$ so that the acceptance probabilities read

$$\mathcal{A}(\mathbf{A} \to F_{ijk\ell}\mathbf{A}) = \min \left\{ 1, \frac{\sum_s k_s R_s}{\sum_s k_s R_s + 2(k_i - k_k)(k_j - k_\ell)} \left[1 + \frac{\Delta_{ijk\ell}|\Phi_{\mathbf{A}}|}{|\Phi_{\mathbf{A}}|} \right] \right.$$
$$\left. \times \frac{(R_i + k_j - k_\ell)^{-1} + (R_j - k_k + k_i)^{-1} + (R_k + k_\ell - k_j)^{-1} + (R_\ell + k_k - k_i)^{-1}}{R_i^{-1} + R_j^{-1} + R_k^{-1} + R_\ell^{-1}} \right\}$$
$$(7.88)$$

The terms $\{R_s\}$ can be calculated at the start of the algorithm and updated during the graph dynamics very efficiently, as only the four terms R_i, R_j, R_k, R_ℓ change upon application of move $F_{ijk\ell}$. However, $\Delta_{ijk\ell}|\Phi_{\mathbf{A}}|$, given in (7.68), requires the update of $\mathrm{Tr}\,(\mathbf{A}^3), \mathrm{Tr}\,(\mathbf{A}^4)$. In the next section we propose a sampling protocol that avoids this difficulty.

7.7 Non-uniform sampling from a restricted set of moves: a hybrid MCMC algorithm

Overview: In this section, we introduce a non-uniform sampling of pairs of ordered links which is efficient for sparse and dense graphs and makes the computation of corrections needed to restore detailed balance particularly easy. This entails drawing uniformly at random a first ordered link, followed by drawing a second ordered link uniformly at random from among those links that do not end at the node where the first ordered link starts or at any of its neighbours.

One can discount the calculation of the first line of (7.88), which comes from the likelihood (7.78), by not restricting oneself exclusively to legal moves, but by also allowing for the (non-uniform) sampling of some illegal moves, whose proposal will be regarded as an iteration spent in the current configuration. This leads to a hybrid algorithm that is a combination of non-uniform sampling and switch-and-hold techniques.

In particular, we draw ordered pairs of ordered links $((i, j), (k, \ell))$ among those that satisfy the requirements $A_{i\ell} = 0$ and $\ell \neq i, j$, as explained above, but we will no longer reject out of hand the pairs for which $A_{kj} = 1$ or $k = i, j$. These will instead be rejected via the move acceptance probability, which will be set to zero for any illegal move, so that illegal moves will count as iteration steps. The advantage is that we no longer need to correct the acceptance probability for the out-of-hand rejection of illegal moves among the moves (non-uniformly) picked and we only need to correct the acceptance probabilities for the bias introduced by our non-uniform sampling of (possibly illegal) moves.

The likelihood of drawing an ordered pair of ordered links $((i, j), (k, \ell))$, satisfying $A_{i\ell} = 0$ and $\ell \neq i, j$, according to the protocol explained above, is

$$P(((i, j), (k, \ell)); A_{i\ell} = 0; \ell \neq i, j) = \frac{1}{N\langle k \rangle} \frac{1}{R_i} \tag{7.89}$$

It follows that moves bringing the network from configuration $\mathbf{A}$ to $F_{ijk\ell}\mathbf{A}$ are sampled with non-uniform probability

$$P(((i, j), (k, \ell)); A_{i\ell} = 0; \ell \neq i, j) + P(((k, \ell), (i, j)); A_{jk} = 0; j \neq k, \ell)$$
$$+ P(((j, i), (\ell, k)); A_{kj} = 0; k \neq i, j) + P(((\ell, k), (j, i)); A_{i\ell} = 0; i \neq k, \ell)$$
$$= \frac{1}{N\langle k \rangle} \left[R_i^{-1} + R_j^{-1} + R_k^{-1} + R_\ell^{-1} \right] \tag{7.90}$$

but they are not all legal. If they can act on the instantaneous configuration $\mathbf{A}$, they will be accepted with a probability $\mathcal{A}(\mathbf{A} \to F_{ijk\ell}\mathbf{A})$ such that

$$\left[R_i^{-1}(\mathbf{A}) + R_j^{-1}(\mathbf{A}) + R_k^{-1}(\mathbf{A}) + R_\ell^{-1}(\mathbf{A}) \right] \mathcal{A}(\mathbf{A} \to F_{ijk\ell}\mathbf{A})$$
$$= \left[R_i^{-1}(\mathbf{A}') + R_j^{-1}(\mathbf{A}') + R_k^{-1}(\mathbf{A}') + R_\ell^{-1}(\mathbf{A}') \right] \mathcal{A}(F_{ijk\ell}\mathbf{A} \to \mathbf{A}) \tag{7.91}$$

where we have denoted $\mathbf{A}' = F_{ijk\ell}\mathbf{A}$. This leads to the algorithm

$$\mathcal{A}(\mathbf{A} \to F_{ijk\ell}\mathbf{A}) = \min \Big\{ 1,$$
$$\frac{(R_i + k_j - k_\ell)^{-1} + (R_j - k_k + k_i)^{-1} + (R_k + k_\ell - k_j)^{-1} + (R_\ell + k_k - k_i)^{-1}}{R_i^{-1} + R_j^{-1} + R_k^{-1} + R_\ell^{-1}} \Big\} \tag{7.92}$$

whose pseudocode can be found in Algorithm 23 of Appendix J. Let us now explain how the proposed hybrid algorithm speeds up the dynamics when compared to (non-uniform) sampling from the set of legal moves only (illustrated in the previous section) and to uniform sampling from all moves (standard switch-and-hold). As an illustrative example, consider the 'nearly hardcore' graph of Figure 7.7. Suppose the first link is

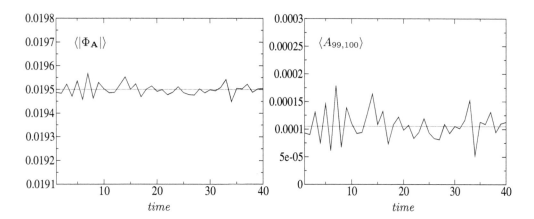

Fig. 7.8 Average mobility $\langle|\Phi_{\mathbf{A}}|\rangle$ (left) and link likelihood $\langle A_{99,100}\rangle$ (right) versus time during graph dynamics preserving the degrees of the near hardcore graph of Figure 7.7 with core size $K = 98$ and total number of nodes $N = 100$. Results are in agreement with predictions (7.94) and (7.93), shown by the dotted line, respectively.

picked from within the 'hardcore' when the graph is in configuration A (left-hand side of the figure). The hybrid algorithm would sample from valid swap candidates: i.e. $(K+1, K+2)$ or $(K+2, K+1)$. Sampling the second link uniformly at random, in contrast, one has only a probability $2/[2+K(K-1)]$ to pick the good link. On average, an $\mathcal{O}(K^2)$ number of iterations would be spent searching for a move which will transition to a new configuration. On the other hand, if one opted for the (non-uniform) sampling from the set of legal moves only, one would need to calculate the $\Delta_{ijk\ell}\mathrm{Tr}(\mathbf{A}^3)$ and $\Delta_{ijk\ell}\mathrm{Tr}(\mathbf{A}^4)$ terms. The number of iterations required for the computation of $\Delta_{ijk\ell}\mathrm{Tr}(\mathbf{A}^4)$ is again $\mathcal{O}(K^2)$ (see Appendix I), and is thus of the same order as those saved on the search for the right move. The hybrid algorithm allows one to save computational time on the search for legal moves without having to pay an equally large price on the computation of the corrections.

Illustration: the nearly hardcore graph. To test the proposed algorithm, we will monitor graph observables that we can calculate exactly. For example, for the near hardcore graph illustrated in Figure 7.7, we can calculate the expectation value of the link $A_{K+1,K+2}$ (which is only present in the state $\mathbf{A}_A$). For the uniform measure that we are targeting $p(\mathbf{A}) = |Z(\mathbf{k})|^{-1} = [K(K-1)+1]^{-1}$, this evaluates to

$$\langle A_{K+1,K+2}\rangle = [1+K(K-1)]^{-1} \tag{7.93}$$

Another quantity we can easily calculate is the average mobility

$$\langle|\Phi_{\mathbf{A}}|\rangle = \frac{K(K-1)(2K-1)}{K(K-1)+1} \tag{7.94}$$

In Figure 7.8 we randomize the near hardcore graph with $N = 100$ nodes and hardcore size $K = 98$, and we plot the link likelihood $\langle A_{99,100}\rangle$ and the mobility $\langle|\Phi_{\mathbf{A}}|\rangle$

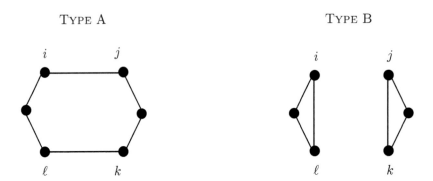

Fig. 7.9 Connected parts of a ring graph with $N = 6$, defined by the degree sequence $k_i = 2$ for $i = 1, \ldots, 6$. Left: A graph of type A, of which there are 60; here $|\Phi_A| = 12$, $\text{Tr}(\mathbf{A}^3) = 0$, $\text{Tr}(\mathbf{A}^4) = 36$. Right: A graph of type B, of which there are 10; here $|\Phi_A| = 18$, $\text{Tr}(\mathbf{A}^3) = 12$, $\text{Tr}(\mathbf{A}^4) = 36$.

during the graph dynamics. These are found in excellent agreement with theoretical predictions, shown as dotted lines.

Illustration: the hexagon graph As a second test, we consider a graph with $N = 6$ nodes and degrees $k_i = 2 \; \forall \; i$. There are only two types of realization of this graph: A, which corresponds to a ring; and B, which corresponds to two disconnected triangles (see Fig. 7.9).

- Type A:

  There are $6!/12 = 60$ such graphs. Here the number of triangles and squares is zero, so $\text{Tr}(\mathbf{A}^3) = 0$ and $\text{Tr}(\mathbf{A}^4) = 2\sum_j k_j^2 - \sum_i k_i = 36$ as it only takes contributions from backtrackings of links and paths of length 2 that start and terminate at the same site. Formula (6.46) then gives $|\Phi_A| = 12$.

- Type B:

  Here we have carried out the edge swap $F_{ijk\ell}$ relative to the ring configuration. The number of type-B graphs is

  $$\frac{1}{2} \binom{6}{3} = 10 \tag{7.95}$$

  Each of these allows for $|\Phi_A| = 2 \cdot 3 \cdot 3 = 18$ edge swaps. There are two triangles, so $\text{Tr}(\mathbf{A}^3) = 2 \cdot 6 = 12$ and $\text{Tr}(\mathbf{A}^4) = 36$, as in configuration A.

If graphs are sampled uniformly, the weight of each graph is $1/70$, and the weighted averages of traces and mobility are:

$$\frac{1}{N}\text{Tr}(\mathbf{A}^3) = \frac{1}{6}\frac{1}{7}12 = \frac{2}{7} \simeq 0.2857$$

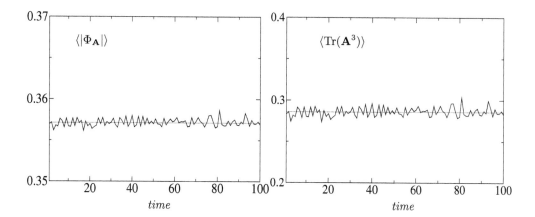

Fig. 7.10 Average mobility $\langle|\Phi_{\mathbf{A}}|\rangle$ (left) trace $\langle\mathrm{Tr}(\mathbf{A}^3)\rangle$ (right) versus time during graph dynamics preserving the degrees of the hexagon graph shown in Figure 7.9. Results are in agreement with theoretical predictions (7.96) shown by the dotted line.

$$\frac{1}{N}\mathrm{Tr}(\mathbf{A}^4) = \frac{1}{6}36 = 6$$

$$\frac{|\Phi_{\mathbf{A}}|}{N^2} = \frac{1}{36}\left(\frac{1}{7}18 + \frac{6}{7}12\right) = \frac{5}{14} \simeq 0.3571 \tag{7.96}$$

In Figure 7.10 we show the evolution of the trace $\langle\mathrm{Tr}(\mathbf{A}^3)\rangle$ and the mobility $\langle|\Phi_{\mathbf{A}}|\rangle$ (averaged over small time windows) during graph dynamics tailored to sampling graphs uniformly, with constrained degrees. These are again found in excellent agreement with theoretical predictions, shown as dotted lines.

We conclude this section by pointing out that the method of sampling links used in the hybrid algorithm can be used to produce graphs with controlled degree correlations, using recipe (7.58) in which for $|\Phi_{\mathbf{A}}|/|\Phi_{F_{ijk\ell;1}\mathbf{A}}|$ one substitutes

$$\frac{P(F_{ijk\ell}|F_{ikj\ell}\mathbf{A})}{P(F_{ikj\ell}|\mathbf{A})} \tag{7.97}$$
$$= \frac{(R_i + k_j - k_\ell)^{-1} + (R_j + k_i - k_k)^{-1} + (R_k + k_\ell - k_j)^{-1} + (R_\ell + k_k - k_i)^{-1}}{R_i^{-1} + R_j^{-1} + R_k^{-1} + R_\ell^{-1}}$$

and R_s given in (7.84). This is what we used to produce Figure 7.4 (panel g, h), Figure 7.5 (panel f) and Figure 7.6 (panel f).

7.8   Extensions to directed graphs

Overview: *For directed graphs, directed edge swaps are not sufficient to guarantee ergodicity, and an additional move has to be introduced, namely a triangle reversion. At each time step, two ordered links are drawn at random upon which either an edge swap or a triangle inversion can be carried out, and either move is carried out with a* *probability which ensures detailed balance with the targeted measure.*

Extension of our randomization protocols to directed graphs is achieved by replacing k with $\mathbf{k} = (k^{\mathrm{in}}, k^{\mathrm{out}})$, allowing repetition of site indices in (7.28) and bearing in mind that $\Pi(\mathbf{k}_a, \mathbf{k}_b) \neq \Pi(\mathbf{k}_b, \mathbf{k}_a)$. However, in contrast to the case of undirected graphs, when self-interactions are forbidden, the directed edge swaps defined above are no longer sufficient to ensure ergodicity, and a new move is required to restore it [151]. The latter can be visualized as reversing a triangle cycle

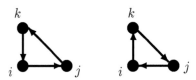

We denote this operation by T_{ijk} and note that its effect on a state $\mathbf{A}$, when it can act on it, is

$$T_{ijk}(\mathbf{A})_{qr} = 1 - A_{qr} \quad \text{for } (q,r) \in \mathcal{S}_{ijk} \tag{7.98}$$
$$T_{ijk}(\mathbf{A})_{qr} = A_{qr} \quad \text{for } (q,r) \notin \mathcal{S}_{ijk} \tag{7.99}$$

where $\mathcal{S}_{ijk} = \{[i,j],[j,k],[k,i]\}$ and $[a,b]$ is non-ordered, i.e. $[a,b] = (a,b)$ if $a < b$ and $[a,b] = (b,a)$ if $a > b$. With two types of move available, edge swaps and triangle reversions, care must be taken to sample evenly from the space of all possible moves[9].

Finally, since for any graph configuration $\mathbf{A}$, the set of graphs $\mathbf{A}'$ generated by application of a single directed edge swap is different from the set of graphs $\mathbf{A}''$ obtained by application of a single triangle reverse, the resulting mobility, for the graph dynamics where both types of moves are allowed, is just the sum of the number of possible moves of each type, i.e. $|\Phi_{\mathbf{A}}| = n_\square(\mathbf{A}) + n_\triangle(\mathbf{A})$. Here $n_\triangle(\mathbf{A})$ is the number of possible triangle reverses that can act on a given configuration $\mathbf{A}$, which can be calculated to be (see [155, 154]):

$$n_\triangle = \frac{1}{3}\mathrm{Tr}(\mathbf{A}^3) - \mathrm{Tr}(\mathbf{A}^{\updownarrow}\mathbf{A}^2) + \mathrm{Tr}(\mathbf{A}^{\updownarrow^2}\mathbf{A}) - \frac{1}{3}\mathrm{Tr}(\mathbf{A}^{\updownarrow^3})$$

in which $A_{ij}^{\updownarrow} = A_{ij} \cdot A_{ji}$.

Shortcut: *In this chapter, we focused on how to generate numerically tailored random graphs with controlled macroscopic structural properties, to serve, e.g. as null models in hypothesis testing. Bias in the generation of random graphs has the potential to invalidate all further statistical analysis performed on the generated networks and has been well documented in the literature. Biases occur when randomizing moves (e.g. hinge swaps or edge swaps) are not drawn uniformly, and no corrections are introduced to restore detailed balance of the graph dynamics with the targeted network distribution.*

---

[9] At the algorithmic level, this is achieved by uniformly sampling pairs of edges and then identifying which type of move they relate to.

We have seen that non-uniform sampling of randomizing moves may make the algorithm much more efficient, and hence it is important to have workable corrections for these algorithms. We have shown an exact approach to generating random graphs from ensembles which share certain topological characteristics with a real network (degree distribution, degree sequences and/or degree correlations), and can therefore serve as a reliable tool for building unbiased null models.

7.9 Solutions of exercises

Exercise 7.1

The pseudocodes for the three sampling protocols are given in Algorithms 16, 17 and 18, respectively. □

Exercise 7.2 We have seen in Exercise 5.1 that we can write

$$p(\mathbf{A}|\mathbf{k},L) = \frac{1}{Z(\mathbf{k},L)} \prod_{i<j} \left[\frac{k_i k_j}{N\langle k\rangle} \delta_{A_{ij},1} + \left(1 - \frac{k_i k_j}{N\langle k\rangle}\right) \delta_{A_{ij},0} \right] \delta_{L,\sum_{i<j} A_{ij}}$$

$$= \frac{1}{Z'(\mathbf{k},L)} e^{\sum_i k_i(\mathbf{A})\log\left[k_i/\sqrt{N\langle k\rangle}\right]} \delta_{L,\sum_{i<j} A_{ij}} \tag{7.100}$$

for large and finitely connected graphs. To sample from this distribution, we use an MCMC algorithm based on hinge flips, with acceptance probabilities $\mathcal{A}[\mathbf{A} \to F_{ijk}\mathbf{A}] = \min\{1, p(F_{ijk}\mathbf{A})/p(\mathbf{A})\}$ and hinge flips sampled uniformly. While hinge flips preserve the hard constraint on the links L, they do not preserve the degree sequence $\mathbf{k}$, and hence we need to work out $Z'(\mathbf{k},L)$, as this may change between configurations $\mathbf{A}$ and $F_{ijk}\mathbf{A}$. Using the Fourier representation of the Kronecker delta, we have

$$Z(\mathbf{k},L) = \sum_{\mathbf{A}} \prod_{i<j} \left[\frac{k_i k_j}{N\langle k\rangle} \delta_{A_{ij},1} + \left(1 - \frac{k_i k_j}{N\langle k\rangle}\right) \delta_{A_{ij},0} \right] \delta_{L,\sum_{i<j} A_{ij}}$$

$$= \int \frac{d\omega}{2\pi} e^{i\omega L} \sum_{\mathbf{A}} \prod_{i<j} \left[\frac{k_i k_j}{N\langle k\rangle} \delta_{A_{ij},1} + \left(1 - \frac{k_i k_j}{N\langle k\rangle}\right) \delta_{A_{ij},0} \right] \prod_{i<j} e^{-i\omega A_{ij}}$$

$$= \int \frac{d\omega}{2\pi} e^{i\omega L} \sum_{\mathbf{A}} \prod_{i<j} \left[\frac{k_i k_j}{N\langle k\rangle} e^{-i\omega} \delta_{A_{ij},1} + \left(1 - \frac{k_i k_j}{N\langle k\rangle}\right) \delta_{A_{ij},0} \right]$$

$$= \int \frac{d\omega}{2\pi} e^{i\omega L} \prod_{i<j} \left[\frac{k_i k_j}{N\langle k\rangle} e^{-i\omega} + \left(1 - \frac{k_i k_j}{N\langle k\rangle}\right) \right]$$

$$= \int \frac{d\omega}{2\pi} e^{i\omega L} \prod_{i<j} e^{\frac{k_i k_j}{N\langle k\rangle}(e^{-i\omega}-1)} \tag{7.101}$$

In the last step, we have assumed that N is large, and used $1+\mathcal{O}\left(\frac{1}{N}\right) = \exp[\mathcal{O}\left(\frac{1}{N}\right)]$. Finally we get

$$Z(\mathbf{k},L) = \int \frac{d\omega}{2\pi} e^{i\omega L} e^{\sum_{i<j} \frac{k_i k_j}{N\langle k\rangle}(e^{-i\omega}-1)}$$

$$= \int \frac{d\omega}{2\pi} e^{i\omega L + N\frac{\langle k \rangle}{2}(e^{-i\omega}-1)} = e^{-L} L^L / L! \tag{7.102}$$

where we used $L = N\langle k \rangle / 2$ and the integral

$$\int \frac{d\omega}{2\pi} e^{i\omega a + be^{-i\omega}} = \int \frac{d\omega}{2\pi} e^{i\omega a} \sum_{n \geq 0} \frac{(be^{-i\omega})^n}{n!} = \sum_{n \geq 0} \frac{b^n}{n!} \int \frac{d\omega}{2\pi} e^{i\omega a - i\omega n}$$

$$= \sum_{n \geq 0} \frac{b^n}{n!} \delta_{n,a} = b^a / a! \tag{7.103}$$

Hence, to leading orders in N, the normalization factor $Z(\mathbf{k}, L)$ only depends on L, which is hard constrained. In addition, from (7.42) we have

$$Z'(\mathbf{k}, L) = Z(\mathbf{k}, L) \prod_{k<\ell} \left(1 - \frac{\langle k \rangle}{N} \frac{W(k_k, k_\ell)}{p(k_k)p(k_\ell)}\right)^{-1} = Z(\mathbf{k}, L) \, e^{L + \mathcal{O}(1)}$$

$$= (L^L / L!) e^{\mathcal{O}(1)} \tag{7.104}$$

Hence, in relevant orders in N, $Z'(\mathbf{k}, L)$ drops out of the acceptance probabilities

$$\mathcal{A}[\mathbf{A} \to F_{ijk}\mathbf{A}] = \min \left\{ 1, \frac{e^{(k_j(\mathbf{A})-1)\log \frac{k_j}{\sqrt{N\langle k \rangle}} + (k_k(\mathbf{A})+1)\log \frac{k_k}{\sqrt{N\langle k \rangle}}}}{e^{k_j(\mathbf{A})\log \frac{k_j}{\sqrt{N\langle k \rangle}} + k_k(\mathbf{A})\log \frac{k_k}{\sqrt{N\langle k \rangle}}}} \right\}$$

$$= \min \left\{ 1, e^{-\log \frac{k_j}{\sqrt{N\langle k \rangle}} + \log \frac{k_k}{\sqrt{N\langle k \rangle}}} \right\} = \min \left\{ 1, \frac{k_k}{k_j} \right\} \tag{7.105}$$

□

Exercise 7.3

First, we define our constraining observables, i.e. the degree sequence and the degree correlation:

$$k_i(\mathbf{A}) = k_i \qquad (\forall i) \tag{7.106}$$

$$W(k, k'|\mathbf{A}) = \frac{1}{N\langle k \rangle} \sum_{ij} A_{ij} \delta_{k_i, k} \delta_{k_j, k'} \qquad (k, k' > 0) \tag{7.107}$$

We aim to show that the random graph ensemble (7.39) maximizes the Shannon entropy

$$S[\mathbf{k}, W] = \sum_{\mathbf{A} \in G} p(\mathbf{A}|\mathbf{k}, W) \log p(\mathbf{A}|\mathbf{k}, W) \tag{7.108}$$

subject to the constraints $W(k, k') = \sum_{\mathbf{A}} p(\mathbf{A})W(k, k'|\mathbf{A})$ for all (k, k') and $k_i = k_i(\mathbf{A})$ for all i. We will limit ourselves to symmetric graphs without self-interactions; derivations for non-symmetric graphs is straightforward.

Extremization of (7.108) with Lagrange multipliers, without enforcing $p(\mathbf{A}) \geq 0$ explicitly, gives

$$\forall \mathbf{A}: \quad \frac{\partial}{\partial p(\mathbf{A})} \left\{ \sum_{\mathbf{A}' \in G} p(\mathbf{A}') \left[\log p(\mathbf{A}') + \Lambda_0 + \sum_{kk'} \lambda(k, k') W(k, k' | \mathbf{A}') \right] \right\} = 0$$

$$\forall \mathbf{A}: \quad \log p(\mathbf{A}) + \Lambda_0 + \sum_{kk'} \lambda(k, k') W(k, k' | \mathbf{A}) + 1 = 0$$

$$\forall \mathbf{A}: \quad p(\mathbf{A}) = \frac{1}{Z} e^{-\sum_{kk'} \lambda(k,k') W(k,k'|\mathbf{A})}$$

$$\forall \mathbf{A}: \quad p(\mathbf{A}) = \frac{1}{Z} e^{-\frac{1}{N\langle k \rangle} \sum_{ij} A_{ij} \lambda(k_i(\mathbf{A}), k_j(\mathbf{A}))} \tag{7.109}$$

with Z such that $\sum_{\mathbf{A} \in G} p(\mathbf{A}) = 1$. As expected for an ensemble of random graphs with maximum entropy, where a set of averages of observables are constrained to assume prescribed values, the result of the extremization gives an exponential family, where the parameters $\{\lambda(k_i, k_j)\}$ are to be calculated from the equations for the constraints. What is left is to show that the exponential family can be reduced to the micro-canonical ensemble (7.39), where degrees are prescribed, by a simple redefinition of the Lagrange multipliers. Let us first rewrite (7.109):

$$\forall \mathbf{A}: \quad p(\mathbf{A}) = \frac{1}{Z} \left(\prod_{i<j} e^{-\frac{1}{N\langle k \rangle} A_{ij} [\lambda(k_i, k_j) + \lambda(k_j, k_i)]} \right) \left(\prod_i \delta_{k_i, k_i(\mathbf{A})} \right) \tag{7.110}$$

$$\forall \mathbf{A}: \quad p(\mathbf{A}) = \frac{1}{Z} \left(\prod_{i<j} \left[e^{-\frac{1}{N\langle k \rangle} [\lambda(k_i, k_j) + \lambda(k_j, k_i)]} \delta_{A_{ij}, 1} + \delta_{A_{ij}, 0} \right] \right) \left(\prod_i \delta_{k_i, k_i(\mathbf{A})} \right) \tag{7.111}$$

We can then redefine our Lagrange multipliers in terms of the ratio $W(k, k')/p(k)p(k')$ via

$$\frac{\langle k \rangle}{N} \frac{W(k, k')}{p(k)p(k')} = \frac{e^{-\frac{1}{N\langle k \rangle} [\lambda(k_i, k_j) + \lambda(k_j, k_i)]}}{1 + e^{-\frac{1}{N\langle k \rangle} [\lambda(k_i, k_j) + \lambda(k_j, k_i)]}}$$

This results in

$$p(\mathbf{A}) = \frac{1}{Z} \prod_{i<j} \left(1 - \frac{\langle k \rangle W(k_i, k_j)}{N p(k_i) p(k_j)} \right)^{-1}$$

$$\times \prod_{i<j} \left[\frac{\langle k \rangle}{N} \frac{W(k_i, k_j)}{p(k_i) p(k_j)} \delta_{A_{ij}, 1} + \left(1 - \frac{\langle k \rangle}{N} \frac{W(k_i, k_j)}{p(k_i) p(k_j)} \right) \delta_{A_{ij}, 0} \right] \cdot \prod_i \delta_{k_i, k_i(\mathbf{A})} \tag{7.112}$$

The first product in (7.112) depends only on the constrained degrees $\{k_i\}$. In fact, to leading order this dependence is only via their average $\langle k \rangle$, via (7.104), so it drops out of the measure, and hence (7.112) can be rewritten as

$$p(\mathbf{A}) = \frac{1}{Z(\mathbf{k}, W)} \prod_{i<j} \left[\frac{\langle k \rangle}{N} \frac{W(k_i, k_j)}{p(k_i) p(k_j)} \delta_{A_{ij}, 1} + \left(1 - \frac{\langle k \rangle}{N} \frac{W(k_i, k_j)}{p(k_i) p(k_j)} \right) \delta_{A_{ij}, 0} \right] \prod_i \delta_{k_i, \sum_j A_{ij}}$$

$$Z(\mathbf{k}, W) = \sum_{\mathbf{A} \in G} \prod_{i<j} \left[\frac{\langle k \rangle}{N} \frac{W(k_i, k_j)}{p(k_i)p(k_j)} \delta_{A_{ij},1} + \left(1 - \frac{\langle k \rangle}{N} \frac{W(k_i, k_j)}{p(k_i)p(k_j)}\right) \delta_{A_{ij},0} \right] \prod_i \delta_{k_i, \sum_j A_{ij}}$$

$$(7.113)$$

which indeed reduces to (7.39), as claimed. □

Exercise 7.4

We can regard the two networks $\mathbf{A}_A, \mathbf{A}_B$, with degree distribution and degree correlations $p_A(k), W_A(k, k')$ and $p_B(k), W_B(k, k')$, respectively, as members of the random graph ensembles $p(\mathbf{A}|p_A, W_A)$ and $p(\mathbf{A}|p_B, W_B)$ respectively,

$$p(\mathbf{A}|p_A, W_A) = \sum_{\mathbf{k}} p(\mathbf{A}|\mathbf{k}, W_A) \prod_i p_A(k_i) \tag{7.114}$$

$$p(\mathbf{A}|p_B, W_B) = \sum_{\mathbf{k}} p(\mathbf{A}|\mathbf{k}, W_B) \prod_i p_B(k_i) \tag{7.115}$$

where $p(\mathbf{A}|\mathbf{k}, W_A), p(\mathbf{A}|\mathbf{k}, W_B)$ are maximum entropy ensembles with constrained degree sequence $\mathbf{k}$ and degree correlations W_A, W_B, respectively. These are ensembles of the type (7.39)

$$p(\mathbf{A}|\mathbf{k}, W) = \frac{w(\mathbf{A}|\mathbf{k}, W)}{Z(\mathbf{k}, W)} \prod_i \delta_{k_i, k_i(\mathbf{A})}$$

$$Z(\mathbf{k}, W) = \sum_{\mathbf{A}} w(\mathbf{A}|\mathbf{k}, W) \prod_i \delta_{k_i, k_i(\mathbf{A})} \tag{7.116}$$

where

$$w(\mathbf{A}|\mathbf{k}, W) = \prod_{k<\ell} \left[\frac{\langle k \rangle}{N} \frac{W(k_k, k_\ell)}{p(k_k)p(k_\ell)} \delta_{A_{k\ell},1} + \left(1 - \frac{\langle k \rangle}{N} \frac{W(k_k, k_\ell)}{p(k_k)p(k_\ell)}\right) \delta_{A_{k\ell},0} \right] \tag{7.117}$$

Hence, we define the distance between networks $\mathbf{A}_A, \mathbf{A}_B$ as the Kullback–Leibler distance between two network ensembles $p(\mathbf{A}|p_A, W_A), p(\mathbf{A}|p_B, W_B)$

$$D_{AB} = \frac{1}{2N} \sum_{\mathbf{A}} p(\mathbf{A}|p_A, W_A) \log \frac{p(\mathbf{A}|p_A, W_A)}{p(\mathbf{A}|p_B, W_B)} + \frac{1}{2N} \sum_{\mathbf{A}} p(\mathbf{A}|p_B, W_B) \log \frac{p(\mathbf{A}|p_B, W_B)}{p(\mathbf{A}|p_A, W_A)}$$

$$= \frac{1}{2} \left[\langle \Omega(\mathbf{A}|p_A, W_A) \rangle_A - \langle \Omega(\mathbf{A}|p_B, W_B) \rangle_A + \langle \Omega(\mathbf{A}|p_B, W_B) \rangle_B - \langle \Omega(\mathbf{A}|p_A, W_A) \rangle_B \right]$$

$$(7.118)$$

where $\langle (\cdot) \rangle_A = \sum_{\mathbf{A}} (\cdot) p(\mathbf{A}|p_A, W_A)$, $\langle (\cdot) \rangle_B = \sum_{\mathbf{A}} (\cdot) p(\mathbf{A}|p_B, W_B)$ and

$$\Omega(\mathbf{A}|p, W) = \frac{1}{N} \log p(\mathbf{A}|p, W) \tag{7.119}$$

Since $p(\mathbf{A}|p, W) \sim e^{-NS[p,W]}$, where $S[p, W]$ is the Shannon entropy, we expect Ω to be $\mathcal{O}(1)$. Indeed, the prefactor $1/N$ has been included in (7.118) to work with $\mathcal{O}(1)$ quantities. Carrying out the summation over $\mathbf{k}$ in (7.114), (7.115), we get

$$p(\mathbf{A}|p, W) = \frac{w(\mathbf{A}|\mathbf{k}(\mathbf{A}), W)}{Z(\mathbf{k}(\mathbf{A}), W)} \prod_i p(k_i(\mathbf{A})) \tag{7.120}$$

where we dropped the indices A, B, as the same manipulations apply to both network ensembles A, B. Inserting in (7.119), we get

$$\Omega(\mathbf{A}|p, W) = \frac{1}{N} \sum_i \log p(k_i(\mathbf{A})) + \frac{1}{N} \log w(\mathbf{A}|\mathbf{k}(\mathbf{A}), W) - \frac{1}{N} \log Z(\mathbf{k}(\mathbf{A}), W)$$

$$= \sum_k p(k|\mathbf{A}) \log p(k) + F_2(\mathbf{A}|\mathbf{k}(\mathbf{A})) - F_1(\mathbf{k}(\mathbf{A}), W) \tag{7.121}$$

with

$$F_1(\mathbf{k}(\mathbf{A}), W) = \frac{1}{N} \log Z(\mathbf{k}(\mathbf{A}), W) \tag{7.122}$$

$$F_2(\mathbf{A}|\mathbf{k}(\mathbf{A}), W) = \frac{1}{N} \log w(\mathbf{A}|\mathbf{k}(\mathbf{A}), W) \tag{7.123}$$

and $p(k|\mathbf{A}) = N^{-1} \sum_i \delta_{k, \sum_j A_{ij}}$. Next we calculate $F_1(\mathbf{k}(\mathbf{A}), W)$ and $F_2(\mathbf{A}|\mathbf{k}(\mathbf{A}), W)$. Using the identity

$$\prod_{i<j} [C_{ij} \delta_{A_{ij},1} + D_{ij} \delta_{A_{ij},0}] = \left(\prod_{i<j} D_{ij} \right) e^{\sum_{i<j} A_{ij}[\log C_{ij} - \log D_{ij}]} \tag{7.124}$$

we find, to leading orders in N,

$$F_2(\mathbf{A}|\mathbf{k}(\mathbf{A}), W) = \frac{1}{N} \sum_{i<j} \log \left(1 - \frac{\langle k(\mathbf{A}) \rangle}{N} \frac{W(k_i, k_j)}{p(k_i|\mathbf{A}) p(k_j|\mathbf{A})} \right)$$

$$+ \frac{1}{N} \sum_{i<j} A_{ij} \log \left[\frac{\langle k(\mathbf{A}) \rangle}{N} \frac{W(k_i, k_j)}{p(k_i|\mathbf{A}) p(k_j|\mathbf{A})} \right] + \mathcal{O}(N^{-1})$$

$$= -\frac{\langle k(\mathbf{A}) \rangle}{2} + \frac{\langle k(\mathbf{A}) \rangle}{2} \log \left[\frac{\langle k(\mathbf{A}) \rangle}{N} \right]$$

$$+ \frac{\langle k(\mathbf{A}) \rangle}{2} \sum_{kk'} W(k, k'|\mathbf{A}) \log \left[\frac{W(k, k')}{p(k|\mathbf{A}) p(k'|\mathbf{A})} \right] + \mathcal{O}(N^{-1})$$

$$= \frac{\langle k(\mathbf{A}) \rangle}{2} \Big(\log[\langle k(\mathbf{A}) \rangle / N] - 1 \Big)$$

$$+ \frac{\langle k(\mathbf{A}) \rangle}{2} \sum_{kk'} W(k, k'|\mathbf{A}) \log \left[\frac{W(k, k')}{p(k|\mathbf{A}) p(k'|\mathbf{A})} \right] + \mathcal{O}(N^{-1}) \tag{7.125}$$

which depends on $\mathbf{A}$ only through $p(k|\mathbf{A})$ and $W(k, k'|\mathbf{A})$.

We can work out F_1 by using the Fourier representation for Kronecker deltas, performing the summation over graph configurations and using Taylor expansion of the exponential $1 + x = \exp[x + \mathcal{O}(x^2)]$, which leads, to leading orders in N, to

$$F_1(\mathbf{k}(\mathbf{A}), W) = \frac{1}{N} \log \sum_{\mathbf{A}'} w(\mathbf{A}'|\mathbf{k}(\mathbf{A}), W) \delta_{\mathbf{k}(\mathbf{A}), \mathbf{k}(\mathbf{A}')}$$

$$= \frac{1}{N} \log \int \frac{d\boldsymbol{\Omega}}{(2\pi)^N} e^{i\boldsymbol{\Omega}\cdot\mathbf{k}(\mathbf{A})} \exp\left[\sum_{i<j} \frac{\langle k(\mathbf{A}) \rangle}{N} \frac{W(k_i, k_j)}{p(k_i|\mathbf{A})p(k_j|\mathbf{A})} \left(e^{-i(\Omega_i + \Omega_j)} - 1 \right) \right]$$

(7.126)

Introducing the order parameter

$$R(k|\mathbf{A}, \boldsymbol{\Omega}) = \frac{1}{N} \sum_i \delta_{k,k_i(\mathbf{A})} e^{-i\Omega_i}$$

(7.127)

and inserting unity in the representation $1 = \int \prod_k dR(k|\mathbf{A})\, \delta[R(k|\mathbf{A}) - R(k|\mathbf{A}, \boldsymbol{\Omega})]$ with δ-functions written in integral form

$$1 = \int \prod_k \frac{dR(k|\mathbf{A})d\hat{R}(k|\mathbf{A})}{2\pi/N} e^{iN \sum_k R(k|\mathbf{A})\hat{R}(k|\mathbf{A}) - i\sum_i \hat{R}(k_i|\mathbf{A})e^{-i\Omega_i}}$$

(7.128)

allows us to write, with the shorthand $\{dRd\hat{R}\} = \prod_k \frac{dR(k|\mathbf{A})d\hat{R}(k|\mathbf{A})}{2\pi/N}$

$$F_1(\mathbf{k}(\mathbf{A}), W) = \frac{1}{N} \log \int \{dRd\hat{R}\} e^{N\Xi[R,\hat{R}]}$$

(7.129)

where

$$\Xi[R, \hat{R}] = i \sum_k R(k|\mathbf{A})\hat{R}(k|\mathbf{A}) - \frac{\langle k(\mathbf{A}) \rangle}{2} + \frac{\langle k(\mathbf{A}) \rangle}{2} \sum_{kk'} R(k|\mathbf{A})R(k'|\mathbf{A}) \frac{W(k, k')}{p(k|\mathbf{A})p(k'|\mathbf{A})}$$

$$+ \sum_k p(k|\mathbf{A}) \log \frac{(-i\hat{R}(k|\mathbf{A}))^k}{k!}$$

(7.130)

In the limit $N \to \infty$, expression (7.129) can be calculated by steepest descent (see Appendix D)

$$\lim_{N \to \infty} F_1(\mathbf{k}(\mathbf{A}), W) = \operatorname{extr}_{R,\hat{R}} \Xi[R, \hat{R}]$$

(7.131)

Differentiation of Ξ gives the following saddle point equations:

$$-iR(k|\mathbf{A}) = \frac{kp(k|\mathbf{A})}{\hat{R}(k|\mathbf{A})}, \qquad -i\hat{R}(k|\mathbf{A}) = \langle k(\mathbf{A}) \rangle \sum_{k'} R(k'|\mathbf{A}) \frac{W(k, k')}{p(k|\mathbf{A})p(k'|\mathbf{A})}$$

(7.132)

so we deduce that at the saddle point

$$\sum_{kk'} R(k|\mathbf{A})R(k'|\mathbf{A}) \frac{W(k, k')}{p(k|\mathbf{A})p(k'|\mathbf{A})} = 1$$

(7.133)

and

$$F_1(\mathbf{k}(\mathbf{A}), W) = -\langle k(\mathbf{A}) \rangle + \sum_k p(k|\mathbf{A}) \log \left[\frac{kp(k|\mathbf{A})}{R(k|\mathbf{A})} \right]^k - \sum_k p(k|\mathbf{A}) \log k!$$

(7.134)

where R is the solution to the self-consistency equation

$$R(k|\mathbf{A}) = \frac{kp(k|\mathbf{A})}{\langle k(\mathbf{A}) \rangle \sum_{k'} R(k'|\mathbf{A}) \frac{W(k,k')}{p(k|\mathbf{A})p(k'|\mathbf{A})}} \qquad (7.135)$$

In conclusion, $F_1(\mathbf{k}(\mathbf{A}), W)$ depends on $\mathbf{k}(\mathbf{A})$ only through $p(k|\mathbf{A})$, so it is a function of $p(k|\mathbf{A})$ and $W(k,k')$ only. Inserting in (7.121), we have

$$\Omega(\mathbf{A}|p, W) = \sum_k p(k|\mathbf{A}) \log p(k) + \frac{\langle k(\mathbf{A}) \rangle}{2} \left(\log \frac{\langle k(\mathbf{A}) \rangle}{N} + 1 \right)$$

$$+ \frac{\langle k(\mathbf{A}) \rangle}{2} \sum_{kk'} W(k,k'|\mathbf{A}) \log \frac{W(k,k')}{p(k|\mathbf{A})p(k'|\mathbf{A})}$$

$$- \sum_k p(k|\mathbf{A}) \log \left[\frac{kp(k|\mathbf{A})}{R(k|\mathbf{A})} \right]^k + \sum_k p(k|\mathbf{A}) \log k! \qquad (7.136)$$

Hence we find

$$D_{AB} = \frac{1}{2} \left\{ \sum_k p_A(k|\mathbf{A}) \log \left[\frac{p_A(k)}{p_B(k)} \right] + \sum_k p_B(k|\mathbf{A}) \log \left[\frac{p_B(k)}{p_A(k)} \right] \right.$$

$$+ \frac{\langle k_A \rangle}{2} \sum_{kk'} W_A(k,k'|\mathbf{A}) \log \left[\frac{W_A(k,k')}{W_B(k,k')} \right] + \frac{\langle k_B \rangle}{2} \sum_{kk'} W_B(k,k'|\mathbf{A}) \log \left[\frac{W_B(k,k')}{W_A(k,k')} \right]$$

$$\left. - \sum_k p_A(k)k \left[\log \left[\frac{R_{BA}(k)}{R_{AA}(k)} \right] + \log \left[\frac{R_{AB}(k)}{R_{BB}(k)} \right] \right] \right\} \qquad (7.137)$$

where

$$R_{AB}(k) = \frac{kp_B(k)}{\langle k_B \rangle \sum_{k'} R_{AB}(k') \frac{W_A(k,k')}{p_B(k)p_B(k')}}$$

$$R_{BA}(k) = \frac{kp_A(k)}{\langle k_A \rangle \sum_{k'} R_{BA}(k') \frac{W_B(k,k')}{p_A(k)p_A(k')}}$$

$$R_{AA}(k) = \frac{kp_A(k)}{\langle k_A \rangle \sum_{k'} R_{AA}(k') \frac{W_A(k,k')}{p_A(k)p_A(k')}}$$

$$R_{BB}(k) = \frac{kp_B(k)}{\langle k_B \rangle \sum_{k'} R_{BB}(k') \frac{W_B(k,k')}{p_B(k)p_B(k')}} \qquad (7.138)$$

The last two equations are clearly solved by $R_{AA}(k) = p_A(k)$ and $R_{BB}(k) = p_B(k)$, while the remaining two must be solved numerically. Hence, we can work out D_{AB} as

$$D_{AB} = \frac{1}{2} \left\{ \sum_k p_A(k|\mathbf{A}) \log \frac{p_A(k)}{p_B(k)} + \sum_k p_B(k|\mathbf{A}) \log \frac{p_B(k)}{p_A(k)} \right.$$

$$
+ \frac{\langle k_A \rangle}{2} \sum_{kk'} W_A(k,k'|\mathbf{A}) \log \frac{W_A(k,k')}{W_B(k,k')} + \frac{\langle k_B \rangle}{2} \sum_{kk'} W_B(k,k'|\mathbf{A}) \log \frac{W_B(k,k')}{W_A(k,k')}
$$

$$
+ \frac{\langle k_A \rangle}{2} \sum_{kk'} W_A(k,k') \left[\log \frac{R_{AA}(k)}{R_{BA}(k)} + \log \frac{R_{AA}(k')}{R_{BA}(k')} \right]
$$

$$
+ \frac{\langle k_B \rangle}{2} \sum_{kk'} W_B(k,k') \left[\log \frac{R_{BB}(k)}{R_{AB}(k)} + \log \frac{R_{BB}(k')}{R_{AB}(k')} \right] \Bigg\}
$$

$$
= \frac{1}{2} D(p_A \| p_B) + \frac{1}{2} D(p_B \| p_A) + \frac{\langle k_A \rangle}{4} D(W_A \| W_B) + \frac{\langle k_B \rangle}{4} D(W_B \| W_A)
$$

$$
+ \frac{\langle k_A \rangle}{4} \sum_{kk'} W_A(k,k') \log \frac{p_A(k)p_A(k')}{R_{BA}(k)R_{BA}(k')}
$$

$$
+ \frac{\langle k_B \rangle}{4} \sum_{kk'} W_B(k,k') \log \frac{p_B(k)p_B(k')}{R_{AB}(k)R_{AB(k')}} \tag{7.139}
$$

where $D(p_A \| p_B)$ and $D(W_A \| W_B)$ are the Kullback–Leibler distances between the degree distributions p_A and p_B and the joint distributions $W_A(k,k')$ and $W_B(k,k')$, respectively. Alternatively, we can write

$$
D_{AB} = \frac{1}{2} D(p_A \| p_B) + \frac{1}{2} D(p_B \| p_A) + \frac{\langle k_A \rangle}{4} D(W_A \| W_{BA}) + \frac{\langle k_B \rangle}{4} D(W_B \| W_{AB})
$$

$$
\tag{7.140}
$$

with

$$
W_{BA}(k,k') = \frac{W_B(k,k')R_{BA}(k)R_{BA}(k')}{p_A(k)p_A(k')}
$$

$$
W_{AB}(k,k') = \frac{W_A(k,k')R_{AB}(k)R_{AB}(k')}{p_B(k)p_B(k')} \tag{7.141}
$$

representing the joint distributions of degrees of connected nodes in an ensemble of the type (7.120) that would have been obtained for the hybrid combination $\{p_A, W_B\}$ and $\{p_B, W_A\}$, respectively. The last two terms in (7.139) reflect interference between the constraints imposed by prescribed degree statistics and those imposed by prescribed degree correlations. See [71,171] for applications of network distances to protein interaction networks.   □

Exercise 7.5

One may represent the genomic and evolutionary properties of each node i by a node label or vector $\mathbf{x}^{(i)} = (x_1^{(i)}, \ldots, x_q^{(i)})$, where each component represents a node property. We could assume as a null model a random graph with the same number of nodes, same node labels and same number of links as in the real network, where the likelihood of having a link between two nodes is given by the empirical frequency

$$
W(\mathbf{x},\mathbf{y}) = \frac{\sum_{ij} A_{ij} \delta_{\mathbf{x}^{(i)},\mathbf{x}} \delta_{\mathbf{x}^{(j)},\mathbf{y}}}{\sum_{ij} A_{ij}} \tag{7.142}
$$

with which two nodes with the same labels are seen to interact in the real network $\mathbf{A}$. This leads to the null model (see also Section 10.3)

$$p(\mathbf{A}|\mathbf{x}, W) = \frac{1}{Z(\mathbf{x}, W, L)} \prod_{i<j} \left[\frac{\langle k \rangle}{N} \frac{W(\mathbf{x}^{(i)}, \mathbf{x}^{(j)})}{p(\mathbf{x}^{(i)})p(\mathbf{x}^{(j)})} \delta_{A_{ij},1} + \left(1 + \frac{\langle k \rangle}{N} \frac{W(\mathbf{x}^{(i)}, \mathbf{x}^{(j)})}{p(\mathbf{x}^{(i)})p(\mathbf{x}^{(j)})} \right) \right]$$

$$\times \delta_{L, \sum_{i<j} A_{ij}}$$

$$= \frac{1}{Z'(\mathbf{x}, W, L)} e^{\sum_{ij} A_{ij} L_{ij}} \delta_{L, \sum_{i<j} A_{ij}} \tag{7.143}$$

by repeating the same steps as in (7.42), where

$$L_{ij} = \log \frac{\langle k \rangle W(\mathbf{x}^{(i)}, \mathbf{x}^{(j)})}{Np(\mathbf{x}^{(i)})p(\mathbf{x}^{(j)}) - \langle k \rangle W(\mathbf{x}^{(i)}, \mathbf{x}^{(j)})} \tag{7.144}$$

To generate graphs from this ensemble, we can use repeated applications of hinge flips F_{rst}, starting from a graph with the same labels and total number of links and accepting them with probabilities

$$\mathcal{A}[\mathbf{A} \to F_{rst}\mathbf{A}] = \min\left\{ 1, e^{L_{rt} - L_{rs}} \right\} \tag{7.145}$$

where we used the fact that $Z'(\mathbf{x}, W, L)$ depends only on L, as in Exercise 7.2, and hence it drops out of the acceptance probabilities. Indeed, we have

$$Z'(\mathbf{x}, W, L) = Z(\mathbf{x}, W, L) \prod_{i<j} \left(1 - \frac{\langle k \rangle}{N} \frac{W(\mathbf{x}^{(i)}, \mathbf{x}^{(j)})}{p(\mathbf{x}^{(i)})p(\mathbf{x}^{(j)})} \right)^{-1}$$

$$= Z(\mathbf{x}, W, L) e^{\sum_{i<j} \frac{\langle k \rangle}{N} \frac{W(\mathbf{x}^{(i)}, \mathbf{x}^{(j)})}{p(\mathbf{x}^{(i)})p(\mathbf{x}^{(j)})}} = Z(\mathbf{x}, W, L) e^L \tag{7.146}$$

where we used $L = N\langle k \rangle / 2$ and

$$\frac{\langle k \rangle}{N} \sum_{i<j} \frac{W(\mathbf{x}^{(i)}, \mathbf{x}^{(j)})}{p(\mathbf{x}^{(i)})p(\mathbf{x}^{(j)})} = \frac{\langle k \rangle}{2N} \sum_{\mathbf{xy}} \sum_{ij} \delta_{\mathbf{x}, \mathbf{x}^{(i)}} \delta_{\mathbf{y}, \mathbf{y}_j} \frac{W(\mathbf{x}, \mathbf{y})}{p(\mathbf{x})p(\mathbf{y})}$$

$$= \frac{N\langle k \rangle}{2} \sum_{\mathbf{xy}} \frac{W(\mathbf{x}, \mathbf{y})}{p(\mathbf{x})p(\mathbf{y})} p(\mathbf{x})p(\mathbf{y}) = L$$

and repeating the steps in (7.101)

$$Z(\mathbf{x}, W, L) = \int \frac{d\omega}{2\pi} e^{i\omega L} e^{\frac{\langle k \rangle}{N} \sum_{i<j} \frac{W(\mathbf{x}^{(i)}, \mathbf{x}^{(j)})}{p(\mathbf{x}^{(i)})p(\mathbf{x}^{(j)})} (e^{-i\omega} - 1)} = e^{-L} \frac{L^L}{L!} \tag{7.147}$$

□

Part IV

Graphs defined by algorithms

8

Network growth algorithms

Introduction. This chapter looks at how networks can be generated via growth processes. This is a fundamentally different approach to graph generation compared to that followed in the algorithms described so far, which all start from a given seed graph which is then randomized via an appropriately defined Markov chain process. A growth algorithm starts from a small seed network or even a 'tabula rasa' network without any links, and then adds edges and/or nodes according to some predefined rules. We look at this from the point of view of how network growth processes can be defined to capture some relevant features of actual physical growth processes, as well as from the point of view of how network growth rules can be analyzed to predict the expected topological properties of the resulting networks.

8.1 Configuration model

Definition of the algorithm. Suppose every node i in an empty N-node network is issued with k_i half-edges (also referred to as *stubs*). Pairing these half-edges will by construction create a nondirected network[1] where the nodes have the required degrees $\{k_1, \ldots, k_N\}$. This is the idea at the heart of the configuration model; the N degrees might have been prescribed directly, or they could have been drawn from some degree distribution. Algorithm 27 provides an implementation of this idea. Figure 8.1 illustrates the connection between pairing stubs randomly and the frequency with which we would expect to see every potential topology.

If the half-edges are joined indiscriminately, loops or double edges may be created. Hence, this process might not result in a simple network, as is illustrated in Figure 8.2. Various adaptations have been proposed for those cases where it is essential to generate a simple network. The probabilistically ideal protocol is to abandon the run as soon as a double or self-link is created, and then restart the growth process from scratch. It has been shown [128] that, if N is large, the configuration model algorithm has a strictly positive chance of creating a simple graph, so long as the degree distribution has a finite second moment. This criterion can be enforced by careful choice of degree distribution, or by limiting the maximum allowed degree [44, 128].

Real-world networks are often both large, and have fat-tailed degree distributions. For example, the prescribed degree distribution may for large k obey a power-law $p(k) \propto k^{-\gamma}$ with γ typically in the range $(2, 3)$. The fatter the tail of this distribution, the higher the chance that the algorithm will fail because of double or self-links. Ultra-high-degree nodes have many separate opportunities to create undesirable self-links or

---

[1]Provided the sum of the degrees is an even number.

Generating Random Networks and Graphs. First Edition. A.C.C. Coolen, A. Annibale, and E.S. Roberts. © A.C.C. Coolen, A. Annibale, and E.S. Roberts 2017. Published in 2017 by Oxford University Press. DOI: 10.1093/acprof:oso/9780198709893.001.0001

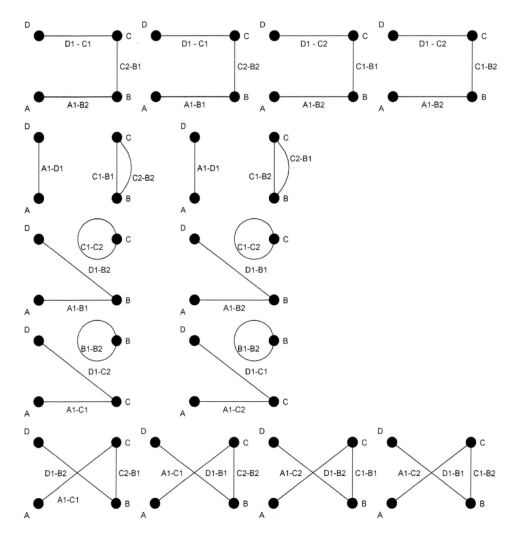

Fig. 8.1 In this figure, we illustrate the issues related to the possible bias in the configuration model. Consider an example where nodes A and D have degree 1, and nodes C and B have degree 2. The diagram above enumerates all possible ways to connect the stubs (A1, B1, B2, C1, C2, D1). If we do not see the labels of the stubs, there are five possible configurations (one per row of the diagram). The configurations in row 1 and row 5 are the only two which correspond to simple graphs. If we were to run the algorithm many times, rejecting non-simple graphs, then configurations 1 and 5 would be seen the same number of times. This is as we would expect from a process that we call *unbiased*. If we did not mind whether the network was simple or not, we would find that configurations 2, 3 and 4 only appeared half as frequently as configurations 1 and 5. The process is still unbiased from the point of view of connecting stubs, but not in the sense of each possible topology appearing the same number of times. This is a general observation that can be shown with simple combinatorics considering permutations of stubs at each node.

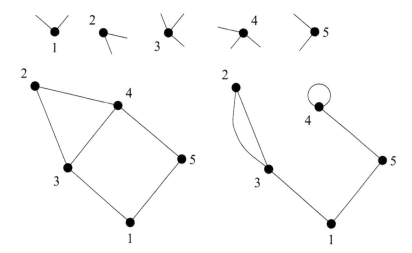

Fig. 8.2 Illustration of the stub-joining method (also known as the configuration model) for generating a graph with a specified degree sequence. Here we have $N = 5$ nodes and prescribe the degree sequence $\mathbf{k} = (2, 2, 3, 3, 2)$. The first step is to allocate to each node the correct number of 'half-edges', as in the top row. Then these half-edges are paired to realize a graph. This is described in Algorithm 27. However, this protocol may generate a graph which has loops or double edges (as in the graph on the bottom right).

double links. If repeatedly restarting makes the algorithm unacceptably slow, some authors (e.g. [127]) recommend a 'backtracking' approach. In this variant, when a double link or self-link is created, the algorithm reverts to an earlier configuration (i.e. undoes the last few moves before the breach) and then continues. However, as described in [101]:

> For the efficient variant, where non-feasible edges are rejected and the construction of a graph continues, there exists a bias ..... We find that this bias does not disappear with growing system size. This becomes also visible, e.g. for scale-free graphs when measuring quantities like the graph diameter.

In [101] the existence of this bias is demonstrated analytically, and also by working through a simple example. The intuition behind it can be understood quite easily and is illustrated in Figure 8.3. Hence, the backtracking approach should never be used to generate a network which is meant to form the basis for any further statistical tests, e.g. as a null model. It is, however, valid to use first the backtracking method to generate an example network with a given degree sequence, and then to use as a second step a degree-preserving graph randomization algorithm (see Chapter 7) to

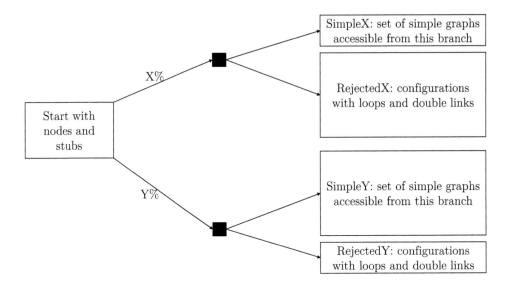

Fig. 8.3 An illustration of why backtracking-based approaches in the configuration model do not sample all possible graphs with the required degree sequence uniformly. We start on the left-hand side of the diagram with a collection of nodes and edges. We run the algorithm partially indicated by the two branches labelled $X\%$ and $Y\%$ – where the labels indicate the probability of following that particular branch. The algorithm will terminate at one of the right-hand boxes. The second and fourth boxes indicate illegal configurations with loops and/or double edges. Suppose that if we end up with one of these configurations, we choose to backtrack to an intermediate point (indicated by the solid black square). This then directly implies that the probability of sampling a configuration from the set SimpleX is $X\%$, and the probability of sampling a configuration from the set SimpleY is $Y\%$. For example, we may have a situation where the majority of the simple graphs lie in SimpleY – but if $Y\% < X\%$, then SimpleX would be visited more frequently.

erase the bias generated by the backtracking execution of the configuration model.

Variations on the basic theme. An alternative way of adjusting the algorithm to create simple graphs is the so-called 'erased' configuration model. Algorithm 27 is first executed and allowed to run to conclusion. Then double links are merged and loops are deleted. Obviously this would reduce the overall number of links in the network, and some of the individual node degrees. Hence, the degree sequence/distribution will no longer match the original one; it does not work as a hard constraint. In [32] it is shown that this construction nonetheless asymptotically gives the correct degree distribution so long as the degree distribution has a finite mean. The fatter the tail of the distribution, the more likely it is to find ultra-high-degree nodes. These nodes will be allocated a large number of stubs, and hence have many opportunities to create self-links and double links. Erasing these materially deforms the degree distribution – more than for a distribution which is more closely gathered around its mean. For a power-law degree distribution, i.e. $p(k) \propto k^{-\gamma}$ for large k, the smaller the exponent, the fatter will be the tail. An exponent $\gamma > 2$ gives a finite mean. Most real-world networks fit distributions with a finite mean, so the erased configuration model can be theoretically justified as a way of imposing a degree sequence as a soft constraint. The construction does not give information about how likely each configuration is to be generated, so the process should be considered biased.

From [45] one learns how to adapt the configuration model to generating directed networks. In this case, each node is allocated a certain number of inbound and outbound stubs. The matching stage pairs in-stubs with out-stubs, but otherwise the results and procedures are broadly analogous to the undirected case. It is also possible to impose additional conditions on how links are paired.

Exercise 8.1 Generate a nondirected simple graph with $N = 10$ nodes and with the degree sequence $\mathbf{k} = (1, 1, 6, 4, 2, 3, 1, 1, 2, 4)$. Is this possible? If not, why not?

Shortcut: *The configuration model allocates a half-edge to each node, based on a target degree sequence or degree distribution. Joining half-edges in pairs realizes a graph with the desired degree sequence, but potentially also creates loops or double edges. If the graph is required to be simple, one should re-run the algorithm from the beginning until a simple graph is achieved. Less computationally demanding alternatives are to erase the loops and double edges, or to backtrack several steps and restart when illegal edges are proposed. However, these alternative approaches are biased and may deform the topology, so they should not be used to generate networks employed as the basis for statistical tests.*

8.2   Preferential attachment and scale-free networks

Definition of preferential attachment. The popular preferential attachment algorithm [3] for generating nondirected graphs begins with a small number of seed vertices. At every step, a new vertex is added with a certain number of nondirected edges that link the new vertex to existing vertices i with probability $k_i / \sum_j k_j$.

Algorithm 24 gives a simple preferential attachment subroutine where each node is born with one link [3]. The next generalization would be to add $m > 1$ links with every new node. If m is chosen to be greater than 1, then it is in principle possible for this algorithm to generate double links. In the examples that follow, we presume that $m = 1$, which means that there will be no double links (unless there were some already present in the seed graph). If m is set greater than 1, then it is up to the programmers to decide whether they wish to add an extra step to the pseudocode above to forbid double links, or tolerate the risk of having double links (which will be small if m is small compared to m_0).

Variations on this theme include rewiring edges or deleting edges. Generalization to directed networks is straightforward. The main models are listed in Tables 8.1 and 8.2 together with references which provide further details about their practical implementation and performance. A hybrid approach – giving greater control of the topological features of the ensemble – would be to generate a starting network with a preferential attachment growth scheme and then to carry out a degree-preserving randomization process on it, as described in Chapter 7. This procedure would then erase the topological artefacts created by the growth process, and result in a network with the desired degree distribution, but otherwise maximally random.

Expected degree distribution of a network generated by preferential attachment. Below we demonstrate that, provided the network is sufficiently large (i.e. the algorithm is run for sufficiently long), the preferential attachment scheme described in Algorithm 24 is expected to generate a network with a power-law degree distribution. In Algorithm 24, a node is born at each time step t. It forms a link to one of the existing nodes with a probability proportional to the degree of the existing node. For easier algebra, we assume that the algorithm starts at time $t = 2$ from a seed network of two nodes connected by a double link[2]. We can write the probability $p(i|t)$ of a newborn node t at time step t linking to node i as $p(i|t) = k_i / \sum_{j=1}^{t-1} k_j$, where k_i is the degree of node i at the beginning of time step t. Since we know that at step $t - 1$ the network will have $t - 1$ links distributed between $t - 1$ nodes, this probability equivalently can be written as $p(i|t) = k_i/2(t-1)$. Hence, the probability that node i will have degree k at time step t can be written as

$$p(k, i, t) = \left[\frac{k-1}{2(t-1)}\right] p(k-1, i, t-1) + \left[1 - \frac{k}{2(t-1)}\right] p(k, i, t-1) \qquad (8.1)$$

The first term gives the probability of node i having gained a link, increasing its degree from $(k-1)$ to k. The second term gives the probability that node i already had degree k in the previous time step, and remained unchanged. This is valid for $i < t$ since node t is born at time step t and will definitely have degree 1.

To obtain a formula for the degree distribution, we sum over all nodes and normalize (remembering that at time step t the network will have t nodes).

$$p(k, t) = \frac{1}{t} \sum_{i=1}^{t-1} \left\{ \left(1 - \frac{k}{2(t-1)}\right) p(k, i, t-1) + \frac{k-1}{2(t-1)} p(k-1, i, t-1) \right\} + \frac{1}{t}\delta_{k1}$$

where the last term reflects the fact that at each step a new node of degree 1 is added to the graph. Rewriting the summations using the standard relationship $p(k) = \frac{1}{t} \sum_{i=1}^{t} p(k_i = k)$, we can see that

$$p(k, t) = \frac{t-1}{t}\left(1 - \frac{k}{2(t-1)}\right) p(k, t-1) + \frac{t-1}{t}\frac{k-1}{2(t-1)} p(k-1, t-1) + \frac{\delta_{k1}}{t}$$

$$= \frac{t-1}{t} p(k, t-1) - \frac{k}{2t} p(k, t-1) + \frac{k-1}{2t} p(k-1, t-1) + \frac{\delta_{k1}}{t} \qquad (8.2)$$

---

[2]This assumption is not essential, but just keeps the formulae more compact.

Exercise 8.2 Show that equation (8.2) is correctly normalized, i.e. that $\sum_{k\geq 0} p(k,t) = 1$ for all $t \geq 0$. What do you expect the average degree to be after t steps?

Equation (8.2) rearranges to

$$2t\,(p(k,t) - p(k,t-1)) = (k-1)p(k-1,t-1) - (k+2)p(k,t-1) + 2\delta_{k1} \quad (8.3)$$

To turn the difference equation into a differential equation, we introduce a real-valued time variable $\tau = \epsilon t$ for some very small but positive ϵ.

$$2\tau\frac{(p(k,\tau) - p(k,\tau-\epsilon))}{\epsilon} = (k-1)p(k-1,\tau-\epsilon) - (k+2)p(k,\tau-\epsilon) + 2\delta_{k1} \quad (8.4)$$

Take the limit of $\epsilon \to 0$ (assuming that it exists[3]) to find that

$$2\tau\frac{\partial p(k,\tau)}{\partial \tau} = (k-1)p(k-1,\tau) - (k+2)p(k,\tau) + 2\delta_{k1} \quad (8.5)$$

The aim of this calculation is to find the degree distribution that $p_\tau(k)$ converges to after the algorithm has been allowed to run for a very long time. When $p_\tau(k)$ converges, then $\partial p_\tau(k)/\partial \tau = 0$ and the stationary solution will satisfy

$$(k-1)p(k-1) - (k+2)p(k) + 2\delta_{k1} = 0 \quad (8.6)$$

which is solved by

$$p(0) = 0, \qquad p(k>0) = \frac{4}{k(k+1)(k+2)} \quad (8.7)$$

Hence, Algorithm 24 will converge to a power-law degree distribution, where $p(k) \propto k^{-3}$ for large k. In Figure 8.4 we show distribution (8.7) together with other conventional degree distributions that have the same average degree (for comparison).

Exercise 8.3 Calculate the average degree $\langle k \rangle = \sum_{k>0} kp(k)$ for the distribution (8.7), and verify that the result agrees with the answer to the previous exercise. Use the identity $\sum_{k\geq 2}[(k+1)(k-1)]^{-1} = 3/4$ (see, e.g. [82]), together with normalization of $p(k)$, to deal with the degree summation.

Scale-free networks. The term scale-free is normally used to denote networks with a degree distribution which (at least for large k) behaves in leading order as $p(k) \propto k^{-\gamma}$ for some exponent γ, generally in the range $2 < \gamma \leq 3$. In this regime, the average degree $\langle k \rangle$ is finite, but $\langle k^2 \rangle$ (and hence the variance of $p(k)$) is not[4]. Reference [113]

---

[3] At this point, we require that the number of iterations $t = \tau/\epsilon$ of the algorithm is sufficiently large for this argument to apply, since we need $\epsilon \to 0$ but also $\tau \to \infty$ (in subsequent steps).

[4] It should be noted that the divergence of $\langle k^2 \rangle$ is, however, only very slow. For large but finite values of N, one will in practice even in scale-free networks find a finite value for $\langle k^2 \rangle$.

MODEL VARIANT	DESCRIPTION	REFERENCES
Barabási–Albert model	Begin with a small number of seed vertices. At every step, a new vertex is added with m edges that link the vertex to existing vertices with degree-weighted probability $k_i / \sum_j k_j$	[3]
+ Edge shuffling	At each step, the process chooses between adding a node (and associated links), adding links to existing nodes, or rewiring. This gives more parameters to tune the model, and reduces the strong correlation between node degree and node age.	[4]
+ Add edges between existing vertices	At each step, with some probability, either a new node is added, or an edge is added between two existing vertices. This can be used to influence degree–degree correlations.	[39, 105]
+ Triad formation	This process combines adding new (preferentially attached) nodes and creating triads. The aim is to more closely mimic clustering in real networks.	[90]
+ Deletions	Edges (or even nodes) can be removed. The probability of removal can itself be weighted – for example, by vertex degree. The aim is to soften some of the characteristic topological artefacts from the Barabási–Albert preferential attachment scheme.	[57, 33, 79]
+ Fitness	Every node is given a fitness parameter η_i. New nodes attach to existing nodes with a probability proportional to $\eta_i k_i$. Network properties will depend strongly on the fitness parameter distribution. This model can generate richer topologies.	[21, 29]
+ Ageing of sites	The likelihood of a new vertex to linking an existing node is positively weighted by the degree of this node, and negatively weighted by its age. For example, Algorithm 26.	[61, 84]

Table 8.1 A non-exhaustive list of variations on the standard preferential attachment growth algorithms for producing simple nondirected networks. Each aims to influence the structural features of the resulting networks, via heuristic adaptations of the dynamical rules.

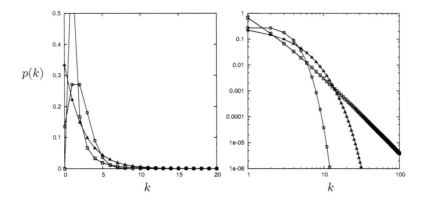

Fig. 8.4 Degree distribution (8.7) resulting from the standard preferential attachment algorithm (Algorithm 24) (connected squares), shown for comparison together with the degree distributions $p(k) = e^{-2}2^k/k!$ of an Erdös–Rényi graph (3.1) (connected circles) and $p(k) = \frac{1}{2}(2/3)^k$ (connected triangles). All three distributions have the same average degree $\langle k \rangle = 2$. The panel on the right show the same distributions, but in a log–log plot. Although in the preferential attachment algorithm most nodes (a fraction 2/3) have degree $k = 1$ (i.e. are disconnected), the width of $p(k)$ still diverges because of the small number of nodes with very high degree.

provides an interesting discussion on the common usage of the term *scale-free*, and the extent to which additional topological properties are implied (beyond the power-law degree distribution). The label scale-free seems to have grown to be associated with an observation of fractal properties in the topology – that a connected subgraph has typically qualitatively similar topology to the whole network. The study of the generation of scale-free/power-law networks has substantially diverged from the study of generating networks with a specified degree distribution. There is now extensive literature devoted specifically to scale-free networks. This is because many authors have promoted the scale-free topology as the one which is most closely aligned to real-world networks.

The authors of [70], who focused on scale-free network topologies, specifically studied the World Wide Web, but subsequent authors claimed the same topology for social networks, citation networks and protein–protein interaction networks. Within model-fitting tolerances, the scale-free random graph has been widely accepted as a description for many complex systems – although there are other fat-tailed degree distributions that may also be good models for real networks. Also, the particular example studied in [3] was the Internet; at every step a new web page is added, and the web page hyperlinks to a certain number of other web pages, with a higher propensity to link to popular pages. The same model makes sense for a citation network: highly cited papers are more likely to be cited by a new paper, since they will be thought to have greater gravitas and credibility. Algorithm 24 indeed successfully produces the power-law degree distribution, and many of the qualitative features that are found in such real networks. This has motivated a subsequent generation of researchers to approach the problem of generating scale-free graphs not simply as a technical exercise,

MODEL VARIANT	DESCRIPTION	REFERENCES
Directed graphs	One specifies the number of outbound and inbound links which are attached to every newborn node. The other ends of these links are attached to existing nodes in the same way as in undirected algorithms.	[105, 27]
Vertex copying	At each time step, a randomly drawn vertex is duplicated, inheriting some or all of the connections of its parent.	[106, 176, 142, 96, 172, 137, 74, 48]
Random-walk methods	One picks at random a first node, which is then connected to a second node that is the outcome of a length l random walk on the instantaneous network, starting from the first node.	[175, 161]
Polya's urn model	This model predates preferential attachment. The simplest version refers to distributing a certain number of balls in buckets. The pertinent variant creates a new bucket containing one ball at every time step, and places a ball in an existing bucket. While this does not model a network explicitly, it allows the prediction of the degree distribution of a network generated by preferential attachment – and the benefit of an alternative well-developed set of tools to analyze the process.	[46, 51]
Geometric preferential attachment	Vertices are placed sequentially in a metric space. The probability that a link forms between a new and an existing vertex is driven by the metric distance between the vertices and the degree of the existing vertex. See also Sections 9.2 and 10.3.	[73] and citing articles

Table 8.2 Variations on growth algorithms that are somewhat further away from the standard preferential attachment algorithm.

but as a way of proposing and testing models for how the network may have come to have its specific topology.

Shortcut: *Preferential attachment is an algorithm where a new node is added at every step, and a link is drawn between the new node and existing nodes with a probability proportional to the number of links that the existing node already has. Preferential attachment can be shown to typically generate networks with power-law degree distributions $p(k) \propto k^{-\gamma}$. Another term for a network with a power-law degree distribution* *is a scale-free network. Real-world networks' degree distributions can often be fitted to a power-law, which makes preferential attachment a popular choice as a random graph model for real-world processes.*

8.3   Analyzing growth algorithms

The bias in growth algorithms. The downside of approaching model generation via heuristic growth algorithms is that it is generally not easy to determine the statistical properties of the resulting ensemble. This is in contrast to the Markov Chain Monte Carlo (MCMC) randomization algorithms of previous chapters which were *designed* to produce graph instances according to a user-defined microscopic probability distribution, so we knew exactly the individual likelihoods of all possible graph realizations. For instance, the classic preferential attachment model produces graphs with an expected power-law degree distribution with $p(k) \propto k^{-3}$, and typically produces a network with non-trivial degree correlations ([28] and [62]). This is generally phrased as a benefit of the model – provided one seeks to produce scale-free graphs and since real-world networks also typically have non-trivial degree correlations. However, the direct implication is that this model does not reach every network with the desired degree distribution with equal probability; it exhibits a consistent bias towards graphs with certain degree correlations. It is likely that there are other topological features which may be favoured by a particular choice of growth algorithm, and this may lead to misleading conclusions if these random networks are subsequently used to evaluate the statistical significance of topological features observed in real networks. The root of the problem is this: while one can calculate the asymptotic form of $p(k)$ for protocols such as preferential attachment, it is generally not (yet) possible to express analytically the precise microscopic probabilities $p(\mathbf{A})$ with which individual graphs will asymptotically be generated. Hence, we cannot even quantify properly the extent and the nature of the bias of the generated graphs, let alone adapt the growth algorithms such that this bias is removed.

As an example, consider a variant of preferential attachment, as discussed in [62]. When a new node is added, instead of attaching to two randomly chosen nodes with a probability weighted linearly with the number of existing links to that node, the node is now instead linked to both ends of a randomly chosen edge (see Figure 8.5). Picking randomly chosen edges is, in substance, similar to choosing nodes with a probability weighted by the number of links that node has. The resulting degree distribution is almost identical to the distribution achieved by a preferential attachment algorithm, so this alternative algorithm could reasonably be promoted as a null model for a scale-free network. However, it is clear that the networks produced will be highly clustered – and hence will be very different in their topology from networks produced by classical preferential attachment – or from a typical network with a power-law degree distribution.

Another convincing example that illustrates why growth algorithms need to be treated with caution was provided by the succinctly named paper "Are randomly grown graphs really random?" [41]. In their model, at each time step a vertex is added, and two randomly picked vertices are connected. The model is expressly designed to evoke the classic Erdös–Rényi random graph model. However, the authors show that

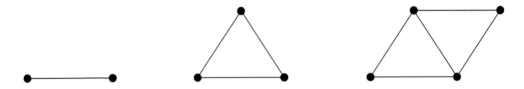

Fig. 8.5 An alternative preferential attachment protocol [62], to illustrate the variability in outcomes generated via variations on conventional growth algorithms. Linking to both ends of a randomly chosen edge (instead of linking to a randomly chosen node) again produces a fat-tailed degree distribution similar to the outcome of conventional preferential attachment, but results in highly clustered networks.

the topological properties of a graph grown in this way are fundamentally different from their Poissonian counterparts (see Exercise 8.4 for a demonstration of why this process would not result in a Poissonian degree distribution). In particular, older vertices tend to not only have more connections overall, but are also more likely to link to other older, highly connected vertices. This is equivalent to observing that there is a strong positive degree–degree correlation induced by the growing process.

Sometimes the strength of the intuition behind a growth algorithm may be sufficient to make it an interesting problem to study the algorithm on its own terms – without expecting or claiming it to produce graphs that are representative of any particular topology, and without seeking to ascertain the individual graph probabilities. For example, the duplication network growth algorithm [176, 142, 96, 172, 137, 74, 48] adds new vertices by copying an existing vertex, inheriting some of its links and acquiring some random new links. Its interest is less as a method of generating a network, and more as a way to validate a proposed mechanism for how observed protein interaction networks may have evolved during evolution. Paralogues are pairs of genes that derive from the same ancestral gene. The hypothesis is that the complexity of modern protein interaction networks developed as certain genes randomly mutated and became incorporated into a genome, together with many of the properties inherited from the original gene. This process obviously cannot be observed in real time, but simulating a network evolution model equipped with similar growth rules can give a sense of whether the proposed mechanism is consistent with observed topologies.

Growth algorithms in practice. For practical applications, preferential-attachment-type algorithms are programmed into many graph manipulation packages, and hence they can provide an easy way of generating a random network with topological properties that are broadly aligned to a real complex network. As discussed in [107], to achieve the power-law distribution, the weighting of the attachment probability has to be asymptotically linear in k (the number of existing connections of the node) in leading order. There is a wealth of variations on the basic method, targeting either a more tailored topology, or to capture known features or mechanisms of the system being modelled. Examples include combining preferential attachment with rewiring, to soften the strong link between the age of a node and its degree. It is also possible to design more complicated expressions to define the probability of a new link forming, incorporating some manner of *fitness* criteria [21], reflecting inhomogeneity among the nodes. This is discussed further in Section 10.2.

While it is generally not realistic to expect an explicit formula for the probabilities $p(\mathbf{A})$ with which individual network instances $\mathbf{A}$ will be generated in network growth algorithms, it is obviously important to predict as much information as we can about their outcomes. There are several mathematical schemes for determining analytically the mean topological properties of the resulting ensembles of graphs. We will concentrate below on the master equation approach, following closely the exposition in [62] – as this scheme benefits from being both intuitive and flexible. Earlier we used this approach to demonstrate that Algorithm 24 (the simplest preferential attachment scheme) will typically generate a random graph with power-law degree distribution. The basic steps of the approach are to express the transition probabilities of states via a master equation, as per standard techniques for analyzing Markov chains, but defined at a more coarse-grained level, as opposed to the level of the adjacency matrices $\mathbf{A}$. Sometimes these equations can be solved by inspection. Alternatively, the continuum limit can be employed to turn discrete difference equations into differential equations. This is not a fully rigorous step, and typically requires the graph size N to be very large, but tends to give solutions which are asymptotically very close to the true ones. It is a satisfactory choice for practical applications. Focusing on determining the degree distribution, the standard steps to solution are as follows:

1. Write a difference equation for the probability $p(k, i, t)$ of node i having degree k at time $t + 1$, given the degree probabilities at time t.
2. Determine boundary conditions based on the starting configuration or seed vertices chosen.
3. Sum over all nodes in existence at that point (being $i = 1 \ldots t$), to obtain an equation for the expected degree distribution $p(k, t) = t^{-1} \sum_{i=1}^{t} p(k, i, t)$.
4. Introduce a duration $0 < \epsilon \ll 1$ of individual iterations, and define the new non-integer time variable $\tau = \epsilon t$.
5. Take the continuum limit, i.e. $\epsilon \to 0$ and $t \to \infty$ for fixed τ, to find an expression for the rate of change $dp(k, \tau)/d\tau$ of the average degree distribution.
6. Determine the stationary state, i.e. the solution of $dp(k, \tau)/d\tau = 0$.

This procedure should result in an expression which relates the input parameters of the growth algorithm and the resulting limiting degree distribution, in principle allowing one to be determined from the other (i.e. the right choice of preferential attachment function to the desired degree distribution or vice versa).

This is a general scheme for analyzing a Markov chain, and although mostly applied to the calculation of degree statistics, it can in principle be adapted to more sophisticated quantities such as degree–degree correlations. However, the calculations become rapidly more complicated, more messy and in need of further assumptions.

Shortcut: *Growth algorithms, such as preferential attachment and its variations, will generally induce non-trivial topological artefacts, such as degree–degree correlations and other biases, which can be very difficult to quantify. It is typically not possible to* *find analytical formulae for the detailed probabilities with which individual graphs will be generated. The most natural use of a growth algorithm is therefore to model network growth processes for their own sake, rather than using these algorithms for generating statistical null models.*

Exercise 8.4 This exercise inspects the *indifferent attachment* algorithm (adapted from [62]). Use the master equation approach to calculate the asymptotic degree distribution $p(k)$ for the following nondirected network growth algorithm. Start from a pair of nodes with two links between them at $t = 2$, then add a new node at every step, which attaches *at random* with uniform probabilities to one of the existing nodes.

Exercise 8.5 Imagine generalizing in the preferential attachment algorithm the probability that a newborn node at time t links to node i to $p(i|t) = k_i^a / \sum_{j=1}^{t-1} k_j^a$. The case $a = 1$ reproduces the conventional algorithm, and for $a = 0$ we recover that of the previous exercise. What would you expect to observe for $a > 1$, or for $a < 0$?

8.4   Solutions of exercises

Exercise 8.1

There are $\sum_{i=1}^{10} k_i = 25$ stubs. This is an odd number, so the stubs cannot be matched in pairs, and it is impossible to construct a graph with the stated requirements.   □

Exercise 8.2

The initial degree distribution is $p(k, 2) = \delta_{k,2}$, which obviously satisfies $\sum_k p(k, 2) = 1$. Summing the right-hand side of equation (8.2) gives us

$$\sum_k p(k, t) = \frac{t-1}{t} \sum_k p(k, t-1) + \frac{1}{t}$$

$$= \frac{t-1}{t} \left[\frac{t-2}{t-1} \sum_k p(k, t-2) + \frac{1}{t-1} \right] + \frac{1}{t} \tag{8.8}$$

Continuing in this vein for $t - 2$ steps, we ultimately reach the expression

$$\sum_k p(k, t) = \frac{t - (t-2)}{t} \sum_k \delta_{k,2} + \sum_{r=1}^{t-2} \frac{1}{t} = \frac{2}{t} + \frac{t-2}{t} = 1 \tag{8.9}$$

Hence, the degree distribution is always normalized. The average degree can be found via a simple argument. At every step the network gains a link and a node, with the starting configuration having two links and two nodes. Hence, at time t the network will have t nodes, t edges and $\sum_{i=1}^t k_i = 2t$. Hence, $\langle k \rangle = t^{-1} \sum_{i=1}^t k_i = 2$.   □

Exercise 8.3

We use some simple manipulations to cast our calculation in a form where the summation identity provided can be employed:

$$\langle k \rangle = \sum_{k>0} \frac{4k}{k(k+1)(k+2)} = \sum_{k>0} \frac{4(k+1)}{k(k+1)(k+2)} - \sum_{k>0} \frac{4}{k(k+1)(k+2)}$$

$$= \sum_{k>0} \frac{4}{k(k+2)} - \sum_{k\geq 0} p(k) = \sum_{k>1} \frac{4}{(k-1)(k+1)} - 1$$

$$= 3 - 1 = 2 \tag{8.10}$$

This is indeed the same value found in the previous exercise. □

Exercise 8.4

This is basically an exercise in careful adaptation of the equivalent master equation derivation for the preferential attachment model. At step $t-1$ our network still has $t-1$ links and $t-1$ nodes, but the probability of selecting node i for linking to the new vertex is now $p(i|t) = 1/(t-1)$. Hence, we have

$$p(k, i, t) = \frac{1}{t-1} \, p(k-1, i, t-1) + \left(1 - \frac{1}{t-1}\right) p(k, i, t-1) \tag{8.11}$$

We sum over all nodes to find $p(k, t) = t^{-1} \sum_{i=1} p(k, i, t)$:

$$p(k, t) = \frac{1}{t} \sum_{i=1}^{t-1} \left\{ \frac{t-2}{t-1} \, p(k, i, t-1) + \frac{1}{t-1} \, p(k-1, i, t-1) \right\} + \frac{1}{t} \delta_{k1}$$

$$= \left(1 - \frac{2}{t}\right) p(k, t-1) + \frac{1}{t} p(k-1, t-1) + \frac{1}{t} \delta_{k1} \tag{8.12}$$

Equivalently:

$$t[p(k, t) - p(k, t-1)] = p(k-1, t-1) - 2p(k, t-1) + \delta_{k1} \tag{8.13}$$

We next define a real-valued time $\tau = \epsilon t$, and obtain

$$\tau \frac{p(k, \tau) - p(k, \tau - \epsilon)}{\epsilon} = p(k-1, \tau - \epsilon) - 2p(k, \tau - \epsilon) + \delta_{k1} \tag{8.14}$$

Taking the limits $\epsilon \downarrow 0$ and $t \to \infty$ for fixed τ then gives

$$\tau \frac{\partial}{\partial \tau} p(k, \tau) = p(k-1, \tau) - 2p(k, \tau) + \delta_{k1} \tag{8.15}$$

whose stationary solutions $p(k, \tau) = p(k)$ for all τ are to be solved from

$$p(k-1) - 2p(k) + \delta_{k1} = 0 \tag{8.16}$$

For $k=0$ this gives $p(0) = 0$, as before (which makes sense, since it is by construction still impossible to find disconnected nodes). For all $k \geq 2$ we have $p(k) = \frac{1}{2} p(k-1)$, giving $p(k) = 2^{-(k-1)}p(1)$. Finally, for $k = 1$ our equation (8.16) becomes $p(1) = \frac{1}{2}$. The conclusion is therefore

$$p(0) = 0, \qquad p(k > 0) = 2^{-k} \tag{8.17}$$

which is also normalized correctly. As a consequence of removing the preference for attaching to high-degree nodes in the growth algorithm, the power-law degree statistics have been replaced by an exponential degree distribution, and the resulting network is no longer scale-free. □

Exercise 8.5

Full solutions for this generalized algorithm can be found in papers such as [107, 105], providing in-depth studies of the topologies expected with specific sub- and super-linear preferential attachment functions. The added complication in doing master equation derivations is that for $a \notin \{0, 1\}$ one now has to keep track of the denominators in the probabilities $p(i|t)$, which are no longer trivial functions of t.

By intuition we can predict that super-linear preferential attachment, weighting the preference still further towards high-degree nodes via values $a > 1$, will enhance the imbalance between the few high-degree nodes and the many low-degree ones even further, creating ultimately a tendency for the network to condense – with the large majority of links attaching to a super-hub. For $a \in (0, 1)$ (sub-linear preferential attachment), it was found that the degree distribution will be a stretched exponential [107]. The limit $a \to 0$ is the case studied in Exercise 8.4 – where we have already demonstrated that an exponential-type degree distribution will result. For $a < 0$ the new vertices will link preferably to the existing nodes with the lowest degrees, and here one expects to find very narrow degree distributions.

There is, however, a paradox in these observations. In real situations, it is quite plausible that the preference function might be idiosyncratically damped or amplified. Nonetheless, we see power-law degree distributions everywhere. Reference [104] described the pre-asymptotic behaviour of a super-linear preferential attachment function, in order to demonstrate how networks which are degenerate in the asymptotic limit may still demonstrate rich and interesting topologies for finite sizes. □

9
Specific constructions

Introduction. In this chapter, we discuss classes of network-generating models which are neither based on carefully tailored Markov chains, nor on growing networks according to specific iterative wiring recipes. Instead they are defined mechanistically and with the objective of investigating specific phenomena. Similar to growth algorithms it is usually difficult to give precise formulae for the probabilities with which individual network realizations will be generated.

9.1 Watts–Strogatz model and the 'small world' property

Overview: The Watts–Strogatz algorithm is designed to generate graphs which show a high degree of clustering (i.e. contains many triangles) and short average path lengths. The strength of the algorithm is its transparency. The downside is that it cannot easily be tailored to model more complex systems, so it is essentially an illustrative model.

Motivation and definition. Watts and Strogatz [183] proposed a model that was designed to capture the *small world* property. This is the observation that real social networks typically have shorter average path lengths than Erdös–Rényi graphs (as defined in (3.1)). This is a concept sometimes colloquially referred to as 'six degrees of separation', to reflect Milgram's famous experiment [125], which counted the average number of steps required to pass a letter between two randomly chosen American citizens. The Watts–Strogatz networks are generated in a way which results in high clustering and a relatively low average shortest path length between any two nodes.

   The Watts–Strogatz prescription begins with a row of nodes arranged in a line. Every node is connected to K nearest neighbours (e.g. two or three neighbours either side of it). Periodic boundary conditions – linking the start and end of the line so that the end nodes also have the correct number of neighbours – gives a circular graph as shown on the left-hand side of Figure 9.1. This initial graph has high clustering (provided $K > 1$), but long average path lengths. A message can only move $K/2$ positions around the circle at each step, so it may take a long time to pass a message between nodes on opposite sides of the circle. Reference [183] proposed adding shortcuts into the network by randomly rewiring a small fraction p of the links between immediate neighbours on the circle, to go instead to a randomly chosen node in the network. The result is Algorithm 29. This will typically create long-range links that connect nodes across the circle, as shown by the second and third graphs in Figure 9.1. The result is a radical reduction of the average shortest path length, while retaining the clustered nature of the network. Figure 9.2 shows how for low values of the rewiring probability

Generating Random Networks and Graphs. First Edition. A.C.C. Coolen, A. Annibale, and E.S. Roberts. © A.C.C. Coolen, A. Annibale, and E.S. Roberts 2017. Published in 2017 by Oxford University Press. DOI: 10.1093/acprof:oso/9780198709893.001.0001

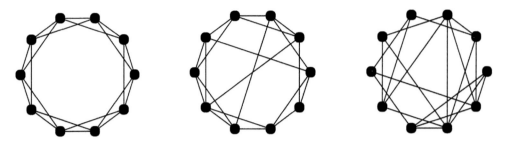

Fig. 9.1 Watts–Strogatz model for different values of the rewiring probability p. From left to right, a regular ring graph with short-range links only (the seed graph, corresponding to $p = 0$), a partially rewired ring graph and an Erdös–Rényi random graph (from [183]).

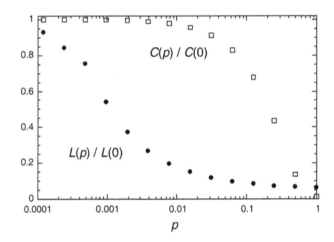

Fig. 9.2 Relationship between the value of the rewiring probability p and the topology of the final graph. The top line shows the proportional reduction in the clustering coefficient, as defined in (2.2). The bottom line shows the proportional reduction in shortest path length between pairs of distinct nodes, averaged over all such pairs. This shows the success of the Watts–Strogatz algorithm in generating networks which simultaneously show high clustering and low path lengths – similar to real-world networks. Figure reproduced from [183].

p, the clustering coefficient remains almost unchanged, but the average shortest path length drops rapidly.

Exercise 9.1 Calculate the clustering coefficients $C_i(\mathbf{A})$, as defined in (2.2), for the seed graph of the Watts–Strogatz model with $K = 2$ and $N > 2K + 1$, as shown on the left of Figure 9.1.

Drawing on the analogy between the Watts–Strogatz construction of a random graph and Milgram's message passing experiment, the regular starting lattice can be

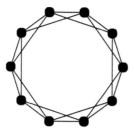 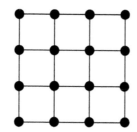

Fig. 9.3 A Watts–Strogatz algorithm can take different regular lattices as a starting point. Here we show lattices with dimension $d = 1$ (ring) and $d = 2$ (square), but one could equally start with higher-dimensional and/or non-cubic lattices.

seen as representing acquaintance based on geographical proximity: you are likely to know lots of people who live in your immediate local area. The additional shortcut links represent acquaintances on another basis (e.g. friends).

What Watts and Strogatz found interesting about Milgram's experiment was not merely that short paths existed between pairs of randomly chosen American citizens. They observed that it was not obvious that individuals passing the message one step at a time would be able to identify a path to the intended recipient within a reasonable number of steps. The intermediate message carriers were not able to see the overall network topology, but in fact they did know about node attributes of the targeted recipients and their social circle. In the particular case of Milgram's experiment, the intended recipient was (in one case) a stockbroker from Boston, and (in another case) the wife of a student in Cambridge, MA. Hence, the intermediate message carriers chose the next step of the path based on people in their social circle who may have encountered the target professionally – as well as based on trying to shrink the geographic distance.

Variations and extensions. The characteristic feature of a Watts–Strogatz type protocol for building networks is to have a two-stage/bi-modal approach. In stage one, short-range links are placed deterministically. In the original paper by [183], stage one connected nearest-neighbour nodes on a ring. This stage will typically deliver a highly clustered graph. Stage two is a stochastic process that typically generates long-range links, whose purpose is to make the model more realistic and to deliver a radically reduced mean shortest path length.

Probably the simplest variation on the Watts–Strogatz model is, after starting from the conventional seed graph, to add random links without removing any of the short-range ones, as described in Algorithm 28. Excluding the rare occurrences where the new links happen by accident to connect nodes that are already connected in the seed graph, this is equivalent to superimposing Erdös–Rényi graphs upon the seed graph. In contrast to the original version, the final graph will consequently have more links than the seed graph, and the average degree $\langle k \rangle$ will increase monotonically with the number of random long-range links that are introduced. Repeating the measurement

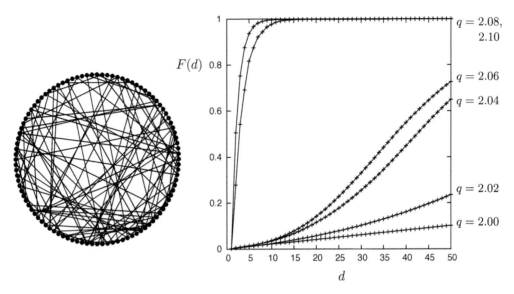

Fig. 9.4 The fraction $F(d)$ of node pairs (i, j) that are found to have a distance $d_{ij} \leq d$, shown versus d, measured in 'small world' graphs created by superimposing an Erdös-Rènyi graph with average degree $q-2$ upon a ring with nearest-neighbour interactions only (i.e. a $K = 1$ seed graph). The combined graph, whose topology will be as shown on the left, has expected average degree $\langle k \rangle = q$. For $q = 2$ we would have only short-range links. Here $N = 1000$ and $\langle k \rangle \in [2.0, 2.1]$. One observes a relatively sharp transition – at around $q = 2.07$, where there is still only a small fraction of long-range links – to graphs with the short typical distances that characterize the small-world effect.

of shortest paths (as was done, e.g. in Figure 3.1 for the Erdös–Rényi graphs) results in Figure 9.4. Clearly, the phenomenology of this alternative construction, in terms of clustering and path length statistics, is identical to that of the original.

A ring, which formed the skeletal architecture for the seed graph in [183], is the one-dimensional version of a regular lattice with periodic boundary conditions. The protocol of the Watts–Strogatz model can be generalized immediately to other regular lattices (see, e.g. [103] and later papers referencing it). The starting point of the algorithm could be a two-dimensional lattice of nodes arranged in a grid as shown in Figure 9.3, or a non-cubic lattice (e.g. triangular, hexagonal, etc.). The only essential ingredient is that in the seed graph all links are still strictly short range, to ensure that a high degree of clustering is built in.

Taking the next logical step, [103] developed an algorithm starting from a regular hierarchical graph. The analogy is that the leaves of the tree may represent different professions. Hence, in Milgram's experiment someone may have passed the letter for the stockbroker to an acquaintance working in financial services, assuming that they are more likely to know someone in the same field. In [103] the hierarchical regular graph represents the intrinsic groupings in the population. Nodes are connected to one another based on a strictly decreasing function $f(h(x, y))$, where $h(x, y)$ gives the

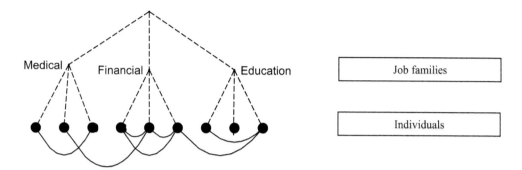

Fig. 9.5 In this variant on the Watts–Strogatz model, based on the work of [103], long-range links are formed depending on how far back any two given nodes in a hierarchical tree had a common ancestor.

'height' or level at which nodes x and y last had a common ancestor. This means that nodes in closely related professions are more likely to be linked than nodes from more distant professions. This idea is illustrated in Figure 9.5. This is a special case of the hidden variable construction presented in Section 10.3. A closer analogy to the Watts–Strogatz algorithm would be to start with all the nodes within each smallest subgrouping connected to one another in cliques. This is the same as assuming that everyone within a profession/social grouping is inter-acquainted. On can then proceed with adding, deleting and rewiring links driven by probability $f(h(x, y))$, as before.

Another extension which can make the Watts–Strogatz model more realistic is to weight the rewiring probability/new link probability with the distance between the two proposed nodes, based either on their distance in the graph, or on an alternative geometric distance definition (of which the construction in Figure 9.5 could in fact be seen as an example), or any other underlying hidden variable (e.g. professional field).

9.2   Geometric graphs

Overview: In geometric random graphs, nodes are put into one-to-one correspondence with points in a geometric space, and the probability of two nodes being connected depends on the distance between their associated points [145]. For example, unit disk graphs are defined by setting a threshold such that any two nodes within a certain distance of each other are connected. The interest is in the relationship between the chosen geometric space, the rule used to calculate link probabilities in terms of physical locations of nodes and the features of the resulting graph.

Definitions. We have already encountered the simplest example of a geometric graph earlier, acting as the seed graph for the Watts–Strogatz algorithm. The graph on the left in Figure 9.3 can be seen as having its nodes i associated with evenly spaced points $\mathbf{x}_i$ on a periodic line (the line being the simplest non-trivial geometric space), and nodes

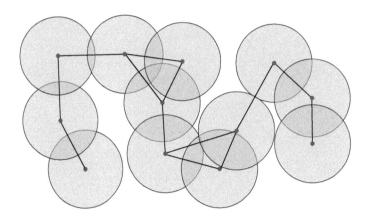

Fig. 9.6 A geometric graph is constructed in two stages. First, nodes are placed randomly in a chosen space (black dots in the figure). A transmission range is specified, which can be understood as the diameter of the shaded circles. All those node pairs whose shaded circles intersect are connected by a link.

(i, j) are connected if their locations are less than a specified cut-off distance[1] apart. In geometric graphs, which by construction are normally simple and nondirected, one generalizes this construction. The choice made for the metric and the underlying geometric space, i.e. the recipe for calculating distances between points, has a strong influence on the topological characteristics of the graphs being generated.

The simplest way to generate a random geometric N-node graph, once the geometry has been chosen (i.e. a space $S \subseteq \mathbb{R}^d$ of d-dimensional points with a distance function $D(\mathbf{x}, \mathbf{x}')$ assigned to any two points $\mathbf{x}, \mathbf{x}' \in S$), is the following: first distribute N vectors $\mathbf{x}_i \in S$ randomly over S, with $\mathbf{x}_i = (x_{i1}, \ldots, x_{id})$; then link nodes that are within a distance of 1 from each other

$$\text{for all } (i, j): \quad A_{ij} = \theta\big(1 - D(\mathbf{x}_i, \mathbf{x}_j)\big) \tag{9.1}$$

Using a step function, $\theta(z > 0) = 1$ and $\theta(z < 0) = 0$. The resulting graphs are sometimes called *unit disk graphs*. The protocol is illustrated in Figure 9.6. For a distance definition $D(\mathbf{x}, \mathbf{x}')$ on S to be acceptable – for S to be a genuine geometric space – it must satisfy the following conditions:

1. $D(\mathbf{x}, \mathbf{x}') \geq 0$ for all $\mathbf{x}, \mathbf{x}' \in S$
2. $D(\mathbf{x}, \mathbf{x}') = 0$ if and only if $\mathbf{x} = \mathbf{x}'$
3. $D(\mathbf{x}, \mathbf{x}') = D(\mathbf{x}', \mathbf{x})$ for all $\mathbf{x}, \mathbf{x}' \in S$
4. $D(\mathbf{x}, \mathbf{x}') \leq D(\mathbf{x}, \mathbf{x}'') + D(\mathbf{x}'', \mathbf{x}')$ for all $\mathbf{x}, \mathbf{x}', \mathbf{x}'' \in S$

This gives a relatively simple protocol for generating geometric random graphs, in which the randomness lies only in the placement of the N position vectors $\{\mathbf{x}_i\}$ within

---

[1]This graph is strictly speaking not random, but one can see deterministic graph constructions as special cases of random graphs, where all link probabilities happen to be either 0 or 1.

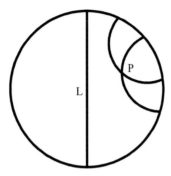

Fig. 9.7 One way of visualizing hyperbolic geometry is via the Poincaré hyperbolic disk. Lines start and end at the perimeter of the (open) circle. As we approach the edge of the circle, the space is defined to expand to infinity. A straight line in the hyperbolic space is defined as the intersection of the disk and another circle meeting the perimeter orthogonally. Parallel lines are lines which do not meet. The sketch shows two of the many possible straight lines which pass through point P and are parallel to L. The multiplicity of such lines is a key difference between Euclidean and hyperbolic space.

the space S. Given these positions, the wiring of the graph is deterministic. The corresponding pseudocode for the case where the space S is a d-dimensional cube is given in Algorithm 30.

Exercise 9.2 Choose the set S to be the interval $[-1, 1]$, equipped with the standard Euclidean distance definition $D(x, x') = |x - x'|$. Distribute N nodes randomly and uniformly on this interval, and connect those node pairs (i, j) that have $D(x_i, x_j) < \Delta/N$. Calculate the average degree distribution of the resulting graph in the limit where $N \to \infty$ for fixed Δ.

Unit disk graphs are particularly evocative of broadcast networks [95]. Nodes represent transmitters (e.g. mobile phone masts) and are placed in $\mathbb{R}^2$ (i.e. on the map/ground). The transmission range [49] of individual stations is represented by the (possibly rescaled) radius of the circle based on the Euclidean distance formula $D(\mathbf{x}_i, \mathbf{x}_j) = [\sum_{\ell \le d}(x_{i\ell} - x_{j\ell})^2]^{\frac{1}{2}}$. It may in addition be preferred to generate such networks under further constraints [160]. These constraints may, for example, require that the whole space S is monitored, preferring configurations with fewer nodes or incorporating topological constraints which are known to increase attack tolerance. Constraints may reflect specific elements of the problem at hand such as physical geography or geology restricting the placement of transmitters. Alternatively, a growth model may be more realistic than placing nodes randomly in S. In real communications networks, it is essential that the graph is connected, so that a message can be passed between any two given nodes. The growth algorithm can be designed to ensure this if every node is placed subject to being within the transmission range of an existing node. In [138] one finds several growth models with constraints; [112] provided a

Fig. 9.8 Illustration from [184] of a triangular tessellation in hyperbolic space. The fast-expanding nature of hyperbolic space is visible in the rapid increase in the number of triangles that make up each concentric circle of the pattern. This pattern can also be seen as representing a hierarchical network, rooted at the centre of the disk. This give some intuition for why several authors have described hyperbolic spaces as 'continuous trees' (or, conversely, trees as 'discrete hyperbolic spaces'). This implies that hyperbolic space is the natural geometry within which to define real-world graphs which have a broadly hierarchical topology, such as the Internet.

thorough survey of methods (see, e.g. Section 7 of [112]). Authors such as [19] look at typical connectivity properties of ensembles of geometric graphs, in order to deduce the required physical properties of the nodes, such as the transmission range required to ensure that the network typically remains connected.

Hyperbolic graphs. We have described above how a graph can be generated by placing points in a flat space and joining points within a certain threshold distance of each other. The same procedure could be applied placing nodes in a curved space. The two main flavours of curved space are spherical geometry (positive curvature – placing nodes on the surface of a sphere) and hyperbolic geometry (negative curvature). The essence is that distances in hyperbolic geometries grow much faster when points move away from each other than in normal Euclidean space. This turns out to be a useful feature when embedding a fast-expanding graph such as, for example, a regular $k-tree$ (see Figure 9.8). A Euclidean space is characterized by the axiom that (in a plane),

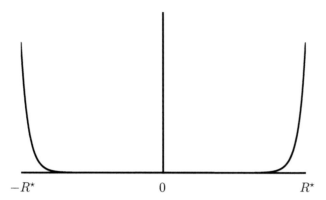

Fig. 9.9 A sketch of the density of nodes (vertical axis) shown as a function of the spatial coordinate that defines a cross section of a Poincaré hyperbolic disk. The first step to defining a hyperbolic geometric graph is by scattering nodes uniformly in the hyperbolic space. In polar coordinates (r, ϕ), this can be achieved by sampling the angular coordinate ϕ uniformly from $[0, 2\pi)$, and sampling the radial coordinate from $[0, R^\star)$ proportional to e^r.

given a point x and a line L, there is a unique line which is parallel to L and passes through x. Conversely, on a hyperbolic geometry, there must be at least two lines which pass through x but never intersect L. Geometric graphs defined on hyperbolic spaces have particularly ignited interest, because they generate networks which have high clustering, small diameter and fat-tailed degree distributions. More information on hyperbolic spaces can be found in the accessible textbook [8].

There are several alternative models for understanding hyperbolic space; we will describe the approach based on the Poincaré hyperbolic disk model, which is illustrated in Figure 9.7. First, the nodes need to be placed at random on an open disk of radius $R^\star \gg 1$ such that the (polar) angle coordinate ϕ is sampled uniformly from $(0, 2\pi]$ whereas the radial coordinate is sampled according to the density $\rho(r) \sim e^r$ for $r \in [0, R^\star)$ (sketched in Figure 9.9). To determine which nodes should be connected to each other, we need to use a formula for hyperbolic distance. For the simplest case of curvature -1, [108] shows that the distance between points (r, ϕ) and (r', ϕ') can be defined as

$$D((r, \phi), (r', \phi')) = \cosh^{-1}\left[\cosh(r)\cosh(r') - \sinh(r)\sinh(r')cos(\Delta\phi)\right] \quad (9.2)$$

where $\Delta\phi = \pi - |\pi - |\phi - \phi'||$. We could complete the construction by following (9.1)

$$\text{for all } (i, j): \quad A_{ij} = \theta\big(1 - D((r_i, \phi_i), (r_j, \phi_j))\big) \quad (9.3)$$

or use another distance-driven connection probability. Figure 9.10 roughly illustrates this. More general schemes for generating hyperbolic graphs are described in [178].

9.3 Planar graphs

Overview: A graph is described as planar if it can be drawn on a plane with no intersecting edges. In general, it is difficult to check for planarity, and the techniques for generating planar graphs are heuristic and bespoke.

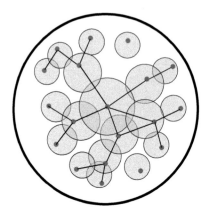

Fig. 9.10 A rough sketch of a hyperbolic geometric graph on a Poincaré disk. As before (see 9.6) nodes are placed randomly in the space; equal-sized circles are drawn around each node; node pairs whose shaded circles intersect are connected by links. The circles appear to become smaller and off-centre towards the perimeter owing to the faster expansion of hyperbolic space towards the outside of the circle. This is also why there are relatively fewer nodes towards the centre of the disk. Even in this rough sketch, it is already possible to see a hierarchical structure emerging in the induced graph.

It is important to realize that geometric graphs with planar point sets $S \subseteq \mathbb{R}^2$ are not the same as planar graphs. A graph is described as planar if it can be drawn on a plane with no intersecting edges. Hence, planar graphs may or may not be geometric, and geometric graphs may or may not be planar; the two concepts are very different. Determining from its adjacency matrix $\mathbf{A}$ alone whether a given graph is planar is a very challenging mathematical problem, so it is important to find an efficient procedure to generate random planar graphs in order to be able to study them numerically. Some techniques to test planarity are presented in [31] and related articles.

An early technique for generating random planar graphs came from [59]. Pairs of nodes are picked from a generic random graph, and an edge is removed (if present) or created (if absent, and if its creation does not violate planarity). The authors of [24] developed the first polynomial time algorithm based on classifying recursive decompositions of planar graphs and calculating associated coefficients in order to place appropriate weights on the decomposition tree. The resulting algorithm is rather involved, but claims uniformity. The Kuratowski generator is based on the Kuratowski theorem, which states that a graph is planar if and only if no subgraph is present that is a subdivision (see Figure 9.12) of a complete five-node graph (K5) or a complete three node by three node bipartite graph (K3;3). (K5) and (K3;3) are illustrated on the right-hand side of figure 9.11. The generator starts with a non-planar random graph. The algorithm searches for a K5 or K3;3 subgraph and removes one of its edges until no more such subgraphs can be found. This approach is illustrated in Figure 9.13. In [120] one finds an experimental survey of available planar graph generators which discusses some other algorithms.

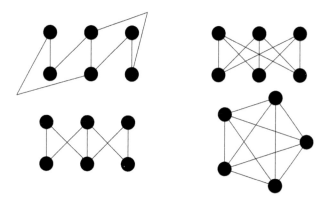

Fig. 9.11 The left-hand pictures show planar graphs. The bottom graph is drawn with intersecting edges, but can be drawn as the top graph (non-intersecting edges) and so is planar. The right-hand pictures show non-planar graphs – they cannot be drawn without intersecting edges. Determining whether a given N-node graph is planar is very difficult if N is large.

9.4   Weighted graphs

Overview: In this section, we review how the configuration model, the preferential attachment model and the Erdös–Rényi model can be extended to generating weighted graphs.

The literature on random graphs normally assumes (imposes) binary links between nodes. The link between nodes i and j is either present or absent. This simplifies the analysis considerably, but clearly does not accurately capture the more nuanced relationships that are observed in real-world scenarios. In this section, we briefly review some modest extensions to standard techniques that allow for weighted edges.

There are two major viewpoints to modelling a weighted network. The first approach constructs a binary network and then ascribes weights to the edges. The most straightforward adaptation of preferential attachment to generating weighted networks was proposed by [187]. The basis for this is a standard preferential attachment scheme: at every time step a new node is born, which connects to m existing nodes with a probability proportional to the degree (number of links) that the target node already has. The additional step is to then assign each of the new links a weight, which is itself proportional to the degree of the target node. In the case of [187], the weights are chosen to be normalized so that for the new node i we have $\sum_j w_{ij} A_{ij} = 1$ and hence

$$w_{ij} = \frac{k_j}{\sum_{s \in \partial_i} k_s} \tag{9.4}$$

where the denominator sums over the degrees of the newly created neighbours of node i. This is found to create a power-law distribution of link weights, with a different exponent from the (power-law) distribution of node degrees.

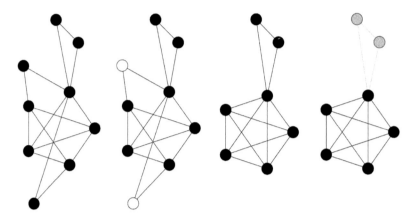

Fig. 9.12 This figure analyzes a graph using Kuratowski's theorem in order to check if it is planar. The original graph is shown on the left. A subdivision of a graph is when a node is place on an edge creating a new network containing one extra node of degree 2. The second graph from the left identifies the two white nodes as constituting a subdivision. The third network from the left removes these nodes. In the final network, it has been shown that the original network contained a subgraph which was a subdivision of K(5), and hence this network was not planar.

The binary edges can be viewed as infrastructure, and the weights can be interpreted as 'traffic' arising from some process occurring on the network. For example, [14] found that passenger capacity between any two airports is roughly proportional to the square root of the product of the 'degrees' of the respective airport nodes. Alternatively, the weights could be assigned independently of the topology referencing underlying node properties. For example, [15] propose a preferential attachment scheme driven by a targeted 'strength' for each node. Strength is defined as the sum of weights of edges starting from that node, and is the weighted network generalization of the concept of degree. In [15], the weights ascribed to edges are allowed to vary as the network evolves. An analogy would be building a new road and consequently observing reduced traffic on previously existing roads.

An alternative viewpoint is to generate networks with integer-weighted links, by viewing these as single/double/triple/multiple links. Examples where this would be a natural model include weighted citation networks (weight set as the number of co-authored papers between two given authors). Integer-weighted links can naturally occur if we choose to relax the requirement that only one link is allowed to pass between any given pair of nodes. The configuration (stub joining) model described in Section 8.1 could be allowed to form multiple edges, or could even be adapted to use target weights as one of the drivers of the algorithm. Reference [80] proposed the integer-weighted version of the Erdös–Rényi model by setting the probability of an edge with (integer) weight w as

$$\text{Prob}(A_{ij} = w) = p^w(1 - p) \quad \forall \, i, j \tag{9.5}$$

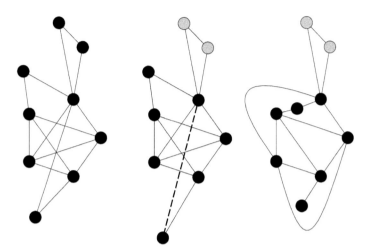

Fig. 9.13 This figure starts with the non-planar graph from Figure 9.12, and then applies the Kuratowski generator approach in order to transform it into a planar graph. The first step identifies the K(5) subgraph (or subdivision of a subgraph). One edge is removed from the K(5) subgraph. The final network demonstrates that the remaining nodes and edges constitute a planar graph.

for some $p \in (0, 1)$. If the aim is to generate a network with total weight $W = \sum_{i<j} w_{ij}$, then p should be chosen as

$$p = \frac{2W}{N(N-1) + 2W} \tag{9.6}$$

Algorithm 9 implements this in pseudocode. A generalization which allows the user to tune the expected strength of each node i is to set

$$\text{Prob}(A_{ij} = w) = (y_i y_j)^w (1 - y_i y_j) \quad \forall \, i, j \tag{9.7}$$

for suitably chosen parameters y_i for each node i.

Exercise 9.3 Calculate the expected average weight per link in a network generated according to (9.5).

9.5   Solutions of exercises

Exercise 9.1

All nodes in the proposed $K = 2$ seed graph have degree $k_i(\mathbf{A}) = 4$, so the clustering coefficient of any node i is $C_i(\mathbf{A}) = 2T_i(\mathbf{A})/k_i(\mathbf{A})[k_i(\mathbf{A}) - 1] = \frac{1}{6}T_i(\mathbf{A})$. Here $T_i(\mathbf{A})$

is the number of distinct triangles that start at site i and end at site i. It is clear from Figure 9.1 (left graph) that each node has exactly three such triangles:

$$i \to i-1 \to i+1 \to i$$

$$i \to i+1 \to i+2 \to i$$

$$i \to i-1 \to i-2 \to i$$

Hence, $C_i(\mathbf{A}) = \frac{1}{2}$ for all i. We require $N > 2K + 1$ to avoid the meaningless case where the seed graph is actually a complete graph on five nodes, which has a clustering coefficient of 1. □

Exercise 9.2

We will set out two different approaches to the solution. The first is more intuitive than formal. The question asks us to calculate the degree distribution of a graph formed by placing N nodes at random in the interval $[-1, 1]$, and connecting two nodes if they are within a distance of Δ/N of each other. Let us start by considering a given node i placed at random within the interval $[-1, 1]$. We know that Δ is fixed, while N is large and tending to infinity; we use this to justify concentrating on the dominant case where node i falls in the range $\left[\frac{\Delta}{N} - 1, 1 - \frac{\Delta}{N}\right]$.

A node j placed uniformly at random in the interval will fall within a distance $\frac{\Delta}{N}$ of node i with a probability of $\frac{1}{2}\frac{2\Delta}{N} = \frac{\Delta}{N}$. A link will be placed between nodes i and j if node j falls within a distance $\frac{\Delta}{N}$ of node i. Repeating this until we reach N nodes is mathematically the same as carrying out $N - 1$ Bernoulli trials (independent coin flips) with success probability Δ/N each time. The degree of node i is the number of successful trials – the number of nodes that fall within a distance $\frac{\Delta}{N}$ of i. It is a standard result [83] that the probability of observing exactly k successful Bernoulli trials follows a Binomial distribution

$$p(k|i) = \binom{N}{k} \left(\frac{\Delta}{N}\right)^k \left(1 - \frac{\Delta}{N}\right)^{N-k} \tag{9.8}$$

which for $N \to \infty$ is known to converge to a Poissonian distribution centred around Δ. So we can conclude that

$$p(k) = e^{-\Delta} \Delta^k / k! \tag{9.9}$$

It is left to the interested reader to calculate results for finite N including the boundary regions.

The second route to solving the problem requires more formal analysis. We note that the positions x_i are each distributed independently according to $p(x_i) = \frac{1}{2}\theta(1 - |x_i|)$. Given the realization $\mathbf{x} = (x_1, \ldots, x_N)$ of all positions, the corresponding graph is defined by $A_{ij} = \theta(\frac{\Delta}{N} - |x_i - x_j|)$, and will have the following degree distribution:

$$p(k|\mathbf{x}) = \frac{1}{N} \sum_{i=1}^{N} \delta_{k, \sum_{j \neq i} \theta(\frac{\Delta}{N} - |x_i - x_j|)}$$

$$= \int_{-\pi}^{\pi} \frac{d\omega}{2\pi} e^{i\omega k} \frac{1}{N} \sum_{i=1}^{N} e^{-i\omega \sum_{j \neq i} \theta(\frac{\Delta}{N} - |x_i - x_j|)} \tag{9.10}$$

(where we used the integral representation of the Kronecker delta symbol). We average over all positions, and take the limit $N \to \infty$:

$$p(k) = \lim_{N \to \infty} \left\langle \int_{-\pi}^{\pi} \frac{d\omega}{2\pi} e^{i\omega k} \frac{1}{N} \sum_{i=1}^{N} e^{-i\omega \sum_{j \neq i} \theta(\Delta/N - |x_i - x_j|)} \right\rangle_{x_1 \dots x_N}$$

$$= \lim_{N \to \infty} \frac{1}{N} \sum_{i=1}^{N} \int_{-\pi}^{\pi} \frac{d\omega}{2\pi} e^{i\omega k} \left\langle e^{-i\omega \sum_{j \neq i} \theta(\Delta/N - |x_i - x_j|)} \right\rangle_{x_1 \dots x_N}$$

$$= \frac{1}{2} \int_{-\pi}^{\pi} \frac{d\omega}{2\pi} e^{i\omega k} \int_{-1}^{1} dx \lim_{N \to \infty} \left(\frac{1}{2} \int_{-1}^{1} dy \, e^{-i\omega\theta(\Delta/N - |x-y|)} \right)^{N-1}$$

$$= \frac{1}{2} \lim_{N \to \infty} \int_{-\pi}^{\pi} \frac{d\omega}{2\pi} e^{i\omega k} \int_{-1}^{1} dx \left(1 + \frac{1}{2}(e^{-i\omega} - 1) \int_{-1}^{1} dy \, \theta(\Delta/N - |x - y|) \right)^{N-1}$$

$$= \frac{1}{2} \lim_{N \to \infty} \int_{-\pi}^{\pi} \frac{d\omega}{2\pi} e^{i\omega k} \int_{-1-\Delta/N}^{1+\Delta/N} dx \left(1 + \frac{\Delta}{N}(e^{-i\omega} - 1) \right)^{N-1}$$

$$= \int_{-\pi}^{\pi} \frac{d\omega}{2\pi} e^{i\omega k} e^{\Delta(e^{-i\omega} - 1)} \tag{9.11}$$

We write the exponential that contains $e^{-i\omega}$ as a power series, and use the integral representation of the Kronecker delta in reverse mode:

$$p(k) = e^{-\Delta} \sum_{\ell \geq 0} \frac{\Delta^\ell}{\ell!} \int_{-\pi}^{\pi} \frac{d\omega}{2\pi} e^{i\omega k - i\omega \ell} = e^{-\Delta} \sum_{\ell \geq 0} \frac{\Delta^\ell}{\ell!} \delta_{k\ell}$$

$$= e^{-\Delta} \Delta^k / k! \tag{9.12}$$

We thus find a Poissonian degree distribution, with average degree $\langle k \rangle = \Delta$. $\quad\square$

Exercise 9.3

The average weight is given by

$$\sum_{w \in \mathbb{N}} w p^w (1 - p) = (1 - p) \left[p + 2p^2 + 3p^3 \dots \right] = (1 - p) \frac{p}{(1 - p)^2} = \frac{p}{1 - p} \tag{9.13}$$

using standard series expansions. If we check ourselves by substituting in the result given by (9.6) in terms of the total weight W, this evaluates as

$$\frac{2W}{N(N - 1)} \tag{9.14}$$

which is as we would expect – although it prompts us to observe that the summation (9.13) includes zero-weighted links (the case where $w = 0$). This means that the average is given over all possible pairs of nodes, rather than present links. There are

$N(N-1)/2$ pairs of nodes; multiplying this by the average calculated in (9.13) gives the total weight $W = \sum_{i<j} w_{ij}$

$$W = \frac{p}{1-p} \times \frac{N(N-1)}{2} \tag{9.15}$$

which, incidentally, is the route to deriving (9.6). Given the total weight of links W, we can choose to average it over the $pN(N-1)/2$ non-zero links in the network to get an average weight per present link of

$$\frac{p}{1-p} \times \frac{N(N-1)}{2} \frac{2}{pN(N-1)} = \frac{1}{1-p} = 1 + \frac{2W}{N(N-1)} \tag{9.16}$$

which is bounded from below by 1 (as we would expect). $\qquad\qquad\square$

Part V

Further topics

10
Graphs on structured spaces

Introduction. So far we have mainly taken the viewpoint of networks and graphs being built of nodes whose properties are defined fully and solely in terms of their links with the other nodes in the graph. Apart from connectivity patterns, nodes are indistinguishable. In this chapter, we subvert that viewpoint by anchoring nodes to an external (geometric or value-driven) structure. Even before creating links, nodes are *defined* to be different, and these differences influence the connectivity of the graph. The outside world is not just to be mimicked in a graph representation, but gives *a priori* meaning to our nodes. Here we explore further examples, such as multiplex networks, and graphs where nodes are assigned fitness values that influence their ability to form links. The techniques for generating graphs on structured spaces are generally extensions of those developed for the more conventional ones. The additional structure tends to be reflected in having more parameters to be varied or controlled in ensembles, and sometimes more computationally demanding algorithms.

10.1   Temporal networks

Overview: Temporal networks are a rich but potentially complicated way to represent datasets in which time plays a crucial role. Such datasets are often naturally represented by networks in which the node and/or link variables evolve over time, i.e. links or nodes can disappear and reappear. Terminology and techniques need to be reviewed in light of the extra time dimension. In most of the relevant models, only the links evolve.

Motivation and definitions. In the simplest type of temporal network, the N nodes remain fixed, but at every time step there is a distinct connectivity representing the contacts/interactions which are actually occurring between these nodes at that time point. The adjacency matrix acquires a time-dependence, and must now be written as $\mathbf{A}(t)$. Time is a valuable extra dimension of information. For example, in cell biology, these are profoundly different scenarios: proteins A, B and C interacting with one another simultaneously (suggestive of a complex?), compared to, in separate time periods, A interacting with B, then B interacting with C, and then C interacting with A (suggestive of a feedback loop?). If node A interacts with node B, then node B interacts with node C – we can conclude that a signal/infection cannot pass from node C to node A. In a static graph, we might have drawn the opposite conclusion, since we would have observed a path between node A and node C.

Most many-variable systems in the real world do have a time dimension, but it is a judgement call for the researcher whether it is worthwhile to model the network as

Generating Random Networks and Graphs. First Edition. A.C.C. Coolen, A. Annibale, and E.S. Roberts. © A.C.C. Coolen, A. Annibale, and E.S. Roberts 2017. Published in 2017 by Oxford University Press. DOI: 10.1093/acprof:oso/9780198709893.001.0001

a temporal network. As observed in [91], an important consideration is whether the timescales of topological evolution of the network are comparable to the timescales of the process being studied. Reference [91] uses the example of the Internet, where information propagates very fast, compared to the rate of new connections being made and broken. Hence, it is usually sufficient to model the Internet as a fixed network.

Temporal networks will in many cases be very sparse and hence difficult to meaningfully statistically analyze. A compromise can be to draw a temporal network as a flat projection (with edges annotated with the time points at which interactions occur) to give an overall sense of the connectivity of the network – and display this alongside a time-series type of chart (see Figure 10.1 for an example). A temporal network can also be approximated by a weighted network, in the sense that we may choose to record only the frequency of interaction between pairs of nodes, and represent these frequencies as weights (or interaction strengths). A temporal network may have a cyclic property, where the interaction patterns depend on the time of day or time of year (examples would include metabolic networks and ecological networks).

Email and telephone records are especially suited for modelling with temporal networks, since they come naturally time-stamped, and time is a crucial variable. A well-known example is the Enron email database [60]. Modelling this as a temporal network gives a fascinating quantitative insight into the differences between normal and stressed communication patterns. The authors of [60] noted that the crisis period was characterized by people communicating more intensively with a broader range of colleagues, bypassing normal hierarchy.

It will be obvious that most of our conventional topological measurements, like shortest path length, diameter, clustering coefficient, degree distribution and so on, will now all become dependent on time. Temporal networks therefore often require new tools and an adapted vocabulary. For example, instead of speaking about the graph being connected, for a temporal network it makes more sense to speak about the *reachability ratio* of a network, which gives the average fraction of vertices that can be reached from node i within some allowed time period [89].

A random model of such a network can be created by defining update rules (for example, rewirings) that code, deterministically or partially probabilistically, how the links change at every time step. Conceptually, the mathematical equations that would describe this situation would be very similar to those describing our earlier Markov Chain Monte Carlo (MCMC) graph randomization processes; the main difference is that we now design the evolution rules to mimic some evolution process observed in the real world (which need not be Markovian, and will usually not equilibrate), instead of designing rules to propel the system towards a specific equilibrium state.

ERGM formulations for generating temporal networks. A relatively simple approach to defining and generating random temporal networks is to construct a time-dependent version of the exponential random graph model, i.e. to generate a *sequence* of networks – one network for each time step – based on a joint probability distribution that is (normally) derived from the maximum-entropy construction, subject to soft constraints in the form of averages. If $\mathbf{A}_T \in G$ for some set G of N-node graphs, for all $t = 1 \ldots T$, and upon writing the joint probability to observe graphs $\{\mathbf{A}(1), \mathbf{A}(2), \ldots, \mathbf{A}(T{-}1), \mathbf{A}(T)\}$

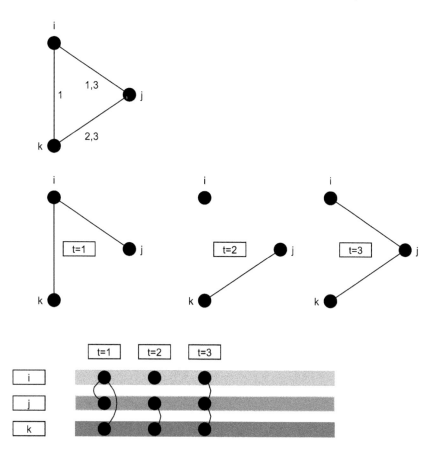

Fig. 10.1 Here we show three alternative but equivalent ways of drawing temporal networks, with stationary node sets but time-dependent links, for time steps $t \in \{1, 2, 3\}$. In the top row, we indicate for each link at which of these time steps it is active. In the middle row, we draw explicitly the instantaneous networks at each time step. In the bottom row, we give a similar representation, but now deformed to create a time line. Clearly, visualization of temporal networks becomes increasingly problematic for large graphs and larger durations.

at time steps $t = 1, 2, \ldots, T$, respectively, as $p(\mathbf{A}(T), \mathbf{A}(T-1), \ldots, \mathbf{A}(2), \mathbf{A}(1))$ (with associated marginal and conditional probabilities defined in the standard manner), we would find within the exponential random graph model (ERGM) picture:

$$p(\mathbf{A}(T), \ldots, \mathbf{A}(1)) = \frac{1}{Z(\boldsymbol{\lambda})} e^{\sum_{\mu=1}^{K} \lambda_\mu \Omega_\mu(\mathbf{A}(T), \ldots, \mathbf{A}(1))} \tag{10.1}$$

$$Z(\boldsymbol{\lambda}) = \sum_{\mathbf{A}(T) \in G} \cdots \sum_{\mathbf{A}(1) \in G} e^{\sum_{\mu=1}^{K} \lambda_\mu \Omega_\mu(\mathbf{A}(T), \ldots, \mathbf{A}(1))} \tag{10.2}$$

Here $\boldsymbol{\lambda} = \{\lambda_1, \ldots, \lambda_K\}$ represent Lagrange parameters associated with our K soft constraints, which are to be solved in the usual manner from the K equations

$$\forall \mu = 1 \ldots K: \quad \Omega_\mu = \langle \Omega_\mu(\mathbf{A}(T), \ldots, \mathbf{A}(1)) \rangle \qquad (10.3)$$

with the average calculated over the joint probabilities (10.1). The numbers $\{\Omega_1, \ldots, \Omega_K\}$ are the values that we need to find on average in our random temporal graphs for the (now multi-time) observables $\Omega_\mu(\mathbf{A}(T), \ldots, \mathbf{A}(1))$. These averages usually relate adjacency matrix entries at different time steps[1]. Without further assumptions or domain knowledge, it is not always self-evident which properties should be constrained and how; this depends on the temporal structure of the data set to be modelled.

Markovian ERGM formulation for generating temporal networks. The previous formulation has the important benefits of being precise (we work with explicit formulae for the microscopic probabilities of generating individual temporal graphs) and of being statistically unbiased (via the maximum-entropy argument). However, the expressions are complex, and solving the equations for the Lagrange parameters $\boldsymbol{\lambda} = (\lambda_1, \ldots, \lambda_K)$ can be painful. One would ideally like to retain the benefit of statistical precision of ERGMs within a computationally convenient and practical framework. Within the so-called temporal ERGM (tERGM) framework, commonly attributed to [86], one assumes that the evolution $\mathbf{A}(1) \to \mathbf{A}(2) \to \ldots$ of the evolving network proceeds as a Markov process. Specifically, it is assumed that

$$p(\mathbf{A}(T), \mathbf{A}(T-1), \ldots, \mathbf{A}(2), \mathbf{A}(1)) = \left[\prod_{t=2}^{T} p(\mathbf{A}(t)|\mathbf{A}(t-1)) \right] p(\mathbf{A}(1)) \qquad (10.4)$$

For this last expression (assuming Markovian evolution) to be consistent with the generic ERGM form (10.1) (assuming maximum entropy, given soft constraints) of the joint probabilities, we see that we must have constraining observables of the form

$$t = 1: \qquad \Omega_1(\cdot) = H(\mathbf{A}(1)), \qquad (10.5)$$
$$t = 2 \ldots T: \qquad \Omega_t(\cdot) = H(\mathbf{A}(t), \mathbf{A}(t-1)) \qquad (10.6)$$

Thus, we have to impose T observables as soft constraints. The first controls the initial probabilities $p(\mathbf{A}(1))$; the remaining $T-1$ constraints control the graph transition probabilities of subsequent time steps. We then find that the factors of (10.1) can be written as

$$p(\mathbf{A}(1)) = \frac{e^{\lambda_1 H(\mathbf{A}(1))}}{Z(\lambda_1)}, \qquad Z(\lambda) = \sum_{\mathbf{A}' \in G} e^{\lambda H(\mathbf{A}')} \qquad (10.7)$$

$$p(\mathbf{A}(t)|\mathbf{A}(t-1)) = \frac{e^{\lambda_t H(\mathbf{A}(t), \mathbf{A}(t-1))}}{Z(\lambda_t, \mathbf{A}(t-1))}, \qquad Z(\lambda, \mathbf{A}) = \sum_{\mathbf{A}' \in G} e^{\lambda H(\mathbf{A}', \mathbf{A})} \qquad (10.8)$$

The Lagrange parameters are to be solved from

$$t = 1: \qquad \Omega_1 = \sum_{\mathbf{A}(1) \in G} \frac{H(\mathbf{A}(1)) \, e^{\lambda_1 H(\mathbf{A}(1))}}{Z(\lambda_1)} \qquad (10.9)$$

---

[1]If we choose observables $\Omega_\mu(\mathbf{A}(T), \ldots, \mathbf{A}(1))$ that each depend only on adjacency matrix entries at *one* time step, the ERGM measure (10.1) will factorize over time, and all adjacency matrices $\{\mathbf{A}(t)\}$ become statistically independent. In this case, we simply find independent ERGM ensembles at each time step, and there is no longer any point in thinking in terms of temporal networks.

$$t = 2\ldots T: \quad \Omega_t = \sum_{\mathbf{A}(1)\ldots\mathbf{A}(t)\in G} H(\mathbf{A}(t), \mathbf{A}(t-1)) \left[\prod_{t'=2}^{t} \frac{e^{\lambda_{t'} H(\mathbf{A}(t'), \mathbf{A}(t'-1))}}{Z(\lambda_{t'}, \mathbf{A}(t'-1))} \right] \frac{e^{\lambda_1 H(\mathbf{A}_1)}}{Z(\lambda_1)}$$

$$(10.10)$$

Choice of observables to act as constraints in tERGMs. Note that our constraints involve only two functions, namely $H(\mathbf{A})$, whose average value at time $t = 1$ is prescribed, and $H(\mathbf{A}', \mathbf{A})$, whose average value is prescribed for joint graph realizations $(\mathbf{A}', \mathbf{A}) = (\mathbf{A}(t), \mathbf{A}(t-1))$ at times $t = 2\ldots T$. If we choose $H(\mathbf{A}', \mathbf{A})$ to be independent of $\mathbf{A}$, we remove the statistical connections between graphs at different times, and return to having separate ERGM formulations for the graph probabilities at each individual time step[2]. More sensible and interesting choices explored in [86] include, e.g. observables that quantify the stability and the transitivity of the graph evolution process. Measuring the stability of the process can be done by calculating how likely it is that a link (or the absence of a link) persists between subsequent time steps:

$$H(\mathbf{A}', \mathbf{A}) = \sum_{ij} \delta_{A'_{ij}, A_{ij}} \tag{10.11}$$

The transitivity measures whether two sides of a triangle being present at time $t - 1$, i.e. having $A(t-1)_{ij} = A(t-1)_{jk} = 1$, makes it more likely that the third side of the triangle will be observed at time t, i.e. that $A(t)_{ik} = 1$. For example, in social networks, two people with a mutual acquaintance are more likely to meet. If set G contains only simple graphs, this can be quantified and built into the graph generation process via the choice

$$H(\mathbf{A}', \mathbf{A}) = \frac{\sum_{ijk}(1-\delta_{ij}) A'_{ik} A_{ij} A_{jk}}{\sum_{ijk}(1-\delta_{ij}) A_{ij} A_{jk}} \tag{10.12}$$

(with the proviso that we set $H(\mathbf{A}', \mathbf{A}) = 0$ if the denominator of this formula is zero). Further worked examples and case studies are provided in [86] and citing articles.

Exercise 10.1 Calculate for $t \in \{2, \ldots, T\}$ the conditional probabilities $p(\mathbf{A}(t)|\mathbf{A}(t-1))$ in the tERGM equations (10.8), including the normalizing constants, for the case where we constrain the transitivity of the process, i.e. for $H(\mathbf{A}', \mathbf{A}) = \sum_{ij} \delta_{A'_{ij}, A_{ij}}$, and where G is the set of simple nondirected N-node graphs. Show that these probabilities can be written as

$$p(\mathbf{A}(t)|\mathbf{A}(t-1)) = \prod_{i<j} \left[\frac{1}{2} + \frac{1}{2}\tanh(\lambda_t)(2\delta_{A_{ij}(t), A_{ij}(t-1)} - 1) \right] \tag{10.13}$$

---

[2] In this simplified case, the graph probabilities $p(\mathbf{A}_t)$ at times $t = 2\ldots T$ need still not be identical, unless we choose $\Omega_2 = \Omega_3 = \ldots = \Omega_T$.

The above Markovian formulation of temporal ERGMs may not always be appropriate. In real temporal networks, timing and duration of contacts is rarely uniform, but often found to be *bursty* [177] – with long quiet periods interspersed with periods of frequent contacts. This is not well captured by assuming that the time evolution of the network proceeds in independent steps. Temporal inhomogeneity, together with other similar correlations and inhomogeneities in real temporal systems, and their intrinsic greater complexity, means that the study of how to generate effective null models of temporal graphs is still developing. A more detailed review can be found in [91].

Sampling algorithms for tERGMs. For numerical algorithms aimed at generating temporal graphs from tERGM ensembles, formulation (10.8) is not the most suitable, and we would instead use

$$p(\mathbf{A}(T), \ldots, \mathbf{A}(1)) = \frac{1}{Z(\boldsymbol{\lambda})} e^{\lambda_1 H(\mathbf{A}(1)) + \sum_{t=2}^{T} \lambda_t H(\mathbf{A}(t), \mathbf{A}(t-1))} \tag{10.14}$$

since normalizing constants always drop out of MCMC sampling processes. The simplest realization of an MCMC sampling process for tERGMs is the direct generalization of the one studied in Section 6.2 for time-independent ERGMs, and is again based on individual 'link flip' moves. The only difference compared to the earlier formulation is that now we need to draw at random not only a candidate node pair (i, j), but also a time label $t \in \{1, \ldots, T\}$. The relevant move set is thus $\Phi = \{F_{ij,t}\}$ of *time-specific* link flips, with $i, j \in \{1, \ldots, N\}$ if $G \subseteq \{0,1\}^N$ and $t \in \{1, \ldots, T\}$, where

$$F_{ij,t} A_{ij}(t) = 1 - A_{ij}(t) \tag{10.15}$$
$$F_{ij,t} A_{rs}(t') = A_{rs}(t') \quad \text{if} \quad (r, s, t') \neq (i, j, t) \tag{10.16}$$

Thus, $F_{ij,t}$ creates a link (i, j) at time t if this link is absent from $\mathbf{A}(t)$, and removes this link if it is present. The move is auto-invertible; i.e. $F_{ij,t}^2 = \mathbf{I}$ for all (i, j, t), and one can go from any initial temporal graph $\{\mathbf{A}(1), \ldots, \mathbf{A}(T)\}$ to any final temporal graph $\{\mathbf{A}'(1), \ldots, \mathbf{A}'(T)\}$ by executing a sequence of elementary moves $F_{ij,t}$. Hence, the set Φ is ergodic in the space of temporal graphs with T time steps. We also need to define the corresponding quantity

$$\Delta_{ij,t}(\mathbf{A}) = \lambda_1 \big[H(F_{ij,t} \mathbf{A}_1) - H(\mathbf{A}_1) \big]$$
$$+ \sum_{\mu=2}^{T} \lambda_\mu \big[H(F_{ij,t} \mathbf{A}(\mu), F_{ij,t} \mathbf{A}(\mu-1)) - H(\mathbf{A}(\mu), \mathbf{A}(\mu-1)) \big] \tag{10.17}$$

This definition simplifies, in terms of the ordinary *time-independent* link flips F_{ij}, to

$$\Delta_{ij,1}(\mathbf{A}) = \lambda_1 \big[H(F_{ij} \mathbf{A}(1)) - H(\mathbf{A}(1)) \big] + \lambda_2 \big[H(\mathbf{A}(2), F_{ij} \mathbf{A}(1)) - H(\mathbf{A}(2), \mathbf{A}(1)) \big]$$
$$\tag{10.18}$$

$$\Delta_{ij,t>1}(\mathbf{A}) = \lambda_t \big[H(F_{ij} \mathbf{A}(t), \mathbf{A}(t-1)) - H(\mathbf{A}(t), \mathbf{A}(t-1)) \big]$$
$$+ \lambda_{t+1} \big[H(\mathbf{A}(t+1), F_{ij} \mathbf{A}(t)) - H(\mathbf{A}(t+1), \mathbf{A}(t)) \big] \tag{10.19}$$

The adaptation to Markovian temporal ERGMs of the standard Metropolis–Hastings MCMC algorithm for ordinary ERGMs is given by Algorithm 7, where we made the

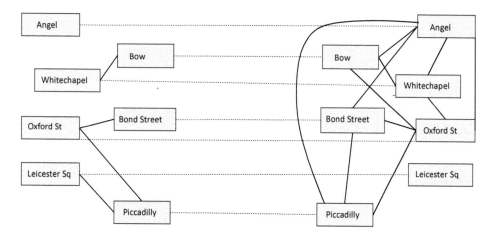

Fig. 10.2 A multiplex network has distinct layers of connections on a given set of nodes. This figure gives a multiplex graph of direct transport connections between London locations. In the left-hand network, two nodes (locations) are linked if there is a direct underground line between them. In the right-hand network, two nodes (locations) are connected if there is a major bus route linking them. The dotted lines connect nodes to their counterparts in the other layer. This is especially useful if the network is drawn without visible node labels.

choice $H(\mathbf{A}_1) = \sum_{i<j} A_{ij}(1)$ and $\Omega_1 = L$, so that the temporal graph at the first time step has a soft-constrained number of links L, and we left unspecified the $T-1$ observables $H(\mathbf{A}(t), \mathbf{A}(t-1)), \forall\, t = 2, \ldots, T$ of networks at two successive time steps.

10.2 Multiplex graphs

Overview: Multiplex networks or graphs are defined in distinct layers, where each layer represents a qualitatively different kind of binary relationship, but each such relationship is formulated for the same common underlying node set.

Motivation and definitions. A multiplex network is a network constructed to consist of layers, where, similar to the situation with temporal graphs, every layer characterizes a different type of pairwise relationship within a common set of nodes. One can see temporal graphs as examples of multiplex networks in which we choose each layer to represent a specific time point of an evolving graph. Here, in contrast, we allow the layers to represent anything we wish. Because the layers can now refer to any type of relationship, one will generally not model such systems using a Markovian assumption to describe link probabilities of successive layers, and it is no longer necessary for any hard topological constraints of different layers to be identical. For example, if the nodes represented people, the layers may correspond to family relationships, professional relationships and social relationships. Alternatively, we could use a multiplex network to model the infrastructure supporting movement of people in a city, with different layers representing different modes of public transport (underground rail, surface rail, buses). The main benefit of defining a multiplex network is the retention of extra information about the different *types* of interactions between two nodes. This is a more detailed representation than projecting to a single variable that summarizes every interaction between the given two nodes. The cost of this extra refinement is substantial extra complexity. In many cases (e.g. cell biology), we may not be able to accurately distinguish the nature of every interaction – hence, the simpler 'flat' network may be more robust, in the sense of having a lower proportion of its data dependent on the judgement of the experimentalist.

To keep the notation compact, we may write a multiplex network of M layers with N nodes each as $\vec{A} = (A^1, A^2, \ldots, A^M)$, where $A^\alpha \in G_\alpha \subseteq \{0, 1\}^N$ for all $\alpha = 1 \ldots M$. The entries of the adjacency matrices now give layer-specific wiring information:

$$A_{ij}^\alpha = 1: \text{ node } i \text{ connects to node } j \text{ in layer } \alpha \tag{10.20}$$
$$A_{ij}^\alpha = 0: \text{ node } i \text{ does not connect to node } j \text{ in layer } \alpha \tag{10.21}$$

An ensemble of random multiplex networks is specified in full by giving the sets $\{G_\alpha\}$ (which unlike those for temporal networks need not be identical, if for instance we wish to impose different constraints on the topologies of the different layers) and a joint probability distribution $p(\vec{A})$ describing the link statistics in all layers.

Successful applications of multiplex networks include transport and communication infrastructures, analysis of epidemics (by modelling more precisely the impacts of different types of interactions between communities [162]) or systemic risk in financial markets. Systemic risk is a term used to describe the risk resulting from the possibility of wholesale collapse of a financial system due to a cascade of interconnected institutions failing. It can be more precisely understood if, rather than modelling the connectivity of a financial market as a conventional graph, one separates interactions among institutions by type (e.g. loan obligations versus insurance obligations, and so on). A fuller survey of the frontiers of research in interconnected networks can be found in [22] and [55]. The authors of [133] present an approach for generating a multiplex network, based on an adaptation of the preferential attachment growth model. Table 10.1 provides references and brief outlines of some common algorithms for generating multiplex networks.

ERGM formulations for generating multiplex networks. Within the transparent and convenient ERGM picture, as described for ordinary graphs in Chapter 4, the joint distribution $p(\vec{A})$ is defined by maximizing the Shannon entropy, subject to a chosen set of K soft constraints (which could refer to observations in a single layer, or in multiple layers), resulting in

$$p(\vec{A}) = \frac{1}{Z(\boldsymbol{\lambda})} e^{\sum_{\mu=1}^{K} \lambda_\mu \Omega_\mu(\vec{A})} \tag{10.22}$$

$$Z(\boldsymbol{\lambda}) = \sum_{A^1 \in G_1} \cdots \sum_{A^M \in G_M} e^{\sum_{\mu=1}^{K} \lambda_\mu \Omega_\mu(\vec{A})} \tag{10.23}$$

Here $\boldsymbol{\lambda} = \{\lambda_1, \ldots, \lambda_K\}$ represent Lagrange parameters associated with our K soft constraints, which are to be solved in the usual manner from

$$\forall \mu = 1 \ldots K: \qquad \Omega_\mu = \langle \Omega_\mu(\vec{A}) \rangle \tag{10.24}$$

DESCRIPTION	REFERENCES
A new node appears in each layer simultaneously, and forms m links within each layer according to the Barabási–Albert preferential attachment algorithm. The degrees of nodes in the layers will be correlated, because the age of the nodes will be the same in every layer.	[98]
A new node appears in each layer simultaneously, and forms m links, preferentially weighted by a linear combination of the degree of the target node in each layer, with links potentally appearing at different times in different layers	[98, 133]
Nodes are born in each layer simultaneously, forming links based on a nonlinear preferential attachment rule.	[134]
A new layer is created at every time step.	[54, 134]
Correlations between layers.	[111, 78] [114, 123]

Table 10.1 Table of some common algorithmic approaches to generating multiplex networks, see [22] for a more detailed review of the field. There are infinite variations on the basic themes of preferential attachment and stub-joining constructions that can be applied to multiplex graphs, but (in contrast to ensemble-based approaches) they all tend to suffer from the handicap that one usually has no access to the precise graph probabilities $p(\vec{\mathbf{A}})$, and therefore one cannot quantify or control the built-in biases.

with the average calculated over the probabilities (10.22). The numbers $\{\Omega_1, \ldots, \Omega_K\}$ are the values we need to find on average in our random multiplex graphs for the observables $\Omega_\mu(\vec{\mathbf{A}})$. Several instances of such ensembles are described in detail in [20].

For example, suppose that we wished to constrain only the average total number of links L^α in each layer of a nondirected multiplex graph. We would now have $K = M$, and the relevant observables would involve only the link variables of one layer at a time, $\Omega_\mu(\vec{\mathbf{A}}) \to \Omega_\alpha(\mathbf{A}^\alpha) = \sum_{i<j} A_{ij}^\alpha$. Insertion into (10.22) would give factorizing joint probabilities $p(\vec{\mathbf{A}})$, describing uncorrelated graphs in each layer,

$$p(\vec{\mathbf{A}}) = \prod_{\alpha=1}^{M} p_\alpha(\mathbf{A}^\alpha), \qquad p_\alpha(\mathbf{A}^\alpha) = \prod_{i<j} \frac{e^{\lambda_\alpha A_{ij}^\alpha}}{1 + e^{\lambda_\alpha}} \qquad (10.25)$$

and the Lagrange parameters are to be solved from $\sum_{\mathbf{A}^\alpha \in G_\alpha} \Omega_\alpha(\mathbf{A}^\alpha) p_\alpha(\mathbf{A}^\alpha) = L^\alpha$. This case thereby reduces to having simply M independent Erdös–Rényi ensembles (3.1), one for each layer, and each with its own layer-specific link probability $p_\alpha = 2L^\alpha/N(N-1)$. Here, and in fact in all cases where the chosen observables refer to link variables of one specific layer only, one always finds distributions $p(\vec{\mathbf{A}})$ that factorize over the M layers, and the concept of multiplex networks becomes pointless since we can simply treat each layer separately using conventional tools.

Correlations between layers and multilinks. The interesting and more relevant scenarios are those where at least some of the observables $\Omega_\mu(\vec{\mathbf{A}})$ contain variables from multiple layers, so that $p(\vec{\mathbf{A}})$ no longer factorizes over α. We may wish to incorporate correlations between links in different layers, or between more complex topological motifs (e.g. triangles). Suppose that the multiplex network represents transport routes. We may observe, for example, that whenever there is a flight route between cities, there is always also a direct train route and a direct coach route. Hence, we would conclude there are correlations within the layers of the multiplex. The authors of [20] propose to simplify the bookkeeping of link correlations in nondirected multiplex graphs by defining *multilinks*. A multilink is a vector $\vec{m}_{ij}(\vec{\mathbf{A}}) = (A_{ij}^1, \ldots, A_{ij}^M) \in \{0,1\}^M$ that collects all link variables in $\vec{\mathbf{A}}$ that relate to the node pair (i,j). Specifying the $\frac{1}{2}N(N-1)$ multilinks is just a slightly different way of grouping the entries of $\vec{\mathbf{A}}$, but it can be useful to define link correlations between layers. For example, we can build such correlations into multiplex ERGMs (10.22) by prescribing as the average number $L_{\vec{m}}$ of each flavour of multilink $\vec{m}$:

$$\text{for all } \vec{m} \in \{0,1\}^M : \quad \Omega_{\vec{m}}(\vec{\mathbf{A}}) = \sum_{i<j} \delta_{\vec{m}, \vec{m}_{ij}(\vec{\mathbf{A}})} \tag{10.26}$$

Clearly, $\sum_{\vec{m}} \Omega_{\vec{m}}(\vec{\mathbf{A}}) = \frac{1}{2}N(N-1)$. Inserting this specific choice of observables into equation 10.22 gives the following ensemble:

$$p(\vec{\mathbf{A}}) = \frac{1}{Z[\{\vec{m}\}]} e^{\sum_{i<j} \lambda[\vec{m}_{ij}(\vec{\mathbf{A}})]} \tag{10.27}$$

$$Z[\{\vec{m}\}] = \sum_{\mathbf{A}^1 \in G_1} \cdots \sum_{\mathbf{A}^M \in G_M} e^{\sum_{i<j} \lambda[\vec{m}_{ij}(\vec{\mathbf{A}})]} \tag{10.28}$$

and the 2^M Lagrange parameters $\{\lambda[\vec{m}]\}$ are to be solved from

$$\text{for all } \vec{m} \in \{0,1\}^M : \quad L_{\vec{m}} = \left\langle \sum_{i<j} \delta_{\vec{m}, \vec{m}_{ij}(\vec{\mathbf{A}})} \right\rangle \tag{10.29}$$

One observes that the distribution $p(\vec{\mathbf{A}})$ now factorizes over all node pairs (i,j), and can be simplified to

$$p(\vec{\mathbf{A}}) = \prod_{i<j} \frac{e^{\lambda[(A_{ij}^1, \ldots, A_{ij}^M)]}}{\sum_{\vec{m}' \in \{0,1\}^M} e^{\lambda[\vec{m}']}} \tag{10.30}$$

We can now calculate the Lagrange parameters $\{\lambda[\vec{m}]\}$ explicitly, and find $\lambda[\vec{m}] = \log L_{\vec{m}}$ (modulo irrelevant $\vec{m}$-independent additive constants, that will drop out of our equations). As a consequence, and upon using the identity $\sum_{\vec{m}} e^{\lambda[\vec{m}]} = \sum_{\vec{m}} L_{\vec{m}} = \frac{1}{2}N(N-1)$, our multiplex graph probabilities simplify further to

$$p(\vec{\mathbf{A}}) = \prod_{i<j} \frac{L_{(A_{ij}^1,\dots,A_{ij}^M)}}{\frac{1}{2}N(N-1)} \tag{10.31}$$

The above approach allows us to impose within the ERGM formalism a target distribution of multilinks. However, it may prove onerous to enumerate all possible multilinks for graphs with many layers.

Sampling algorithms for ERGMs of multiplex graphs. Sampling algorithms for generating multiplex graphs with the ERGM probabilities from (10.22) are largely identical to those inspected earlier for temporal graphs. One replaces all summations over time indices $t = 1 \dots T$ in the latter by summations over layer indices $\alpha = 1 \dots M$. The relevant move set is now $\Phi = \{F_{ij,\alpha}\}$ of *layer-specific* link flips, with $i, j \in \{1, \dots, N\}$ if $G \subseteq \{0,1\}^N$ and $\alpha \in \{1, \dots, M\}$, where

$$F_{ij,\alpha} A_{ij}^{\alpha} = 1 - A_{ij}^{\alpha} \tag{10.32}$$
$$F_{ij,\alpha} A_{rs}^{\beta} = A_{rs}^{\beta} \quad \text{if} \quad (r,s,\beta) \neq (i,j,\alpha) \tag{10.33}$$

As always, we introduce variables to denote the change in the log probability of graphs caused by our elementary moves, which here take the form

$$\Delta_{ij,\alpha}(\vec{\mathbf{A}}) = \sum_{\mu=1}^{K} \lambda_\mu \left[\Omega_\mu(F_{ij,\alpha}\vec{\mathbf{A}}) - \Omega_\mu(\vec{\mathbf{A}}) \right] \tag{10.34}$$

The adaptation to nondirected multiplex ERGMs with M layers and N nodes of the standard Metropolis–Hastings MCMC algorithm for ordinary ERGMs can then be written in the form of algorithm 8.

10.3   Networks with labelled nodes

Overview: Real-world networks often exhibit structural features that can be attributed to their nodes having intrinsic characteristics that are not of a geometric nature. One example is the stochastic block model, which can be extended in various ways to provide realistic models of community structure in graphs. Other examples are graphs where individual nodes have a latent (i.e. hidden) intrinsic fitness. All these models are described by ensembles with structural similar equations.

generalizations of stochastic block models. Large complex systems are more easily understood if they can be described in terms of smaller connected subsystems (modules, groups or families). For example, in social networks, identifying the household as the natural building block of society makes analysis simpler, and is meaningful. In other applications, groupings can also be identified, but the same structure could have been just as accurately described as a correlation between some attributes of connected nodes. For example, people whose surnames begin with the letter 'T' may find themselves adjacent on lists of names, but are not a group in any functional way. Being in a functional group may or may not correspond to some topological signature. There is a distinction between choosing to use a null model with a certain group structure, and seeking to create a model of the actual functional modules of the system. In Chapter 4, we introduced a simple stochastic block model, where the nodes were divided into non-overlapping families, and soft constraints specified the connectivity within a family and between families. Some of the limitations of this model include the assumption of non-overlapping communities and the assumption of independence of links. Nonetheless, this model is easy to analyze, can be adapted to the desired parameters for inter- and intra-community connectivity, and gives a useful benchmark for community detection algorithms.

A natural extension of the stochastic block model is to define a degree distribution for the number of links that each node has, within a community and between communities. One benefit of the stochastic block model representation is that it is possible to take a multi-scale view of the system – binary connections between individual nodes – or 'zoom out' to view the binary or weighted connections between communities. Mixed membership stochastic block models are presented in [2]. Vertices can belong to multiple groups, and two vertices are more likely to be connected if they have more than one group in common. A correction may be desired, in order to counteract the possibly unphysical implied property that the area of overlap between two communities should have a higher average density of edges than an area that falls in just a single community. This construction can be modelled via a simple modification of the standard stochastic block model (SBM) probabilities (4.88). Let M be the number of desired communities in a simple nondirected N-node graph. We assign to each node i a binary vector $\vec{q}_i = (q_{i1}, \ldots, q_{iM})$, in which each $q_{im} \in \{0,1\}$ denotes whether (1) or not (0) node i belongs to community m. We can then define the following probabilities for the adjacency matrices:

$$p(\mathbf{A}) = \prod_{i<j} \left[P_{\vec{q}_i, \vec{q}_j} \delta_{A_{ij},1} + \left(1 - P_{\vec{q}_i, \vec{q}_j}\right) \delta_{A_{ij},0} \right] \qquad (10.35)$$

The function $P_{\vec{q}, \vec{q}'}$ must be monotonically increasing with the number of common nonzero entries in the two vectors $\vec{q}$ and $\vec{q}'$, e.g. $P_{\vec{q}, \vec{q}'} = [p_0 + p_1 \sum_{m \leq M} q_m q'_m]/N$, with $p_0, p_1 > 0$. All node pairs (i, j) with $i < j$ still have statistically independent links, so generating graphs according to (10.35) is trivial.

Determining the appropriate parameters and model to correspond to a given realworld network that shows evidence of structure in the form of weakly interaction sub-networks is essentially the problem of community detection. This is a very active and varied field of which a full survey is beyond the scope of this text.

Exercise 10.2 Protein–protein interaction networks are known to feature *hubs* (very highly connected nodes) and a dissortative structure. Hubs are comparatively unlikely to link to other hubs. This is believed to be an important topological feature which contributes to the stability of the network. Define a stochastic block model that has these features, and support your claim by calculating the correlations between the degrees of connected nodes.

While the stochastic block model is a useful tool for representing and analysing data, it is not always necessary to combine it with a network representation. For example, [56] and [109] identify the groups as colours, and require there to be no edges between nodes of the same colour, i.e. $P_{qq} = 0$ for all q in formula (4.88). This is a restatement of the famous four colour problem. The four colour theorem states that with four colours only it is possible to colour in a political map so that no two adjacent countries have the same colour. It can be transformed into a network problem by placing a node in every country, and linking the nodes if the countries share a border. This is an example of using the stochastic block model as an analytical trick, without it actually reflecting any underlying communities.

The examples discussed above are all described by conventional graphs. All nodes are in principle of the same 'type' (albeit distinguishable by their labels). One could also imagine stochastic block models for bipartite graphs, in which for instance an enty $A_{ij} = 1$ would be interpreted as species i eating food type j. The basic principle for working with the relational tables/stochastic block matrix would remain the same – to determine some meaningful groupings, and then to use that as a basis to define a plausible null model which respects the structural features observed in the network.

Graphs with hidden node labels. As before, we associate intrinsic parameters $\mathbf{x} = (x_1, \ldots, x_N)$ with all N vertices in a graph, and assume that these drive the pairwise connection probabilities. However, in contrast to the stochastic block model, these N parameters now need no longer have discrete values. They are sometime called *hidden* or *latent* variables (e.g. fitness, gender, age) to emphasize that they are intrinsic to the nodes, and cannot be observed directly from the topology of the network. It will not be a surprise that such graphs can again be modelled via ERGMs. In addition, there have also been algorithm-driven approaches, including a preferential attachment model driven by hidden variables.

The ERGM route for building hidden variable ensembles for simple nondirected graphs is easy. We maximize as usual the Shannon entropy (3.12) of the the probabilities $p(\mathbf{A})$, for simple nondirected N-node graphs with $\mathbf{A} \in G = \{0,1\}^{\frac{1}{2}N(N-1)}$, here subject to a soft constraint on the value of a feature of the form $\Omega(\mathbf{A}) = \sum_{i<j} A_{ij} H(x_i, x_j)$, for some a symmetric function $H(x, x')$ of the hidden node features. Inserting this choice for $\Omega(\mathbf{A})$ into (3.18) gives

$$p(\mathbf{A}|\mathbf{x}) = \frac{e^{\sum_{i<j} A_{ij} H(x_i,x_j)}}{Z(\mathbf{x})}, \qquad Z(\mathbf{x}) = \sum_{\mathbf{A} \in G} e^{\sum_{i<j} A_{ij} H(x_i,x_j)} \qquad (10.36)$$

The normalization factor evaluates as

$$Z(\mathbf{x}) = \prod_{i<j} \left[\sum_{A \in \{0,1\}} e^{AH(x_i,x_j)} \right] = \prod_{i<j} (1 + e^{H(x_i,x_j)}) \qquad (10.37)$$

Hence, we obtain

$$p(\mathbf{A}|\mathbf{x}) = \prod_{i<j} \left[\frac{e^{A_{ij} H(x_i,x_j)}}{1+e^{H(x_i,x_j)}} \right] = \prod_{i<j} \left[\frac{e^{H(x_i,x_j)}}{1+e^{H(x_i,x_j)}} \delta_{A_{ij},1} + \frac{1}{1+e^{H(x_i,x_j)}} \delta_{A_{ij},0} \right]$$

$$= \prod_{i<j} \left[W(x_i, x_j)\delta_{A_{ij},1} + (1 - W(x_i, x_j))\delta_{A_{ij},0} \right] \tag{10.38}$$

in which $W(x, x') = [1 + e^{-H(x,x')}]^{-1}$. The links are generated independently, but the probabilities $W(x_i, x_j)$ of individual links (i, j) being present need not be uniform. The function $W(x, x')$ controls the relation between the latent features of individual nodes and their likelihood of being connected, and thereby drives the topology.

The chosen form of $\Omega(\mathbf{A})$ does not obviously link to topological observables, although the analogy between (10.38) and expression (7.28) gives a clue. If we take the labels $\{x\}$ to be discrete, we can define the total number of links which connect nodes with a given combination of hidden attributes x and x', which is given by

$$L(x, x'|\mathbf{A}) = \sum_{i \neq j} A_{ij}\delta_{xx_i}\delta_{x'x_j}\left(1 - \frac{1}{2}\delta_{xx'}\right) \tag{10.39}$$

The last factor corrects for double counting when $x = x'$. Its ensemble average is

$$L(x, x') = \sum_{\mathbf{A} \in G} p(\mathbf{A}|x) \sum_{i \neq j} A_{ij}\left(1 - \frac{1}{2}\delta_{xx'}\right)\delta_{xx_i}\delta_{x'x_j}$$

$$= \sum_{\mathbf{A} \in G} \sum_{i \neq j} A_{ij}\left(1 - \frac{1}{2}\delta_{xx'}\right)\delta_{xx_i}\delta_{x'x_j} \prod_{r<s}\left[W(x_r, x_s)\delta_{A_{rs},1} + (1 - W(x_r, x_s))\delta_{A_{rs},0} \right]$$

$$= \sum_{i<j} \sum_{A_{ij} \in \{0,1\}} A_{ij}\left(1 - \frac{1}{2}\delta_{xx'}\right)\left[\delta_{xx_i}\delta_{x'x_j} + \delta_{xx_j}\delta_{x'x_i}\right]$$

$$\times \left[W(x_i, x_j)\delta_{A_{ij},1} + (1 - W(x_i, x_j))\delta_{A_{ij},0} \right]$$

$$= \sum_{i<j} \left(1 - \frac{1}{2}\delta_{xx'}\right)\left[\delta_{xx_i}\delta_{x'x_j} + \delta_{xx_j}\delta_{x'x_i}\right] W(x_i, x_j)$$

$$= \sum_{i \neq j} \left(1 - \frac{1}{2}\delta_{xx'}\right)\delta_{xx_i}\delta_{x'x_j} W(x_i, x_j)$$

$$= W(x, x')\left[\left(1 - \frac{1}{2}\delta_{xx'}\right)\sum_{ij}\delta_{xx_i}\delta_{x'x_j} - \frac{1}{2}\delta_{xx'}\sum_i \delta_{xx_i}\right] \tag{10.40}$$

Writing this in terms of $n(x) = \sum_i \delta_{x,x_i}$, the number of nodes with label x, gives

$$L(x, x') = W(x, x')\left[(1 - \frac{1}{2}\delta_{xx'})n(x)n(x') - \frac{1}{2}\delta_{xx'}n(x)\right]$$

$$= W(x, x')\left[n(x)n(x')(1 - \delta_{xx'}) + \frac{1}{2}n(x)[n(x) - 1]\delta_{xx'}\right] \tag{10.41}$$

As is clear already from (10.38), the factor $W(x, x')$ controls the fraction of potential node pairs (i, j) with $(x_i, x_j) = (x, x')$ that are connected.

It follows that one can generate maximum-entropy graphs that share with some observed N-node nondirected real-world network both the statistics of the latent variables and the statistics of links between each possible flavour of node pairs in the following way. One measures the numbers $\{n(x), L(x, x')\}$ for all (x, x'), and subsequently generates all links independently according to (10.38), with the choice

$$x \neq x': \quad W(x, x') = \frac{L(x, x')}{n(x)n(x')}, \qquad W(x, x) = \frac{L(x, x')}{\frac{1}{2}n(x)[n(x)-1]} \qquad (10.42)$$

The resulting protocol for generating graphs is summarized in the pseudocode given in Algorithm 6. In the case where the hidden variables are not discrete, one needs to carry out binning and smoothing in order to use the above formulae. This is particularly important if the real network from which the parameters are taken is not large.

Preferential attachment with fitness parameters. In [21] the Barabási–Albert preferential attachment scheme is adapted to allow for a fitness parameter, as summarized by Algorithm 25. The original preferential attachment algorithm was described in Section 8.2. For the hidden variable variant, at every step a new node is added, and it connects to a node i already present with a probability proportional to the fitness parameter x_i and the number of existing links that node i has. In the classic Barabási–Albert scheme, the attachment probability is weighted by the degree alone, which creates an unrealistically strong correlation between the age of the node and its number of links.
Stub joining with fitness parameters. A hidden variable variant of the configuration (or stub-joining) model is proposed in a series of papers by the authors of [168], where the hidden variables are defined as a properties of the links rather than of the nodes. Specifically, there are a number (say K) of allowed link colours. Each node is given a certain number of half-edges of each colour; these numbers are generated randomly, but according to an imposed specified joint probability $p(m_1, \ldots, m_K)$ that gives the likelihood that a node has m_1 half-edges of colour 1, m_2 half-edges of colour 2, and so on. In addition, a $K \times K$ *colour preference matrix* $\boldsymbol{T}$ is defined, such that we expect to observe $n_{ab} = \langle m_a \rangle T_{ab} \langle m_b \rangle$ edges in the final network that are coloured a on one side and coloured b on the other side (where $\langle m_a \rangle$ gives the mean number of stubs of colour a per node). Reference [168] then describes the stub-joining procedure more fully. As always with the configuration model, it is delicate to ensure that the networks that are generated are simple (no loops or double edges) and unbiased (i.e. statistically representative of the ensemble of allowed graphs). If we chose to make $\boldsymbol{T}$ a diagonal matrix (so that stubs can only join to stubs of the same colour), then we could view the network as having K coloured layers (i.e. the red layer consists of only red links between nodes, and so on). Hence, this is an alternative way of defining a *multiplex* graph (as was defined in Section 10.2); 'colours' would be rewritten as 'interaction types' – for example contacts via phone, email or face-to-face. However, since the colour layers are then fully disconnected and independent, the concept of multiplex graphs would be overkill – we would simply have K conventional graphs, and can treat those with standard tools.

More variations. The work of [38] and [77] further illuminates the interplay between degree distributions, degree–degree correlations and hidden variable correlations. Specifically, [38] disputes that preferential attachment is a natural mechanism to describe a network such as the Internet, because new nodes (webpages) do not know (or care) how many existing hyperlinks point to any particular webpage. The decision to link to an existing page should logically be driven by fitness (i.e. quality of content on the page). The authors of [38] reinforce their arguments by showing that a hidden variable scheme for generating networks can be tuned to closely reproduce the topology resulting from a preferential attachment scheme. Their network is constructed statically: each node i is allocated a fitness x_i, drawn at random from some given distribution $\rho(x)$. Some symmetric function $f(x_i, x_j)$ converts the fitness parameter into a connection probability. The authors of [38] note that if the fitness $\rho(x)$ follows a power-law distribution, then this will induce power-law degree distribution. Moreover, for an appropriate choice of the function $f(x_i, x_j)$, non-power-law fitness distributions can also result in power-law distributed degrees. Their example uses an exponential distribution of the fitness parameter, and linking two nodes if the sum of their fitness is greater than a threshold (i.e. $f(x_i, x_j) = \theta\,[x_i + x_j - z(N)]$, where θ is a Heaviside function).

Graph ensembles of the form (10.38) can generate many different types of tailored random graphs. Choosing the hidden variable assigned to a node to be its degree retrieves the ensemble with imposed degree–degree correlations. We can obtain graphs with modular or community structure, upon choosing x_i to represent a cluster label and setting $W(x, x') = a\delta_{x,x'} + b$, with $a, b > 0$. Geometric random graphs can be generated when the latent variables x_i are chosen to represent coordinates in physical space, and $W(x, x')$ is a decreasing function of the distance $|x - x'|$ between the physical locations of the relevant nodes.

10.4   Relations and connections between models

In the last few sections, we have discussed several routes for making random graph models more realistic by relaxing the assumptions that, for example, nodes are homogeneous (e.g. graphs with community structure), or that the links are homogeneous (e.g. the multiplex graphs), or that the connection probabilities are driven by purely topological observables (e.g. the hidden variable ensembles). For any given problem, there will be more than one valid and satisfactory way to model it. For example, if one's study is specifically focused on local properties of the individual (such as fitness), it would make sense to start the analysis from that viewpoint, and to then expect which topological properties appear in the network. In contrast, if we have substantial data on statistical features of the overall network, but incomplete knowledge of the features of the individual nodes (as in, e.g. protein interaction networks) – it would make more sense to concentrate on ensuring that the null models accurately capture the typically observed topology. The practitioner will also need to make a choice of which type of graph generation method to follow – a method driven by an algorithm, or a method designed to achieve a given probability measure. The advantages and disadvantages discussed in Chapters 4 and 8 equally apply to graphs on structured spaces.

Figure 10.3 illustrates some of the interconnections between the various approaches and definitions. For example, there is an arrow from 'Hidden variables' to 'Stochastic block model' because any random graph ensemble that can be described as a stochastic block model can be described via a hidden variable ensemble, where the hidden variables are drawn from a finite set of discrete values (i.e. colours). Similarly, graphs where the connection probability is driven by the metric distance between nodes, can

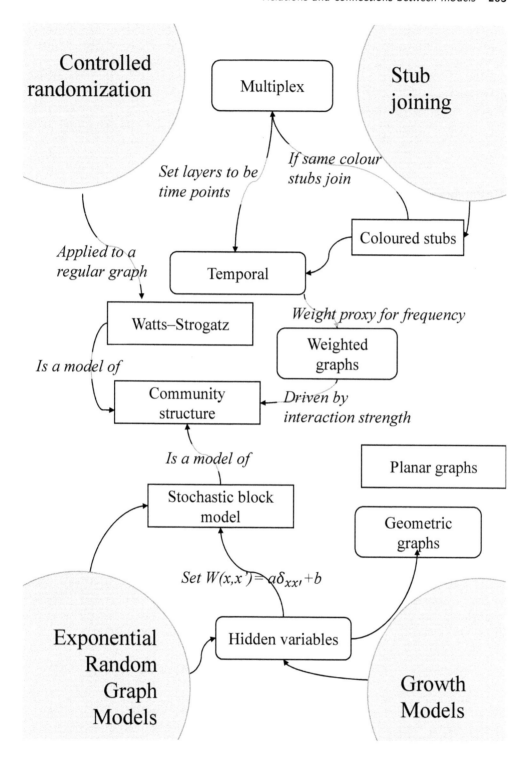

Fig. 10.3 Some commonalities and conceptual relationships between different approaches/terminologies for defining and generating graphs with an underlying structure.

also be expressed as a hidden variable ensemble, where the hidden variable is the coordinate of the node in metric space. Suppose we follow the scheme of [168], and give every node a certain number of coloured stubs, plus a probability of stubs connecting based on their colours. The same topology could be achieved by viewing the number of stubs of each colour as a 'hidden' property of each node. If the ensemble was defined so that stubs only connected to others of the same colour – allowing multiple links so long as they are of different colours – then we retrieve a multiplex graph. This underlines the flexibility of the basic techniques and available approaches.

10.5   Solutions of exercises

Exercise 10.1

We work out the relevant expression and use factorization over node pairs (i, j). For simple nondirected graphs, we will have $H(\mathbf{A}', \mathbf{A}) = 2 \sum_{i<j} \delta_{A'_{ij}, A_{ij}}$, and hence

$$p(\mathbf{A}(t)|\mathbf{A}(t-1)) = \frac{e^{2\lambda_t \sum_{i<j} \delta_{A_{ij}(t), A_{ij}(t-1)}}}{\sum_{\mathbf{A}' \in G} e^{2\lambda_t \sum_{i<j} \delta_{A'_{ij}, A_{ij}(t-1)}}}$$

$$= \prod_{i<j} \left[\frac{e^{2\lambda_t \delta_{A_{ij}(t), A_{ij}(t-1)}}}{\sum_{A' \in \{0,1\}} e^{2\lambda_t \delta_{A', A_{ij}(t-1)}}} \right] = \prod_{i<j} \left[\frac{e^{2\lambda_t \delta_{A_{ij}(t), A_{ij}(t-1)}}}{1 + e^{2\lambda_t}} \right]$$

$$= \prod_{i<j} \left[\frac{\delta_{A_{ij}(t), A_{ij}(t-1)} e^{\lambda_t} + (1 - \delta_{A_{ij}(t), A_{ij}(t-1)}) e^{-\lambda_t}}{2 \cosh(\lambda_t)} \right]$$

$$= \prod_{i<j} \left[\frac{1}{2} \delta_{A_{ij}(t), A_{ij}(t-1)} [1 + \tanh(\lambda_t)] + \frac{1}{2} (1 - \delta_{A_{ij}(t), A_{ij}(t-1)}) [1 - \tanh(\lambda_t)] \right]$$

$$= \prod_{i<j} \left[\frac{1}{2} + \frac{1}{2} \tanh(\lambda_t) (2 \delta_{A_{ij}(t), A_{ij}(t-1)} - 1) \right] \tag{10.43}$$

$\square$

Exercise 10.2

We need to construct a stochastic block model graph ensemble in which high-degree nodes connect preferably with low-degree ones. There are many ways to do this. Here we will construct an SBM of the form (10.35). We assign to each node i a binary variable $q_i \in \{0, 1\}$ indicating whether (1) or not (0) node i prefers to have many links. We then define the following probabilities for the adjacency matrices:

$$p(\mathbf{A}) = \prod_{i<j} \left[P_{q_i, q_j} \delta_{A_{ij}, 1} + (1 - P_{q_i, q_j}) \delta_{A_{ij}, 0} \right] \tag{10.44}$$

To build in the tendency for nodes with $q_i = 1$ to have more links, as well as the dissortativity, we need to choose four numbers $P_{q, q'}$ such that (i) connections are preferably formed between high-degree nodes (those with $q_i = 1$) and low-degree nodes (those with $q_j = 0$), and (ii) nodes with $q_i = 1$ should have more links than those with $q_i = 0$. This, e.g. can be achieved by the following choice, with $b > 0$:

$$P_{q,q'} = \frac{b}{N}(1 - \delta_{qq'}) = \begin{cases} b/N & \text{if } (q, q') = (1, 0) \text{ or } (0, 1) \\ 0 & \text{otherwise} \end{cases} \tag{10.45}$$

We will then find the following average degrees, where we use the shorthand $f = N^{-1}\sum_i \delta_{q_i,1}$ for the fraction of high-degree nodes:

$$\langle k_i(\mathbf{A}) \rangle = \sum_{j \neq i} \langle A_{ij} \rangle = \frac{b}{N} \sum_{j \neq i}(1 - \delta_{q_i, q_j})$$

$$q_i = 0: \quad \langle k_i(\mathbf{A}) \rangle = bf, \qquad q_i = 1: \quad \langle k_i(\mathbf{A}) \rangle = b(1 - f) \tag{10.46}$$

Nodes with $q_i = 1$ will thus indeed have a higher average degree than those with q_i if we choose $f < \frac{1}{2}$. The overall average degree is

$$\langle k \rangle = fb(1 - f) + (1 - f)bf = 2(1 - f)fb \tag{10.47}$$

According to (2.10), to verify that degree–degree correlations of connected nodes are indeed negative, we need to show that

$$\sum_{k,k'} W(k, k'|\mathbf{A})kk' < \left(\sum_{k,k'} W(k, k'|\mathbf{A})k\right)^2 \tag{10.48}$$

where the function $W(k, k'|\mathbf{A})$ is defined in (2.9). We will take N to be large, for simplicity, and calculate the average of $W(k, k'|\mathbf{A})$:

$$W(k, k') = \lim_{N \to \infty} \langle W(k, k'|\mathbf{A}) \rangle$$

$$= \lim_{N \to \infty} \frac{1}{N\langle k \rangle} \sum_{i,j=1}^{N} \left\langle \delta_{k,k_i(\mathbf{A})} A_{ij} \delta_{k',k_j(\mathbf{A})} \right\rangle \tag{10.49}$$

Its moments are

$$\sum_{k,k'} W(k, k')k = \lim_{N \to \infty} \frac{1}{\langle k \rangle N} \sum_i \langle k_i^2(\mathbf{A}) \rangle = \lim_{N \to \infty} \frac{1}{\langle k \rangle N} \sum_i \sum_{rs} \langle A_{ir} A_{is} \rangle$$

$$= \lim_{N \to \infty} \frac{1}{\langle k \rangle N} \sum_i \left[\sum_{rs} \langle A_{ir} \rangle \langle A_{is} \rangle + \sum_r [\langle A_{ir} \rangle - \langle A_{ir} \rangle^2] \right]$$

$$= \lim_{N \to \infty} \frac{1}{\langle k \rangle N} \sum_i \left[\sum_{rs} P_{q_i,q_r} P_{q_i,q_s} + \sum_r P_{q_i,q_r} + \mathcal{O}\left(\frac{1}{N}\right) \right]$$

$$= \frac{bf(1 - f)}{\langle k \rangle}(b + 2) = 1 + \frac{1}{2}b \tag{10.50}$$

$$\sum_{k,k'} W(k, k')kk' = \lim_{N \to \infty} \frac{1}{\langle k \rangle N} \sum_{ij} \langle k_i(\mathbf{A}) A_{ij} k_j(\mathbf{A}) \rangle = \lim_{N \to \infty} \frac{1}{\langle k \rangle N} \sum_{ij} \sum_{rs} \langle A_{ri} A_{ij} A_{js} \rangle$$

$$= \lim_{N\to\infty} \frac{2}{\langle k\rangle N} \sum_{i<j} \sum_{rs} \Big[\delta_{rj}\delta_{is}\langle A_{ij}\rangle + \delta_{rj}(1-\delta_{is})\langle A_{ij}\rangle\langle A_{js}\rangle$$

$$+ (1-\delta_{rj})\delta_{is}\langle A_{ri}\rangle\langle A_{ij}\rangle + (1-\delta_{rj})(1-\delta_{is})\langle A_{ri}\rangle\langle A_{ij}\rangle\langle A_{js}\rangle \Big]$$

$$= \lim_{N\to\infty} \frac{2}{\langle k\rangle N} \sum_{i<j} P_{q_i,q_j} \Big(1 + \sum_{s\neq i} P_{q_j,q_s} \Big) \Big(1 + \sum_{r\neq j} P_{q_r,q_j} \Big)$$

$$= \lim_{N\to\infty} \frac{1}{\langle k\rangle N} \sum_{i\neq j} P_{q_i,q_j} \Big(1 + \langle k_i(\mathbf{A})\rangle + \mathcal{O}\Big(\frac{1}{N}\Big) \Big) \Big(1 + \langle k_j(\mathbf{A})\rangle + \mathcal{O}\Big(\frac{1}{N}\Big) \Big)$$

$$= \lim_{N\to\infty} \frac{1}{\langle k\rangle N} \sum_{i\neq j} P_{q_i,q_j} + \lim_{N\to\infty} \frac{2}{\langle k\rangle N} \sum_{i\neq j} P_{q_i,q_j} \langle k_j(\mathbf{A})\rangle$$

$$+ \lim_{N\to\infty} \frac{1}{\langle k\rangle N} \sum_{i\neq j} P_{q_i,q_j} \langle k_i(\mathbf{A})\rangle\langle k_j(\mathbf{A})\rangle$$

$$= 1 + \lim_{N\to\infty} \frac{2}{\langle k\rangle N} \sum_{j} \langle k_j(\mathbf{A})\rangle^2 + \lim_{N\to\infty} \frac{1}{\langle k\rangle N} \sum_{i\neq j} P_{q_i,q_j} \langle k_i(\mathbf{A})\rangle\langle k_j(\mathbf{A})\rangle$$

$$= 1 + b + bf(1-f) \tag{10.51}$$

Hence

$$\sum_{kk'} W(k,k')kk' - \Big(\sum_{kk'} W(k,k')k \Big)^2 = 1 + b + bf(1-f) - \Big(1 + \frac{1}{2}b\Big)^2$$

$$= bf(1-f) - \frac{1}{4}b^2 = \frac{1}{4}b\Big[4f(1-f) - b\Big] \tag{10.52}$$

So provided we choose $b > 1$, we will always find that the degrees of connected nodes are negatively correlated (since $4f(1-f) \leq 1$ for all $f \in [0,1]$, and hence the graphs from this SBM are dissortative. $\qquad\square$

11
Applications of random graphs

Introduction. This chapter will review the graph generation techniques presented so far in the context of potential applications. We will refer to some specific case studies, as well as discussing wider considerations relevant to particular applications. The first step of analyzing data using networks techniques is to choose a suitable network representation of the system – to refine, simplify and visualize the raw data. Often the choice is made to reduce to the simplest possible such representation: nondirected networks with homogeneous nodes and binary links. In other problems, it may be necessary to work with more structured or detailed networks in order to retain more information. In each section below, we start by discussing suitable choices of network representation; where possible we cover typical observed topological features. We then focus on why and how random graph models have been used in this application.

11.1   Power grids

Overview: *It is necessary to have robust contingency plans to prevent local power outages triggering major cascading blackouts in a power grid. We review an example where the proposed defensive strategies were tested on tailored random graphs.*

The real systems to be modelled. Power grids are vulnerable to cascading power failure [99] – where failure of one element of the network pushes nearby components beyond their operating capacities, causing them also to fail. Examples of major blackouts include 160 million people in Brazil finding themselves without power in 1999 when a bolt of lightning struck a substation; $6 billion worth of damage in the United States in 2003 when a power line brushed against overgrown trees; 10 million Europeans without power in 2006 as a knock-on effect of a high-voltage line being switched off across the river Ems to allow a cruise ship to pass.

One defensive strategy is 'intentional islanding', which splits a network into subnetworks. The aim is for every subnetwork to have independent generating capacity, and relatively few links to other subnetworks. If the threat of a cascading power failure is identified, these links can be cut, and the subnetwork or 'island' can be isolated, thus containing the instability.

Network representations. The authors of [153] described how to construct and use relevant null models for power grids and their instabilities. Their aim was to study the effect of specific network modifications on the performance of network partitioning schemes – including the effect of increasing network size, and the effect of having to cover differently shaped regions.

Generating Random Networks and Graphs. First Edition. A.C.C. Coolen, A. Annibale, and E.S. Roberts. © A.C.C. Coolen, A. Annibale, and E.S. Roberts 2017. Published in 2017 by Oxford University Press. DOI: 10.1093/acprof:oso/9780198709893.001.0001

A power grid is typically modelled in this context as a weighted undirected graph on N nodes (see Figure 11.1). The nodes are heterogeneous – they may represent generators or subnetworks (loads). The weight w_{ij} corresponds here to the capacity of the transmission lines between nodes i and j, and is in this application strictly non-negative. In [153] the weights were defined more specifically as the ratio

$$w_{ij} = \frac{\text{number of grid lines between } i \text{ and } j}{\text{normalized geographic distance between } i \text{ and } j} \tag{11.1}$$

The network model of [153] was built to match the total number of power lines in the real network, as well as the average geographic edge length between any two nodes. They achieved the former aim by applying the configuration model. Specifically, each node was given a random number of half-edges summing to the correct total number of edges. Then these half-edges were paired and connected – permitting double links but not allowing self-links. It would be a natural extension to distribute the half-edges so that the degree of each node in the null-model network was the same as in the real network.

The paper does not specify whether the algorithm was restarted if a self-link was created, or whether the choice was made to backtrack a few steps only. In this case, however, this would not create inaccuracies, since the next step of the procedure was to apply an edge-swapping Markov Chain Monte Carlo (MCMC) algorithm, with the energy function (i.e. minus the logarithm of $p(\mathbf{A})$) specified as the total edge length (i.e. the sum of all the edge lengths in the network). Doing an edge swap MCMC would indeed erase any biases in the application of the configuration model.

The preliminary observations of [153] were that the real (Florida) power networks studied could be partitioned into 'islands' much more effectively than the null models. The authors suggested that this indicated that the real networks had correlations within them which were not present in the null model. The pertinent topological properties of the Florida power grid (and power grids in general) are explored in fuller detail by authors including [186, 139]. For example, the paper [186] criticises [153] for not distinguishing between different types of nodes (generators and substations), and proposes a heuristic correction which removes links between generators, and replaces them with extra links between pairs of generator and load nodes of similar length. The authors of [186] claim that a benefit is that this approach 'uncovers the organisational principles' – although, of course, this comes at the cost of no longer having a demonstrably unbiased null model.

11.2 Social networks

Overview: Network techniques are very well established in the social sciences, where the general culture is towards richer models which capture features such as community structure.

The real systems to be modelled. Networks in social sciences are very well established, although the terminology used can often differ from that in other network application domains. A $p^\star$ model [157] is a generalization of the exponential random graph

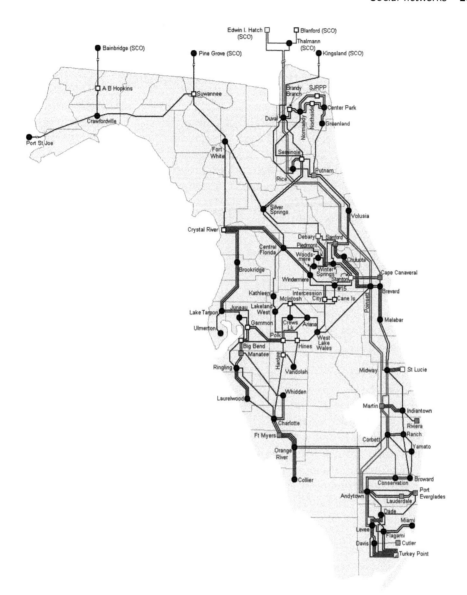

Fig. 11.1 Network representation of the Florida power grid from [186]. Generators are represented as squares and loads as ovals. Thicker lines represent multiple power lines.

models(ERGM) studied in Section 4.1. This is the most widely used class of random graph model in the social sciences. They are popular for their flexibility in being able to incorporate important network features and empirical observables.

The analysis of social networks is typically grounded in the tools of statistics and graph theory. This contrasts in presentation with the branches of network science rooted in condensed matter physics. However, since it is now possible to study very large social networks – for example, using datasets such as Facebook – there is now more common ground than ever with the mindset of condensed matter physics, which imagines the number of nodes to be large and effectively infinite.

Network representations. The general culture in social sciences is towards richer models which can capture more of the empirically observed features of the system. For example, approaches such as the stochastic block model (see Section 4.5) were popularized through their ability to capture human community structure. What we referred to as hidden (node) variables in Section 10.3 would be referred to in social science as *actor attributes* and applied to study whether individuals form relationships based on certain characteristics (e.g. [156]) such as gender, seniority or other attribute. Again, the extra level of complexity is justified by the fact that the model now closely reflects the core of the research question.

As in other applications, clustering is seen as a key property in social networks. Authors such as [182] promote the inclusion of higher-order cliques (e.g. complete subgraphs with 4 or more nodes) into the action term of the exponential random graph model to combat the known degeneracy of the Strauss model, which is the simplest model incorporating a constraint on the total number of links and total number of triangles in a system (see Section 4.4). Social networks are typically relatively small – so some of the theoretical concerns about degeneracy may be less relevant.

Case study in social networks: modelling the spread of disease. Common disease spread models include the SIR model, where a node can be classified as *susceptible* (i.e. liable to fall ill), *infectious* (i.e. ill and liable to pass on the illness to other nodes with which it has contact) and *recovered* (which in this context implies that the node is immune from further infection). An alternative is the SIS model, where a node reverts to being susceptible after it recovers. Traditionally, a disease spread model assumes that the populations are evenly mixed. A disease spread model can be simulated on a network, with the stipulation that an infection can only pass between two nodes which are connected in the network (e.g. two people who meet). A social network is a natural interpretation, and this section will be phrased in that context, although the idea extends in a straightforward way to other applications (e.g. the spread of malicious software through a computer network). A key question is to understand how the topology of a network influences the rate of spread of disease on the network. This question has many practical applications relating to how infections can be contained and whether certain types of communities are more vulnerable to epidemics.

Real–world networks – including social ones – are typically more clustered than is observed in an Erdös–Rényi network, or a random network with some imposed degree distribution. For example, if two people have an common acquaintance, it is clear intuitively that they are more likely to know each other. A model which only controls the degree distribution does not reflect this. Ensembles of networks with clustering, or many short loops, are hard to analyze, because the tools which underpin much of the past and present analytical work on random graph ensembles tend to break down when the networks are not locally tree-like. A more technical discussion on this can be found in Section 4.4 – which uses an exponential random graph construction to generate an

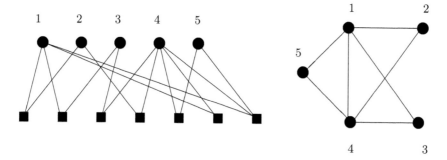

Fig. 11.2 Left figure: A bipartite graph in which the squares represent social groupings (e.g. families) and the circles represent individuals. The links indicate which individuals belong to which family. Right figure: A projection of the information in the bipartite graph, obtained by drawing a relationship link between each pair of nodes that have a social group in common. This construction allows for the ensemble of bipartite graphs to be studied with conventional graph theoretical approaches, and then for the result to be related to an ensemble of highly clustered networks that would have been otherwise difficult to approach analytically. It is a natural and intuitive model in many contexts.

unbiased ensemble of random graphs with a given average degree and given clustering (Strauss Model). However, this model is known to break down for 'realistic' values of the parameters – leading to degenerate clumped networks.

The paper [130] aimed to create a model of highly clustered networks. Its strategy (which was independently reproduced by [1] in the context of modelling the immune system) is to view the social network as a projection of a bipartite network. The two sides of the bipartite network in Figure 11.2 may, for example, represent people and social groupings. The social groupings could be families, or workplaces or social clubs. The ensemble controls the properties of the bipartite network (e.g. the degree distribution in the bipartite network). The usual analytical tools can be applied, since the bipartite network is locally tree-like. The crucial assumption is that people who belong to the same social group all know each other. The bipartite network is projected onto a conventional network that retains only the 'people' nodes, which are then connected according to the rule of 'if two nodes share a common neighbour in the bipartite network, then they are connected in the projection'. This results in a highly clustered network. In [130] SIR-type epidemic processes were studied within this model, and disease spread on the clustered model was compared to that on null models generated by rewiring the clustered model (thereby reverting to the weak clustering of random Erdös–Rényi type graphs). On the basis of this study, it was claimed that clustering decreases the size of epidemics, but also decreases the epidemic threshold, making it easier for diseases to spread. However, this conclusion was challenged by [100], who noted that the construction of [130] controlled the average degree and average clustering, but not the degree distribution. In [100] the results of [130] were attributed to the deformation of the degree distribution, and generally to idiosyncrasies in this construction (which makes no claim to generating an unbiased ensemble).

An alternative approach was proposed in [126], inspired by the configuration model (as described in Chapter 8). Here, each node is given an 'edge degree' and a 'triangle degree'. These quantities correspond to the number of links and triangles respectively that the node participates in. They propose a method for connecting nodes based on these quantities. However, the resulting networks are far from being unbiased representatives of a given average degree/average clustering, and [126] proposes an associated un-clustered null model. These authors found that, with their construction, higher clustering resulted in an increased epidemic threshold.

There are alternative constructions towards generating networks with more realistic clustering, but the contrasting conclusions reached by different authors underline the fact that it is a subtle task to design an algorithm to generate a random graph with several topological properties combined, without inadvertently incorporating biasing the resulting graphs.

11.3   Food webs

Overview: *A network can represent feeding relationships in an ecosystem, and can help answer questions about what makes an ecosystem stable. Some of the topological properties which are especially important to incorporate into random graph models in this context include hierarchy, energy conservation and spatial factors.*

The real systems to be modelled. In ecology, directed graphs can be used to describe food webs: which species eat which species. A node represents a species (e.g. an animal/insect/groups of plants). A directed link from node A to node B means species A eats species B. A historical illustration of this concept is reproduced in Figure 11.3. Nodes in the network can be associated with different roles in the food web. A basal species is a species that eats no other species – a source in a directed network. A top species is a species that is eaten by no other species. This is a sink in the network.

Network representations. On a network representation, a food chain is a path going from a source to a sink. A food chain forms the backbone of an ecosystem, transferring energy from lower to higher organisms. Another topological observable with interesting consequences is how the proportion of top species varies with the size of the system. Food webs are often modelled as tree-like hierarchical directed random graphs. This assumes a strict predator–prey hierarchy. It also implies that the links must be unidirectional. The author of [50] compares the theoretically predicted and observed values for food chain lengths p_{ij}. The simplest version is the homogeneous case ($p_{ij} = p$). Interesting variants would be to have prey-driven or predator-driven models ($p_{ij} = p_j$ and $p_{ij} = p_i$, respectively). In [50] it is suggested that modelling a food web as a random network ignores the finely tuned ecological adaptations which have led to particular predator–prey relationships; the likelihood of a certain predator eating a particular prey must be a function of the total available amount of food. Some manner of energy/mass conservation at each trophic level would be an interesting criteria to incorporate into a null model, though it is not clear how this could be practically implemented.

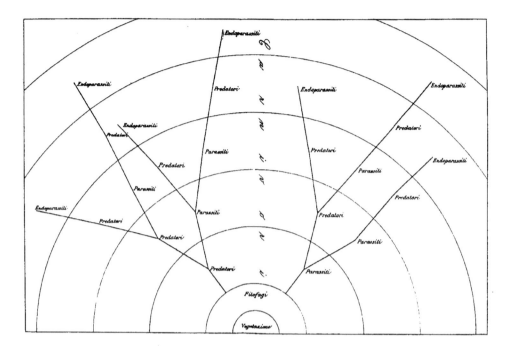

Fig. 11.3 Lorenzo Camerano (1856–1917) was one of the first to draw a *food web* to show the feeding relationships in an ecosystem. *Vegetazione* is vegetation; *Fitofagi* are plant-eating insects; *Predatori* are predators; *Parassiti* are parasites; *Carnivori* are carnivores; *Endoparassiti*, internal parasites such as tapeworms, appear at the top of the feeding chain [43].

In [122] it is observed that a few highly connected species are connected significantly among each other in the food web (i.e. it exhibits assortative degree–degree correlation). They used a rewiring algorithm and proposed their own null model, tailored to the parameters of their problem, which includes correlations. The authors of [121] include a spatial element to the predator–prey equations defined on a food web, to demonstrate the impact of habitat loss on biodiversity. They found qualitatively new consequences, in order to predict which species would be most vulnerable to habitat loss. They considered how the food web structure alters the top species' response to habitat loss under different scenarios.

A classic example of the use of null models in ecology is the paper [185]. The author wished to challenge the (at the time) widely accepted truism that predator/prey ratios within a community were universally roughly constant. He used randomly generated food webs to point out that the apparent constancy was not statistically significant, and could largely be ascribed to scaling between different-sized ecosystems. A more modern version of this observation is by the author of [42], who commented that food webs do not typically have a scale-free degree distribution (unlike many other biological systems). They explain that species specialize and prey on a small set of other species – rather than simply feed on the most abundant food source. That is, ecosystems naturally break up into communities (see Table 1 in [185]).

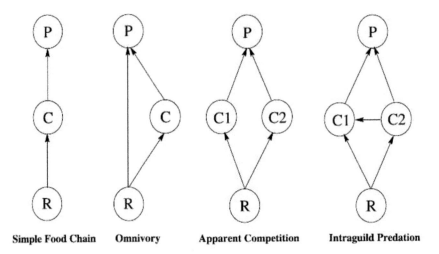

Fig. 11.4 Simple local food web motifs, reproduced from [121]. R denotes the basal species; P denotes the top species. The nodes labelled C (or C1, C2) label carnivores at intermediate levels.

Food chain lengths have been studied in order to determine which environmental factors influence them. This can, for example, help predict which events may lead to some species becoming extinct. Hypotheses include that the lengths of the food chain may depend on how many perturbations the system suffers, whether the ecosystem is in two or three dimensions, or how much energy is available at their base. These are examples of properties that can be tested computationally using tailored random graph models. Other topology-based interesting questions include whether omnivores (i.e. species that feed on more than one level in the system) stabilize (e.g. [148, 69, 117, 92, 118]) or destabilize (e.g. [147]) ecological communities. A related topic is how biodiversity contributes to the functioning and stability of an ecosystem. These might be all be interesting parameters to control or study within a random graph model of a food web.

Food webs present very different modelling challenges compared to other biological networks. Constraints on hierarchy in these systems must form a crucial element of any null model. Considerations of energy conservation, stability and spatial factors are also important. Food webs are an area where quality theoretical predictions bring real benefits in avoiding the need for potentially damaging 'experiments'. We see the effects of dis-balanced food webs in both extinction and pestilence. Hence, it is highly desirable to develop our understanding of the properties and evolution of these kinds of networks.

11.4   World Wide Web

Overview: The Internet is an interesting network, which benefits from being both organically formed and accessible to cheap data collection [143]. One practical application of random graphs in this context is to benchmark and improve the performance of search engines.

The real system to be modelled. The World Wide Web is a popular candidate to be modelled as a random graph process. The most widely seen approach (e.g. [110]) takes web pages to be the nodes of the graph, and draws a directed link from node A to node B if there is a hyperlink leading from web page A to web page B. The World Wide Web is interesting because it is a very large network that is formed organically, largely free of external organizing influences. We can observe features in the World Wide Web that are similar to features observed in networks based on the natural world. However, in contrast to most natural networks, data on interconnections within the World Wide Web is easy to harvest. Moreover, the network is rapidly evolving.

Network representations. The Erdös–Rényi random graph (3.1) does not adequately capture the features seen in the World Wide Web graph. In particular, the World Wide Web has a tendency for hubs (i.e. a higher probability for ultra-highly connected nodes than would be observed in an Erdös–Rényi graph); a plausible model of the World Wide Web must at a minimum show a fat-tailed degree distribution. For this reason, the preferential attachment model (Section 8.2) is often used as a model of the World Wide Web. A new web page/node is created at every time step, and it links preferentially to existing web pages with perceived authority (i.e. those with a large number of other pages linking to it). However, the analogy with the preferential attachment algorithm is not perfect. For example, the World Wide Web does not have such a strong correlation between the degree and age of a node. Another candidate for a natural growth method to generate a random graph to model the World Wide Web is a vertex copying scheme [106], where a new node inherits some of the links of a previously existing node. This encourages the formation of locally dense subgraphs, representing special interest groups or communities.

The authors of [110] proposed that in a successful model for the web there should be a correlation between the 'topics' of the in-neighbourhood and out-neighbourhoods of any particular node. Hence, a non-growing construction would naturally follow a hidden variable type of scheme, where there were specified (power-law or similar) degree distributions, plus some correlations driven by the hidden variables. The optimal descriptive model would also involve several different move types, in order to tune the simulated network to observed network properties, and to reduce model artefacts. However, this makes it difficult to define 'unbiased' for the purpose of statistical tests.

Practical applications of studying and simulating the structure of the web graph include improving the performance of search engines (e.g. [110, 102] and references therein). A search engine is required to find a high-quality page on any given topic. As pointed out by the author of [102], there is no reason to suppose that the website of Honda will contain the phrase 'car manufacturer', nor that the website *www.harvard.edu*

will be distinctly richer in the word 'Harvard' than any of the millions of other web pages in which the word 'Harvard' also appears. This makes the case for focusing search engine operation on analyzing the position of the page (node) within the network – rather than on in-depth analysis of the content of the web page. The terminology in the literature is that when a link is created from node (web page) a to node b, then a *confers authority* on b. The search algorithm may be calibrated to give greater weight to authority conferred from other known authoritative sites – for example giving greater prominence to a page which is linked from *www.bbc.co.uk*. Alternatively, it may be programmed to identify motifs which are known to be associated with communities – giving greater prominence to a node/web page which has incoming links from many other websites featuring the searched for keywords. Methods to evaluate new search algorithms generally involve benchmarking against industry standard web pages (e.g. Google), or manually classifying the quality of obtained results. Both approaches have clear drawbacks; well-specified synthetic null models would make the process of designing and testing search algorithms more automatic and more open to innovation. Similar tools can be developed and applied to citation networks and patent networks.

11.5   Protein–protein interaction networks

Overview: *Due to the advance of modern experimental techniques, protein–protein interaction databases (which can be represented as networks) are now extremely large. There are many applications for synthetic network models in this field. For example, a random graph with controlled topological properties can be used to evaluate the significance of an apparently over-represented motif. A growth process can be designed to mimic an evolutionary model.*

The real system to be modelled. As was already discussed in Chapter 1, a dataset of protein–protein interactions can be interpreted as a network by viewing proteins as nodes and a link between two nodes indicating an interaction between the corresponding proteins; see, e.g. Figure 11.5. A classic application of random graph ensembles in molecular biology is the study of motifs (small over-represented subgraphs) in protein–protein interaction networks. The study [164] aimed to show that certain motifs were observed in the real biological network much more frequently than would be expected in a random graph with the same degree distribution – and hence that these motifs were likely to indicate some special biological function or design. This demonstrates one common use of random graph ensembles in molecular biology: to act as a null model to be able to identify atypical and potentially meaningful topological features. However, the choice of constraints and features is obviously a potential source of interpretation bias. Are these motifs still over-represented in the random graph ensemble which matches the degree distribution and the degree–degree correlation of the real network? Would constraining the average number or distribution of short loops make all the other motifs trivial? Are the motifs in fact simply artefacts of the method used to collect or store the data? If we believed the final statement – then it would be more logical to take the distribution of motifs as the constraint (e.g. constrain a random

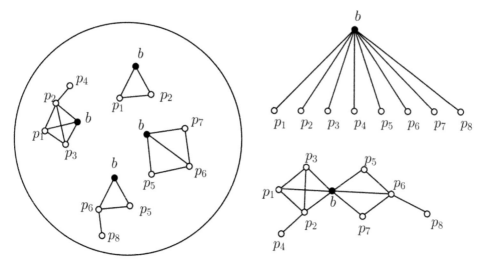

Fig. 11.5 Protein interactions can be identified by using a bait protein **b** which binds to the molecules of interest. The resulting complexes can be analyzed by mass spectroscopy. The resulting networks will typically show clique motifs where the bait 'catches' a protein complex. This can be seen in the illustration above, reproduced from [97]. Alternative experimental methods, such as Yeast2Hybrid, are only able to identify direct interactions, and hence show much fewer clique motifs.

network modelling a mass spectroscopy dataset to have a certain density of cliques) – and to study what topologies result.

Network representations. One approach would be to generate random graphs which match both the degree distribution and the degree–degree correlation of the real networks. This would not perfectly capture the clustering of real networks. As we saw in Section 4.4, creating an unbiased random graph ensemble constrained with a specified distribution of short loops is a difficult problem. An alternative and entirely tangential way to use random graph ensembles in molecular biology is as an evolutionary model. For example, the authors of [176, 142, 96, 172, 137, 74, 48] present growth algorithms for generating a random graph ensemble where one node in the network is copied at every time step, inheriting some of the links of its parents. The expected statistical properties of such an ensemble are unknown up front – they can be deduced. If, however, a certain combination of growth rules and parameters creates a network which is similar to the real network, then it is possible to use this as a basis for speculating about the evolutionary processes that may have formed the network (and how the network may subsequently evolve).

The developing area of synthetic biology hints at the exciting potential of network research to inform very specific biological interventions. For example, the work of [16] (and references therein) echoes the theoretical study of motifs by exploring experimentally the idea that cellular networks are modular. The authors attempt to swap a native module with another module which has apparently the same function. If this is successful, it would allow for biological topologies to be viewed and analyzed as a series of logic gates – and hence for networks designed *in silico* to be implemented *in vivo*. This has profound implications for networks-based drug design. In [75] one finds the design of a computational procedure to create gene networks which are theoretically capable of performing basic tasks. The authors describe this as an 'evolutionary' procedure. It can be understood as essentially an MCMC procedure – albeit a very detailed and complicated scheme incorporating topology and kinetic effects (i.e. reaction/production/degradation rates). Their procedure has two steps. Starting with an ensemble of candidate networks, each network produces a copy. The copy has a small number of mutations. Potential mutations include creating a new gene or interaction or altering a kinetic parameter. The members of the enlarged ensemble are scored on the basis of some fitness function, and the lower-scoring half are removed from the ensemble, thus ending up with an ensemble that has the same number of candidate networks, but that are more suited to performing the specified function. The authors of [75] focused on designing networks which functioned as bistable switches and oscillators. The number of nodes in these schemes is inevitably small given the complexity of the algorithm. Nonetheless it is easy to imagine how streamlined criteria and increased computational power could make this approach more practical.

Networks in molecular biology have gone well beyond being simply a nice tool to visualize large datasets. They are in many ways the natural language to try to understand and find patterns in highly complex information – and a relatively cheap *in silico* alternative to wet lab experiments. Understanding the functioning of cells as a whole allows for the development of drugs that act at several points of the network at once, in order to disrupt a certain pathway (or create a new pathway where function is disrupted). For example, [6] and related articles describe the study of the survivin protein, concluding that

> Despite the fact that survivin is not a traditional drug target – that is not an enzyme or cell-surface molecule – its unique nodal properties imply that even relatively subtle perturbations of its expression, stability or binding . . . could irreversibly impair tumour cell viability.

This is substantially more sophisticated than brute-force or single-target approaches to finding therapeutic drugs. Moreover, it may be possible to predict possible side effects or drugs interactions.

Epilogue

One of the main tasks we set ourselves in writing this book was to present in a unified manner the different strategies, motivations, techniques and notational conventions used by the stakeholding disciplines in this field. We found it to be only partially possible, and this is reflected in the division of the book into parts. We start with random graph ensembles and their MCMC sampling algorithms; here, much can be calculated and proved analytically, and these chapters are therefore more mathematical in spirit. We then move to those graph generation strategies and application-driven network constructions that do not start from a probability distribution; here, one typically finds that much less is known analytically about the detailed probabilities with which graphs will be generated. The later chapters are therefore more descriptive.

A common theme behind all graph generation studies is the desire to generate 'interesting' graphs, i.e. graphs that have non-trivial topological or statistical features that we can somehow control in our generation algorithms. A second common theme is that in practice, whether we like it or not, we usually find ourselves at some stage studying a dynamical process describing the evolution of a graph. This is sometimes very explicit, as with all studies that *define* random graphs directly and solely in terms of a dynamical generation protocol, as with the configuration model or the preferential attachment model. Alternatively, the graph dynamics could enter via the backdoor, as a mechanism for sampling from a specific (static) probability distribution, as with MCMC methods. In either case, it is inevitable that, if we want to have precise control over the topological features of the graphs we are generating, we need to study the mathematical connection between the evolution rules of the processes and the equilibrium state(s) to which they evolve. Establishing this link is sometimes easy, sometimes hard, but always important, especially if the application we have in mind for our random graphs is to serve as 'null models' in some statistical test.

In deciding which material to include in this text, we allowed ourselves to be guided partially by what can already be found in existing textbooks, and by our belief in the importance of understanding the link between graph probabilities and dynamical processes. Given the emphasis in several books on discussions of applications of networks and graphs, we felt justified in only touching that area. We also decided to focus mainly on nondirected graphs – which are the most commonly used ones – but hope and trust that the interested reader will not have too much difficulty adapting and generalizing the algorithms described for nondirected graphs to directed ones.

We initially expected this book to be only about half its present size, but during the writing it became increasingly clear to us how vast and rich the graph generation field actually is. In fact, many interesting problems are still waiting to be solved. We hope and trust that, in spite of its limitations and inevitable bias, this book may serve as a useful and efficient starting point for further research.

Key symbols and terminology

Symbols

$\mathbf{A}$	Adjacency matrix representing a network or graph..
$\mathcal{A}$	Move acceptance probability..
$\langle \cdot \rangle$	Average over given ensemble. .
$\langle \cdot \rangle_{\mathrm{ER}}$	Average over Erdös–Rényi ensemble..
$C_i(\mathbf{A})$	Clustering coefficient..
k_i	Degree of node i..
Δ	Indicates change..
δ_{ab}	Kronecker's delta. Defined to be equal to 1 if $a = b$, and zero otherwise.
∂_i	Set of neighbours of node i..
$F(\lambda)$	Generating function..
F_{ijkl}	Graph randomization move acting on nodes $ijkl$..
G	Space of graphs..
γ	Exponent of power-law degree distribution..
$\langle k \rangle$	Average degree..
L	Number of links..
λ	Lagrange parameters. .
N	Number of nodes in the network..
$\mathbf{\Omega}$	Observed topological feature..
Φ_A	Mobility. Number of legal randomization moves that can act on network $\mathbf{A}$..
$\Pi(k, k')$	Normalized degree–degree correlations..
$p(k)$	Probability of a node having degree k, averaged over ensemble..
$Q(k, k')$	Canonical kernel. .
$q = 1 \ldots Q$	Labels for modules or communities..
S	Number of two-stars..
S	Shannon entropy..
$\mathrm{Tr}(\mathbf{A})$	Trace. Sum of diagonal entries of a matrix..
w_{ij}	Weight of link going from node j to node i..
$W(k, k')$	Probability randomly picked link will connect nodes with degrees k and k'..
$\mathcal{W}$	State transition probability..
x_i	Fitness, or other hidden attribute, of node i..
$Z(\mathbf{\Omega})$	Partition function or normalising constant..

Terminology

Affinity matrix Matrix $\boldsymbol{P}$, with entries $P_{q,q'} \in [0,1]$ that give the probability for any two nodes in group q and q', respectively, to be connected.

Assortativity Pearson correlation between the degrees of connected node pairs, see equation (2.10), page 12.

Canonical Tailored random graphs with soft constraints, page 24.

Canonical kernel $Q^*(k,k') = \Pi(k,k')kk'/\langle k \rangle^2$. This kernel, or kernels related to it by separable transformations, leads to ensembles with average degree correlations given by $\Pi(k,k')$, page 119.

Cliques Small fully connected subgraph.

Clustering coefficient $C_i(\mathbf{A}) \in [0,1]$ Gives the fraction of nodes linked to i that are themselves mutually connected, see equation (2.2), page 10.

Configuration model The stub-joining graph generation algorithm, page 153.

Condensation transition Transition from states where (for example) the triangles are distributed evenly over the graph to states where they are clumped around a small number of nodes.

Dense graph regime Probability of forming a connection is of order 1.

Detailed balance Expression indicating that at equilibrium we would on average expect to observe the same number of transitions $A \to A'$ and $A' \to A$ between any two states, see equation (6.5), page 86.

Edge swap $\{F_{ijk\ell}\}$ – graph randomization move which keeps the degree of each node constant, page 96.

Ensemble average Average of some observable over all the members of a random graph ensemble, see equation (4.2), page 34.

Erdös–Gallai theorem Criterion for a degree sequence to be graphical, see equation (4.23), page 38.

Erdös–Rényi graph ensemble Simplest random graph ensemble, soft-constrained to have L links, or (equivalently) a link formed between any two nodes with some constant probability p, page 20.

Ergodic Process that is able to reach every permitted configuration in a finite number of steps, page 26.

Finite connectivity Scaling regime, where the degrees stay finite even as $N \to \infty$, page 66.

Generalized degrees Count the number of distinct paths of a given length $\ell \geq 1$ that either flow out of, or into a node i, see equation (2.5), page 11.

Geometric random graphs Nodes are placed in a geometric space and the connections are formed based on the distance between two nodes, page 173.

Glauber One possible choice for the acceptance probability of a Markov Chain Monte Carlo process, see equation (6.9), page 86.

Graphical A degree sequence or other set of constraints that can be realized in a network, page 38.

Hamming distance Measure to compare two networks, page 288.

Hidden variable A node attribute that may drive the formation of the network, page 199.

Hinge flip $\{F_{ijk}\}$ – graph randomization move which preserves the total number of links in the graph, page 96.

Hubs Very highly connected nodes.

Kernel Non-negative, real-valued function, with given symmetry and normalization properties, often representing a probability density (or a function related to it). The word *kernel* usually refers to the probability density $W(k, k')$ and to the function $Q(k, k') = W(k, k')/p(k)p(k')$, both symmetric under exchange of k and k'..

Lagrange's method Technique for constrained optimization, see equation (3.13), page 23.

Link flip $\{F_{ij}\}$ – graph randomization move based on creating/destroying individual links, page 96.

Local features Measurements and observations that provide information about the structure of the network in the vicinity of individual nodes, page 9.

Markovian Graph probabilities at the next iteration step depend only on those of the previous step and do not depend on states at earlier stages of the process, page 25.

Maximum entropy ensembles Ensembles in which all graphs **A** that meet our hard constraints are equally probable, see equation (3.14), page 23.

MCMC Markov Chain Monte Carlo. Algorithms to generate random samples from a probability distribution based on a carefully designed Markov chain, page 83.

Mean-field approximation Individual interactions are replaced with averages, considerably simplifying calculations over many-body systems.

Metropolis–Hastings One possible choice for the acceptance probability of a Markov Chain Monte Carlo process, see equation (6.9), page 86.

Microcanonical Tailored random graphs with hard constraints, page 23.

Mobility Number of moves that can be executed on a graph, page 88.

Modularity Topological structure where the network can be divided into groups such that nodes are much more likely to form links within their group rather than outside their group, page 13.

Motifs Small subgraphs which are believed to be significant. They may be considered significant, for example, because they appear in the network much more frequently than we would expect, because they are repeated across different related networks or because they reflect some known functional process.

Multiplex A network defined in layers, where every layer characterizes a different type of relationship within the underlying node set. They are typically used to incorporate information on different types of links between node pairs, page 193.

Nondirected graphs $A_{ij} = A_{ji}$ for all (i, j), so the links carry no arrows.

Null model Control case to test a hypothesis against. In this context generally taken to mean a graph which has certain relevant constraints imposed on it, but is otherwise maximally random, page 6.

Partition function Normalizing constant of the probability distribution of the microscopic configurations of the system given its macroscopic features Ω. In other word, $Z(\Omega)$ is the 'volume' (i.e. number) of the microscopic configurations of the system compatible with macroscopic constraints Ω.

Phase diagram A chart which represents all the possible states of the system. Generally the axes are key parameters, and lines are drawn separating the different phases of the system.

Phase transitions Tendency of some models to generate distinct graphs in extreme phases, page 40.

Planar graph Can be drawn on a plane with no intersecting edges, page 178.

Power law Fat-tailed distribution $p(k) \sim k^{-\gamma}$ where in network applications the exponent γ typically falls between 2 and 3.

Preferential attachment Preferential attachment is an algorithm where a new node is added at every step, and a link is drawn between the new node and existing nodes with a probability proportional to the number of links that the existing node already has, page 157.

Protein interaction networks Each protein is represented by a node, and each pair interaction is represented by a link that connects the two nodes concerned, page 4.

Regular graph Network where every node has the same degree.

Scale-free network Network where the degree distribution (at least for the high-degree nodes) follows $p(k) \propto k^{-\gamma}$ for some exponent γ in the range $2 < \gamma \leq 3$, page 161.

Shannon Entropy Measures the information content of a typical sample from the distribution $p(\mathbf{A})$, defined as $S = -\sum_{\mathbf{A} \in G} p(\mathbf{A}) \log p(\mathbf{A})$, see equation (3.12), page 23.

Stationary states Also known as invariant measure or equilibrium distribution. The probability distribution that a stochastic process equilibrates to.

Steepest descent integration Asymptotic method for evaluating some special kinds of integrals, see equation (D.1), page 237.

Stirling's formula $\log n = n \log n - n + \mathcal{O}(\log n)$ for $n \to \infty$.

Stochastic block model Simple model of community structure where nodes are divided into communities, and given different connection probabilities within and between communities, page 57.

Stochastic Evolution involves random numbers, so that one cannot write deterministic dynamical equations for the graph $\mathbf{A}$ itself, but only equations relating to the *probability* of finding a certain graph at a given time step, page 25.

Strauss Model An ERGM with a controlled average number of links and triangles, page 48.

Stubs Half-edges attached to nodes, page 153.

Temporal network At every time step there is a distinct network representing the contacts/interactions which are actually occurring at that time point, page 187.

Tree-like graph Graph which is connected and has no (short) loops.

Triangle Set of three distinct nodes i, j, k in an undirected simple network such that $A_{ij} = A_{jk} = A_{ki} = 1$.

Two-star model Model where the total number of links and the number of paths of length two is controlled via soft constraints, see equation (4.33), page 41.

Vertex copying Growth algorithm where at each time step, a randomly drawn vertex is duplicated, inheriting some or all of the connections of its parent, see equation (8.7), page 162.

Watts–Strogatz algorithm A two-stage graph generation protocol where nodes are arranged on a line with nearest-neighbour connections, after which long-range links are added at random, giving a network with high clustering and relatively low average path length, page 169.

Weighted graph Links between nodes have weights associated with them - representing features such as traffic or contact frequency, page 179.

List of Algorithms

Appendix A
The delta distribution

Definition. We define the δ-distribution as the probability distribution $\delta(x)$ corresponding to a zero-average random variable x in the limit where the randomness in the variable vanishes. So

$$\int_{-\infty}^{\infty} dx \, f(x)\delta(x) = f(0) \qquad \text{for any function } f$$

By the same token, the expression $\delta(x - a)$ will then represent the distribution for a random variable x with average a, in the limit where the randomness vanishes, since

$$\int dx \, f(x)\delta(x - a) = \int_{-\infty}^{\infty} dx \, f(x + a)\delta(x) = f(a) \qquad \text{for any function } f$$

Formulae for the δ-distribution. A problem arises when we want to write down a formula for $\delta(x)$. Intuitively one could propose taking a zero-average normal distribution and setting its width to zero:

$$\delta(x) = \lim_{\sigma \to 0} p_\sigma(x) \qquad p_\sigma(x) = \frac{1}{\sigma\sqrt{2\pi}} \, e^{-x^2/2\sigma^2} \tag{A.1}$$

This is not a true function in a mathematical sense: $\delta(x)$ is zero for $x \neq 0$ and $\delta(0) = \infty$. However, we realize that $\delta(x)$ only serves to calculate averages; it only has a meaning inside an integration. If we adopt the convention that one should set $\sigma \to 0$ in (A.1) only *after* performing the integration, we can use (A.1) to derive the following properties (for sufficiently well-behaved functions f):

$$\int_{-\infty}^{\infty} dx \, \delta(x)f(x) = \lim_{\sigma \to 0} \int_{-\infty}^{\infty} dx \, p_\sigma(x)f(x) = \lim_{\sigma \to 0} \int \frac{dx}{\sqrt{2\pi}} \, e^{-x^2/2} f(\sigma x) = f(0)$$

$$\int_{-\infty}^{\infty} dx \, \delta'(x)f(x) = \lim_{\sigma \to 0} \int_{-\infty}^{\infty} dx \left\{ \frac{d}{dx}\Big(p_\sigma(x)f(x)\Big) - p_\sigma(x)f'(x) \right\}$$

$$= \lim_{\sigma \to 0} [p_\sigma(x)f(x)]_{-\infty}^{\infty} - f'(0) = -f'(0)$$

The following relation links the δ-distribution to the step function:

$$\delta(x) = \frac{d}{dx}\theta(x) \qquad\qquad \theta(x) = \begin{cases} 1 \text{ if } x > 0 \\ 0 \text{ if } x < 0 \end{cases} \tag{A.2}$$

This is proved by showing that both sides of the equation have the same effect inside an integration:

$$\int_{-\infty}^{\infty} dx \left(\delta(x) - \frac{d}{dx}\theta(x) \right) f(x) = f(0) - \lim_{\epsilon \to 0} \int_{-\epsilon}^{\epsilon} dx \left\{ \frac{d}{dx} \left(\theta(x)f(x) \right) - f'(x)\theta(x) \right\}$$

$$= f(0) - \lim_{\epsilon \to 0} \left[f(\epsilon) - 0 \right] + \lim_{\epsilon \to 0} \int_{0}^{\epsilon} dx \ f'(x) = 0$$

Finally one can use the definitions of Fourier transforms and inverse Fourier transforms to obtain the following integral representation of the δ-distribution:

$$\delta(x) = \int_{-\infty}^{\infty} \frac{dk}{2\pi} e^{ikx} \tag{A.3}$$

Appendix B
Clustering and correlations in Erdös–Rényi graphs

B.1 Clustering coefficients

In this appendix, we show that the average clustering coefficient $C_i = \langle C_i(\mathbf{A}) \rangle$ of any node i in graphs generated from the Erdös–Rényi ensemble (3.1), with the definition of $C_i(\mathbf{A})$ given in (2.2), is

$$\langle C_i(\mathbf{A}) \rangle = p\left[1 - (1-p)^{N-1} - p(N-1)(1-p)^{N-2}\right] \tag{B.1}$$

To prove (B.1) we have to be careful to distinguish between nodes i with $k_i(\mathbf{A}) < 2$, where clustering is impossible (i.e. $C_i(\mathbf{A}) = 0$), and those nodes with $k_i(\mathbf{A}) \geq 2$, where definition (2.2) applies. To handle this conditioning on the degree value, we use the integral representation of the Kronecker δ-symbol (see 2.3):

$$\langle C_i(\mathbf{A}) \rangle = \left\langle \sum_{k \geq 0} \delta_{k,k_i(\mathbf{A})} C_i(\mathbf{A}) \right\rangle = \left\langle \sum_{k \geq 2} \delta_{k,k_i(\mathbf{A})} \frac{\sum_{r \neq s} A_{ir} A_{rs} A_{si}}{k_i(\mathbf{A})(k_i(\mathbf{A}) - 1)} \right\rangle$$

$$= \sum_{k \geq 2} \frac{1}{k(k-1)} \sum_{r \neq s} \left\langle \delta_{k,k_i(\mathbf{A})} A_{ir} A_{rs} A_{si} \right\rangle$$

$$= \sum_{k \geq 2} \frac{2}{k(k-1)} \sum_{r < s} \left\langle \int_{-\pi}^{\pi} \frac{d\omega}{2\pi} e^{i\omega(k - \sum_j A_{ij})} A_{ir} A_{rs} A_{si} \right\rangle$$

$$= \sum_{k \geq 2} \frac{2}{k(k-1)} \sum_{r < s,\ r,s \neq i} \int_{-\pi}^{\pi} \frac{d\omega}{2\pi} e^{i\omega k} \left\langle A_{ir} A_{rs} A_{si} \prod_{j \neq i} e^{-i\omega A_{ij}} \right\rangle$$

$$= \sum_{k \geq 2} \frac{2}{k(k-1)} \sum_{r < s,\ r,s \neq i} \int_{-\pi}^{\pi} \frac{d\omega}{2\pi} e^{i\omega k}$$

$$\times \left\langle A_{rs}\left(A_{ir} e^{-i\omega A_{ir}}\right)\left(A_{is} e^{-i\omega A_{is}}\right) \prod_{j \notin \{i,r,s\}} e^{-i\omega A_{ij}} \right\rangle \tag{B.2}$$

So far we have substituted definitions and rearranged factors such that entries of the adjacency matrix are grouped together. Now we calculate the actual ensemble averages. The distribution $p(\mathbf{A})$ in ensemble (3.1) factorizes over the links – these are distributed independently – which means that our average simplifies to a product of averages:

$$\left\langle A_{rs}\left(A_{ir}e^{-i\omega A_{ir}}\right)\left(A_{is}e^{-i\omega A_{is}}\right)\prod_{j\notin\{i,r,s\}}e^{-i\omega A_{ij}}\right\rangle$$

$$=\langle A_{rs}\rangle\langle A_{ir}e^{-i\omega A_{ir}}\rangle\langle A_{is}e^{-i\omega A_{is}}\rangle\prod_{j\notin\{i,r,s\}}\langle e^{-i\omega A_{ij}}\rangle$$

$$=p(pe^{-i\omega})^2(pe^{-i\omega}+1-p)^{N-3}$$

$$=p^3e^{-2i\omega}(pe^{-i\omega}+1-p)^{N-3} \tag{B.3}$$

Next we use Newton's binomium formula to work out the quantity $(pe^{-i\omega}+1-p)^{N-3}$:

$$\left\langle A_{rs}\left(A_{ir}e^{-i\omega A_{ir}}\right)\left(A_{is}e^{-i\omega A_{is}}\right)\prod_{j\notin\{i,r,s\}}e^{-i\omega A_{ij}}\right\rangle$$

$$=p^3e^{-2i\omega}\sum_{\ell=0}^{N-3}\binom{N-3}{\ell}p^\ell e^{-\ell i\omega}(1-p)^{N-3-\ell}$$

$$=\sum_{\ell=0}^{N-3}\binom{N-3}{\ell}p^{\ell+3}e^{-(\ell+2)i\omega}(1-p)^{N-3-\ell} \tag{B.4}$$

We insert this into our earlier expression for $\langle C_i(\mathbf{A})\rangle$, use the fact that $\sum_{r<s,\ r,s\neq i}1=\frac{1}{2}(N-1)(N-2)$, and do some simple cleaning up:

$$\langle C_i(\mathbf{A})\rangle=\sum_{k>2}\frac{(N-1)(N-2)}{k(k-1)}\int_{-\pi}^{\pi}\frac{d\omega}{2\pi}e^{i\omega k}\sum_{\ell=0}^{N-3}\binom{N-3}{\ell}p^{\ell+3}e^{-(\ell+2)i\omega}(1-p)^{N-3-\ell}$$

$$=\sum_{k>2}\frac{(N-1)(N-2)}{k(k-1)}\sum_{\ell=0}^{N-3}\binom{N-3}{\ell}p^{\ell+3}(1-p)^{N-3-\ell}\int_{-\pi}^{\pi}\frac{d\omega}{2\pi}e^{i\omega(k-\ell-2)} \tag{B.5}$$

At this stage we re-employ the integral representation of the Kronecker delta symbol to get rid of the ω-integral:

$$\langle C_i(\mathbf{A})\rangle=\sum_{k>2}\frac{(N-1)(N-2)}{k(k-1)}\sum_{\ell=0}^{N-3}\binom{N-3}{\ell}p^{\ell+3}(1-p)^{N-3-\ell}\delta_{k,\ell+2}$$

$$=\sum_{\ell=0}^{N-3}\binom{N-3}{\ell}\frac{(N-1)(N-2)}{(\ell+2)(\ell+1)}p^{\ell+3}(1-p)^{N-3-\ell} \tag{B.6}$$

We write explicitly the combinatorial factor, and simplify the fractions where possible:

$$\langle C_i(\mathbf{A})\rangle=\sum_{\ell=0}^{N-3}\frac{(N-3)!}{\ell!(N-3-\ell)!}\frac{(N-1)(N-2)}{(\ell+2)(\ell+1)}p^{\ell+3}(1-p)^{N-3-\ell} \tag{B.7}$$

$$= \sum_{\ell=0}^{N-3} \frac{(N-1)!}{(\ell+2)!(N-3-\ell)!} p^{\ell+3}(1-p)^{N-3-\ell}$$

$$= \sum_{\ell=0}^{N-3} \binom{N-1}{\ell+2} p^{\ell+3}(1-p)^{N-3-\ell}$$

$$= p \sum_{\ell=0}^{N-3} \binom{N-1}{\ell+2} p^{\ell+2}(1-p)^{N-1-(\ell+2)} = p \sum_{\ell'=2}^{N-1} \binom{N-1}{\ell'} p^{\ell'}(1-p)^{N-1-\ell'}$$

$$= p \sum_{\ell'=0}^{N-1} \binom{N-1}{\ell'} p^{\ell'}(1-p)^{N-1-\ell'} - p(1-p)^{N-1} - p(N-1)p(1-p)^{N-2}$$

$$\tag{B.8}$$

We then recognize that Newton's binomial formula can be used to find the sum over ℓ, and proceed to our final result which is indeed (B.1), as claimed:

$$\langle C_i(\mathbf{A}) \rangle = p\left[\left(p+1-p\right)^{N-1} - (1-p)^{N-1} - p(N-1)(1-p)^{N-2}\right]$$

$$= p\left[1 - (1-p)^{N-1} - p(N-1)(1-p)^{N-2}\right] \tag{B.9}$$

B.2 Degree correlations in the sparse regime

Next we work out the ensemble average for the Erdös–Rényi ensemble of the degree–degree correlation kernel (2.9). We start from (3.2) and consider the model only in the 'sparse' or 'finite connectivity' regime, where $\langle k \rangle$ is kept finite and $N \to \infty$.

First, we focus on the large N behaviour of $\bar{k}(\mathbf{A})$. We have already shown in the main text that $\langle \bar{k}(\mathbf{A}) \rangle = \langle k \rangle$. Let us inspect the ensemble fluctuations around this average, for which we require $\langle \bar{k}^2(\mathbf{A}) \rangle$. As before, we use the basic facts that according to (3.2) any two entries A_{ij} and A_{rs} of the adjacency matrix, with $i < j$ and $r < s$, are statistically independent unless $i = r$ and $j = s$, and $\langle A_{ij} \rangle = \langle k \rangle/(N-1)$ for all (i,j):

$$\langle \bar{k}^2(\mathbf{A}) \rangle = \left\langle \left(\frac{2}{N}\sum_{i<j} A_{ij}\right)^2 \right\rangle = \frac{4}{N^2}\sum_{i<j}\sum_{r<s}\langle A_{ij}A_{rs} \rangle$$

$$= \frac{4}{N^2}\sum_{i<j}\sum_{r<s}\left\{\delta_{ir}\delta_{js}\langle A_{ij}^2 \rangle + (1-\delta_{ir}\delta_{js})\langle A_{ij}\rangle\langle A_{rs}\rangle\right\}$$

$$= \frac{4}{N^2}\sum_{i<j}\sum_{r<s}\left\{\delta_{ir}\delta_{js}\frac{\langle k \rangle}{N-1} + (1-\delta_{ir}\delta_{js})\frac{\langle k \rangle^2}{(N-1)^2}\right\}$$

$$= \frac{4\langle k \rangle^2}{N^4}\sum_{i<j}\sum_{r<s}1 + \mathcal{O}(\frac{1}{N}) = \langle k \rangle^2 + \mathcal{O}(\frac{1}{N}) \tag{B.10}$$

Hence, $\langle \bar{k}^2(\mathbf{A}) \rangle - \langle \bar{k}(\mathbf{A}) \rangle^2 = \langle [\bar{k}(\mathbf{A}) - \langle \bar{k}(\mathbf{A}) \rangle]^2 \rangle = \mathcal{O}(N^{-1})$, and for a typical graph realization $\mathbf{A}$ from the ensemble one will find $\bar{k}(\mathbf{A}) = \langle k \rangle + \mathcal{O}(N^{-1/2})$.

It now follows that we may write the ensemble average of (2.9) for $N \to \infty$ as

$$\lim_{N \to \infty} \langle W(k, k'|\mathbf{A}) \rangle = \frac{1}{\langle k \rangle} \lim_{N \to \infty} \frac{2}{N} \sum_{i<j} \langle \delta_{k, k_i(\mathbf{A})} A_{ij} \delta_{k', k_j(\mathbf{A})} \rangle \tag{B.11}$$

With the integral representation of the two Kronecker deltas, and using the invariance of the probability distribution $p(\mathbf{A})$ under index permutations, this can be written as

$$\lim_{N \to \infty} \langle W(k, k'|\mathbf{A}) \rangle = \frac{1}{\langle k \rangle} \int_{-\pi}^{\pi} \frac{\mathrm{d}\omega \mathrm{d}\omega'}{4\pi^2} e^{\mathrm{i}(\omega k + \omega' k')} \lim_{N \to \infty} \frac{2}{N} \sum_{i<j} \left\langle e^{-\mathrm{i}\sum_{\ell}(\omega A_{i\ell} + \omega' A_{j\ell})} A_{ij} \right\rangle$$

$$= \frac{1}{\langle k \rangle} \int_{-\pi}^{\pi} \frac{\mathrm{d}\omega \mathrm{d}\omega'}{4\pi^2} e^{\mathrm{i}(\omega k + \omega' k')} \lim_{N \to \infty} (N-1) \left\langle e^{-\mathrm{i}\sum_{\ell}(\omega A_{1\ell} + \omega' A_{\ell 2})} A_{12} \right\rangle$$

$$= \frac{1}{\langle k \rangle} \int_{-\pi}^{\pi} \frac{\mathrm{d}\omega \mathrm{d}\omega'}{4\pi^2} e^{\mathrm{i}(\omega k + \omega' k')} \lim_{N \to \infty} (N-1) \left\langle e^{-\mathrm{i}A_{12}(\omega + \omega')} A_{12} \right\rangle$$

$$\times \left\langle e^{-\mathrm{i}\sum_{\ell>2}(\omega A_{1\ell} + \omega' A_{2\ell})} \right\rangle$$

$$= \frac{1}{\langle k \rangle} \int_{-\pi}^{\pi} \frac{\mathrm{d}\omega \mathrm{d}\omega'}{4\pi^2} e^{\mathrm{i}(\omega k + \omega' k')} \lim_{N \to \infty} (N-1) e^{-\mathrm{i}(\omega + \omega')} \frac{\langle k \rangle}{N-1}$$

$$\times \left[\frac{\langle k \rangle}{N-1} e^{-\mathrm{i}\omega} + 1 - \frac{\langle k \rangle}{N-1} \right]^{N-2} \left[\frac{\langle k \rangle}{N-1} e^{-\mathrm{i}\omega'} + 1 - \frac{\langle k \rangle}{N-1} \right]^{N-2}$$

$$= \int_{-\pi}^{\pi} \frac{\mathrm{d}\omega \mathrm{d}\omega'}{4\pi^2} e^{\mathrm{i}(\omega k + \omega' k') - \mathrm{i}(\omega + \omega')}$$

$$\times \lim_{N \to \infty} \left[1 + \frac{\langle k \rangle}{N-1} (e^{-\mathrm{i}\omega} - 1) \right]^{N-2} \left[1 + \frac{\langle k \rangle}{N-1} (e^{-\mathrm{i}\omega'} - 1) \right]^{N-2}$$

We may now use $\lim_{n \to \infty} [1 + x/n]^n = e^x$, which gives

$$\lim_{N \to \infty} \langle W(k, k'|\mathbf{A}) \rangle = \left[\int_{-\pi}^{\pi} \frac{\mathrm{d}\omega}{2\pi} e^{\mathrm{i}\omega(k-1) + \langle k \rangle(e^{-\mathrm{i}\omega} - 1)} \right] \left[\int_{-\pi}^{\pi} \frac{\mathrm{d}\omega'}{2\pi} e^{\mathrm{i}\omega'(k'-1) + \langle k \rangle(e^{-\mathrm{i}\omega'} - 1)} \right]$$

$$\tag{B.12}$$

We work out the first factor of (B.12), by writing $\exp[\langle k \rangle e^{-\mathrm{i}\omega}]$ as a Taylor series and again using the integral representation of the Kronecker delta:

$$\int_{-\pi}^{\pi} \frac{\mathrm{d}\omega}{2\pi} e^{\mathrm{i}\omega(k-1) + \langle k \rangle(e^{-\mathrm{i}\omega} - 1)} = e^{-\langle k \rangle} \int_{-\pi}^{\pi} \frac{\mathrm{d}\omega}{2\pi} e^{\mathrm{i}\omega(k-1)} \sum_{\ell \geq 0} \frac{\langle k \rangle^\ell}{\ell!} e^{-\mathrm{i}\ell\omega}$$

$$= e^{-\langle k \rangle} \sum_{\ell \geq 0} \frac{\langle k \rangle^\ell}{\ell!} \int_{-\pi}^{\pi} \frac{\mathrm{d}\omega}{2\pi} e^{\mathrm{i}\omega(k-1-\ell)}$$

$$= e^{-\langle k \rangle} \sum_{\ell \geq 0} \frac{\langle k \rangle^\ell}{\ell!} \delta_{\ell, k-1} = \begin{cases} 0 & \text{if } k = 0 \\ e^{-\langle k \rangle} \frac{\langle k \rangle^{k-1}}{(k-1)!} & \text{if } k > 0 \end{cases} \tag{B.13}$$

We see that both outcomes in the last line, that for $k = 0$ and that for $k > 0$, can be written as $p(k)k/\langle k \rangle$, with $p(k) = e^{\langle k \rangle} \langle k \rangle^k / k!$. The second factor of (B.12) is obtained from the first by replacing $k \to k'$. Hence in combination we find, given that we know from (2.11) that $W(k|\mathbf{A}) = p(k|\mathbf{A})k/\langle k \rangle$ and that for the Erdös–Rényi model in the sparse regime the degree distribution is given by (3.5), that

$$\lim_{N \to \infty} \langle W(k, k'|\mathbf{A}) \rangle = \langle W(k|\mathbf{A}) \rangle \langle W(k'|\mathbf{A}) \rangle \tag{B.14}$$

The degrees in the Erdös–Rényi model in the sparse regime are uncorrelated.

Appendix C

Solution of the two-star exponential random graph model (ERGM) in the sparse regime

---

In this appendix, we calculate the rescaled generating function $f(\alpha, \beta)$ of the two-star model, which is the ERGM of simple nondirected graphs defined by imposing as soft constraints the two observables $(4.32, 4.33)$, or equivalently the first and second moments of the degree distribution, in the sparse regime. We start from definition (4.54). To prepare the sum over graphs in (4.54), we first insert the following representation of unity: $1 = \sum_{\mathbf{k}} \delta_{\mathbf{k}, \mathbf{k}(\mathbf{A})}$, with $\mathbf{k} \in \mathbb{N}^N$ and $\delta_{\mathbf{k}, \mathbf{k}(\mathbf{A})} = \prod_{i \leq N} \delta_{k_i, k_i(\mathbf{A})}$:

$$f(\alpha, \beta) = \frac{1}{N} \log \sum_{\mathbf{k}} \sum_{\mathbf{A} \in G} e^{\sum_{i<j} A_{ij}\left(2\alpha + \beta(k_i + k_j)\right)} \delta_{\mathbf{k}, \mathbf{k}(\mathbf{A})} \tag{C.1}$$

We next write the Kronecker deltas in integral form, using

$$\delta_{nm} = \int_{-\pi}^{\pi} \frac{d\omega}{2\pi} e^{i\omega(n-m)} \tag{C.2}$$

With the shorthand $\boldsymbol{\omega} = (\omega_1, \ldots, \omega_N)$ and $\boldsymbol{\omega} \cdot \mathbf{k} = \sum_{i \leq N} \omega_i k_i$, and by using the symmetry of $\mathbf{A}$ to write $\sum_i \omega_i k_i = \sum_{i<j}(\omega_i + \omega_j)A_{ij}$, we then obtain an expression for $f(\alpha, \beta)$ in which the sum over graphs can be carried out:

$$f(\alpha, \beta) = \frac{1}{N} \log \sum_{\mathbf{k}} \int_{-\pi}^{\pi} \frac{d\boldsymbol{\omega}}{(2\pi)^N} e^{i\boldsymbol{\omega} \cdot \mathbf{k}} \sum_{\mathbf{A} \in G} e^{\sum_{i<j} A_{ij}\left(2\alpha + \beta(k_i + k_j) - i(\omega_i + \omega_j)\right)}$$

$$= \frac{1}{N} \log \sum_{\mathbf{k}} \int_{-\pi}^{\pi} \frac{d\boldsymbol{\omega}}{(2\pi)^N} e^{i\boldsymbol{\omega} \cdot \mathbf{k}} \prod_{i<j} \left[1 + e^{2\alpha + \beta(k_i + k_j) - i(\omega_i + \omega_j)}\right] \tag{C.3}$$

Similar to the dense regime, where to obtain the required scaling of the degrees we had to introduce a specific scaling with N of the parameter β, we find here that for the model to be in the *sparse* regime, i.e. $L/N = \mathcal{O}(1)$, we must ensure that α depends in a specific way on N. To see this, we work out the first equation of (4.55):

$$\frac{L}{N} = \frac{1}{N} \sum_{n<m} \frac{\sum_{\mathbf{k}} \int_{-\pi}^{\pi} d\boldsymbol{\omega} \; e^{i\boldsymbol{\omega} \cdot \mathbf{k}} \frac{e^{2\alpha + \beta(k_i + k_j) - i(\omega_i + \omega_j)}}{1 + e^{2\alpha + \beta(k_i + k_j) - i(\omega_i + \omega_j)}} \prod_{i<j}\left[1 + e^{2\alpha + \beta(k_i + k_j) - i(\omega_i + \omega_j)}\right]}{\sum_{\mathbf{k}} \int_{-\pi}^{\pi} d\boldsymbol{\omega} \; e^{i\boldsymbol{\omega} \cdot \mathbf{k}} \prod_{i<j}\left[1 + e^{2\alpha + \beta(k_i + k_j) - i(\omega_i + \omega_j)}\right]}$$

$$\tag{C.4}$$

This expression shows that if we choose $\alpha = \mathcal{O}(1)$, we will once more find $L/N = \mathcal{O}(N)$ (i.e. dense graphs). In contrast, we will indeed obtain $L/N = \mathcal{O}(1)$ (i.e. our present ERGM will produce sparse graphs) if $e^{2\alpha} = \mathcal{O}(N^{-1})$. Hence, we now set $\alpha = \frac{1}{2}\log(c/N)$. In view of the fact that by constraining $L/N = \frac{1}{2}\langle N^{-1}\sum_i k_i\rangle$ and $(S+L)/N = \frac{1}{2}\langle N^{-1}\sum_i k_i^2\rangle$ we are simply imposing the first and the second moment of the degree distribution, we will put simply $L/N \to \frac{1}{2}\langle k\rangle$ and $(S+L)/N \to \frac{1}{2}\langle k^2\rangle$, upon which the two equations (4.55) become

$$\frac{\langle k\rangle}{c} = \frac{\sum_{\mathbf{k}}\int_{-\pi}^{\pi}\mathrm{d}\boldsymbol{\omega}\; e^{i\boldsymbol{\omega}\cdot\mathbf{k}+\beta(k_1+k_2)-i(\omega_1+\omega_2)+\mathcal{O}(N^{-1})}\prod_{i<j}\left[1+\frac{c}{N}e^{\beta(k_i+k_j)-i(\omega_i+\omega_j)}\right]}{\sum_{\mathbf{k}}\int_{-\pi}^{\pi}\mathrm{d}\boldsymbol{\omega}\; e^{i\boldsymbol{\omega}\cdot\mathbf{k}}\prod_{i<j}\left[1+\frac{c}{N}e^{\beta(k_i+k_j)-i(\omega_i+\omega_j)}\right]}$$

(C.5)

$$\frac{\langle k^2\rangle}{c} = \frac{\sum_{\mathbf{k}}k_1\int_{-\pi}^{\pi}\mathrm{d}\boldsymbol{\omega}\; e^{i\boldsymbol{\omega}\cdot\mathbf{k}+\beta(k_1+k_2)-i(\omega_1+\omega_2)+\mathcal{O}(N^{-1})}\prod_{i<j}\left[1+\frac{c}{N}e^{\beta(k_i+k_j)-i(\omega_i+\omega_j)}\right]}{\sum_{\mathbf{k}}\int_{-\pi}^{\pi}\mathrm{d}\boldsymbol{\omega}\; e^{i\boldsymbol{\omega}\cdot\mathbf{k}}\prod_{i<j}\left[1+\frac{c}{N}e^{\beta(k_i+k_j)-i(\omega_i+\omega_j)}\right]}$$

(C.6)

(where we used, where possible, invariance of terms under permutations of the node indices). Now, provided we keep $c = \mathcal{O}(1)$, all relevant quantities will scale in the desired way with N. Next we expand for large N the factors that resulted from the sum over graphs:

$$\prod_{i<j}\left[1+\frac{c}{N}e^{\beta(k_i+k_j)-i(\omega_i+\omega_j)}\right] = e^{\frac{1}{2}\sum_{i\neq j}\log\left[1+\frac{c}{N}e^{\beta(k_i+k_j)-i(\omega_i+\omega_j)}\right]}$$

$$= e^{\frac{c}{2N}\sum_{ij}e^{\beta(k_i+k_j)-i(\omega_i+\omega_j)}+\mathcal{O}(1)}$$

$$= e^{\frac{1}{2}cN\sum_{\mathbf{k}\mathbf{k}'}\int \mathrm{d}\omega\mathrm{d}\omega' P(k,\omega|\mathbf{k},\boldsymbol{\omega})P(k',\omega'|\mathbf{k},\boldsymbol{\omega})e^{\beta(k+k')-i(\omega+\omega')}+\mathcal{O}(1)}$$

(C.7)

with

$$P(k,\omega|\mathbf{k},\boldsymbol{\omega}) = \frac{1}{N}\sum_{r=1}^{N}\delta_{k,k_r}\delta(\omega-\omega_r)$$

(C.8)

In order to achieve factorization over all variables with node labels i, so that the summations over $\mathbf{k}$ and the integrals over $\boldsymbol{\omega}$ can be done, we next introduce, for each (k,ω), the following factor. Being an integral over a δ-distribution (which we subsequently write in integral form), it is simply equal to one, but inserting it will enable us to transport the dependences on $\mathbf{k}$ and $\boldsymbol{\omega}$ to a more manageable place:

$$1 = \int \mathrm{d}P(k,\omega)\;\delta\left[P(k,\omega)-P(k,\omega|\mathbf{k},\boldsymbol{\omega})\right]$$

$$= \int \frac{\mathrm{d}P(k,\omega)\mathrm{d}\hat{P}(k,\omega)}{2\pi/N}e^{iN\hat{P}(k,\omega)P(k,\omega)-i\sum_{r=1}^{N}\delta_{k,k_r}\delta(\omega-\omega_r)\hat{P}(k,\omega)}$$

(C.9)

Upon inserting this expression for each (k,ω) combination, which in practice is done by first discretizing ω in steps of size Δ which is eventually set to zero, i.e.

$$1 = \int\left[\prod_{k,\omega}\frac{\mathrm{d}P(k,\omega)\mathrm{d}\hat{P}(k,\omega)}{2\pi/N\Delta}\right]e^{\Delta i\sum_{k,\omega}\left[N\hat{P}(k,\omega)P(k,\omega)-\sum_{r=1}^{N}\delta_{k,k_r}\delta(\omega-\omega_r)\hat{P}(k,\omega)\right]}$$

(C.10)

we may replace each occurrence of $P(k, \omega|\mathbf{k}, \boldsymbol{\omega})$ in our above formulae by the new integration variable $P(k, \omega)$. We can then write both the numerators and the denominators of our equations for $\langle k \rangle$ and $\langle k^2 \rangle$ as path integrals over the two functions $\{P, \hat{P}\}$, with the shorthand $\{dPd\hat{P}\} = \lim_{\Delta \to 0} \prod_{k,\omega}[dP(k,\omega)d\hat{P}(k,\omega)\Delta/2\pi]$. Moreover, we will then have obtained full factorization over sites, so that each sum $\sum_{\mathbf{k}}$ becomes the product of N independent summations and each integral $\int d\boldsymbol{\omega}$ the product of N independent integrals. The result is

$$\frac{\langle k \rangle}{c} = c\frac{\int \{dPd\hat{P}\}e^{N\Phi[P,\hat{P}]}\left[\frac{\sum_k \int_{-\pi}^{\pi}d\omega\ e^{i\omega(k-1)-i\hat{P}(k,\omega)+\beta k}}{\sum_k \int_{-\pi}^{\pi}d\omega\ e^{i\omega k-i\hat{P}(k,\omega)}}\right]^2}{\int \{dPd\hat{P}\}e^{N\Phi[P,\hat{P}]}} \tag{C.11}$$

$$\frac{\langle k^2 \rangle}{c} = \frac{\int \{dPd\hat{P}\}e^{N\Phi[P,\hat{P}]}\left[\frac{\sum_k k \int_{-\pi}^{\pi}d\omega\ e^{i\omega(k-1)-i\hat{P}(k,\omega)+\beta k}}{\sum_k \int_{-\pi}^{\pi}d\omega\ e^{i\omega k-i\hat{P}(k,\omega)}}\right]\left[\frac{\sum_k \int_{-\pi}^{\pi}d\omega\ e^{i\omega(k-1)-i\hat{P}(k,\omega)+\beta k}}{\sum_k \int_{-\pi}^{\pi}d\omega\ e^{i\omega k-i\hat{P}(k,\omega)}}\right]}{\int \{dPd\hat{P}\}e^{N\Phi[P,\hat{P}]}}$$
$$\tag{C.12}$$

with

$$\Phi[P, \hat{P}] = i\sum_{k \geq 0} \int_{-\pi}^{\pi} d\omega\ \hat{P}(k,\omega)P(k,\omega) + \log \sum_{k \geq 0} \int_{-\pi}^{\pi} d\omega\ e^{i\omega k - i\hat{P}(k,\omega)}$$
$$+ \frac{1}{2}c \sum_{k,k' \geq 0} \int_{-\pi}^{\pi} d\omega d\omega' P(k,\omega)P(k',\omega')e^{\beta(k+k')-i(\omega+\omega')} + \mathcal{O}\left(\frac{\log N}{N}\right) \tag{C.13}$$

In the limit $N \to \infty$, we can evaluate the path integrals in (C.11,C.12) by the saddle point (or 'steepest descent') method (see Appendix D), and obtain

$$\frac{\langle k \rangle}{c} = \left[\frac{\sum_k \int_{-\pi}^{\pi}d\omega\ e^{i\omega(k-1)-i\hat{P}(k,\omega)+\beta k}}{\sum_k \int_{-\pi}^{\pi}d\omega\ e^{i\omega k-i\hat{P}(k,\omega)}}\right]^2 \tag{C.14}$$

$$\frac{\langle k^2 \rangle}{c} = \left[\frac{\sum_k k \int_{-\pi}^{\pi}d\omega\ e^{i\omega(k-1)-i\hat{P}(k,\omega)+\beta k}}{\sum_k \int_{-\pi}^{\pi}d\omega\ e^{i\omega k-i\hat{P}(k,\omega)}}\right]\left[\frac{\sum_k \int_{-\pi}^{\pi}d\omega\ e^{i\omega(k-1)-i\hat{P}(k,\omega)+\beta k}}{\sum_k \int_{-\pi}^{\pi}d\omega\ e^{i\omega k-i\hat{P}(k,\omega)}}\right] \tag{C.15}$$

where $P(k,\omega)$ and $\hat{P}(k,\omega)$ are now those functions that extremize the $N \to \infty$ limit of (C.13). We note that according to (C.8) the quantity $p(k) = \int_{-\pi}^{\pi} d\omega\ P(k,\omega)$ will be the ensemble average of the degree distribution, i.e. $p(k) = \lim_{N\to\infty}\langle N^{-1}\sum_i \delta_{k,k_i(\mathbf{A})}\rangle$. What remains is to calculate P and $\hat{P}$ by functional differentiation of $\Phi[P, \hat{P}]$. This gives the following two coupled equations:

$$P(k,\omega) = \frac{e^{i\omega k - i\hat{P}(k,\omega)}}{\sum_{k'\geq 0}\int_{-\pi}^{\pi}d\omega'\ e^{i\omega'k'-i\hat{P}(k',\omega')}} \tag{C.16}$$

$$\hat{P}(k,\omega) = ic\ e^{\beta k - i\omega}\sum_{k'\geq 0}\int_{-\pi}^{\pi}d\omega'\ P(k',\omega')e^{\beta k'-i\omega'} \tag{C.17}$$

We conclude from the second equation that $\hat{P}(k,\omega) = i\gamma e^{\beta k - i\omega}$, which leaves us with

$$P(k,\omega) = \frac{e^{i\omega k + \gamma e^{\beta k - i\omega}}}{\sum_{k'} \int_{-\pi}^{\pi} d\omega'\, e^{i\omega' k' + \gamma e^{\beta k' - i\omega'}}}, \qquad \gamma = c \sum_k \int_{-\pi}^{\pi} d\omega\, P(k,\omega) e^{\beta k - i\omega} \quad \text{(C.18)}$$

The denominator on the left is evaluated by writing $\exp[\gamma e^{\beta k' - i\omega'}]$ as a Taylor series:

$$\int_{-\pi}^{\pi} d\omega'\, e^{i\omega' k' + \gamma e^{\beta k' - i\omega'}} = \sum_{\ell \geq 0} \frac{\gamma^\ell e^{\beta \ell k'}}{\ell!} \int_{-\pi}^{\pi} d\omega'\, e^{i\omega'(k'-\ell)} = 2\pi \frac{\gamma^{k'} e^{\beta k'^2}}{k'!} \quad \text{(C.19)}$$

One can in fact evaluate all the remaining integrals over ω in the same way, and find that

$$P(k,\omega) = p(k) \frac{e^{i\omega k + \gamma e^{\beta k - i\omega}}}{2\pi \gamma^k e^{\beta k^2}/k!}, \qquad p(k) = \frac{\gamma^k e^{\beta k^2}/k!}{\sum_{\ell \geq 0} \gamma^\ell e^{\beta \ell^2}/\ell!}, \quad \text{(C.20)}$$

We are then left with one further equation from which to solve γ, plus the two equations (C.11, C.12) from the ensemble constraints, to be solved for c, β, γ and expressed in terms of the imposed values of $\langle k \rangle$ and $\langle k^2 \rangle$. They can be written as, respectively,

$$\gamma \left(1 - ce^\beta \sum_{k \geq 0} e^{\beta k} p(k)\right) = 0 \quad \text{(C.21)}$$

$$\frac{\langle k \rangle}{c} = \left(\frac{1}{\gamma} \sum_{k \geq 0} kp(k)\right)^2, \qquad \frac{\langle k^2 \rangle}{c} = \left(\frac{1}{\gamma} \sum_{k \geq 0} kp(k)\right)\left(\frac{1}{\gamma} \sum_{k \geq 0} k^2 p(k)\right) \quad \text{(C.22)}$$

From this we can deduce that $\gamma = \sqrt{c \langle k \rangle}$, and that (c, β) are subsequently to be solved from any two of the above three equations. If we pick the two moment equations, and if we also write $c = \exp(2\tilde{\alpha})$ (which means that $\tilde{\alpha}$ is related to the original ensemble parameter α via $\alpha = \tilde{\alpha} - \frac{1}{2}\log N$), then we obtain the transparent final result

$$\langle k \rangle = \sum_{k \geq 0} kp(k), \qquad \langle k^2 \rangle = \sum_{k \geq 0} k^2 p(k) \quad \text{(C.23)}$$

with the degree distribution

$$p(k) = \frac{\langle k \rangle^{k/2} e^{\tilde{\alpha} k + \beta k^2}/k!}{\sum_{\ell \geq 0} \langle k \rangle^{\ell/2} e^{\tilde{\alpha} \ell + \beta \ell^2}/\ell!} \quad \text{(C.24)}$$

Appendix D
Steepest descent integration

Steepest descent (or 'saddle point') integration is a method for dealing with integrals of the following type, with $\mathbf{x} \in \mathbb{R}^p$, and for continuous functions $f(\mathbf{x})$ and $g(\mathbf{x})$ of which f is bounded from below, and with $N \in \mathbb{R}$ positive and large:

$$I_N[f, g] = \int d\mathbf{x} \; g(\mathbf{x}) e^{-Nf(\mathbf{x})} \tag{D.1}$$

We first take $f(\mathbf{x})$ to be real-valued; this is the simplest case, for which finding the asymptotic analysis of (D.1) goes back to Laplace. We assume that $f(\mathbf{x})$ can be expanded in a Taylor series around its (assumed unique) minimum $f(\mathbf{x}^\star)$, i.e.

$$f(\mathbf{x}) = f(\mathbf{x}^\star) + \frac{1}{2} \sum_{ij=1}^{p} B_{ij}(x_i - x_i^\star)(x_j - x_j^\star) + \mathcal{O}(|\mathbf{x} - \mathbf{x}^\star|^3) \tag{D.2}$$

$\mathbf{B}$ is the local $p \times p$ curvature matrix, with entries $B_{ij} = \lim_{\mathbf{x} \to \mathbf{x}^\star} \partial^2 f(\mathbf{x})/\partial x_i \partial x_j$. If the integral (D.1) exists, inserting (D.2) into (D.1), followed by transforming $\mathbf{x} = \mathbf{x}^\star + \mathbf{y}/\sqrt{N}$ gives

$$I_N[f, g] = e^{-Nf(\mathbf{x}^\star)} \int d\mathbf{x} \; g(\mathbf{x}) e^{-\frac{1}{2} N \sum_{ij} (x_i - x_i^\star) B_{ij}(x_j - x_j^\star) + \mathcal{O}(N|\mathbf{x} - \mathbf{x}^\star|^3)}$$

$$= N^{-\frac{p}{2}} e^{-Nf(\mathbf{x}^\star)} \int d\mathbf{y} \; g(\mathbf{x}^\star + \frac{\mathbf{y}}{\sqrt{N}}) \; e^{-\frac{1}{2} \sum_{ij} y_i B_{ij} y_j + \mathcal{O}(N^{-\frac{1}{2}}|\mathbf{y}|^3)} \tag{D.3}$$

From this latter expansion, we can obtain two important identities. First, if $\lim_{N \to \infty} p \log(N)/N = 0$ then

$$- \lim_{N \to \infty} \frac{1}{N} \log \int d\mathbf{x} \; e^{-Nf(\mathbf{x})} = - \lim_{N \to \infty} \frac{1}{N} \log I_N[f, 1]$$

$$= f(\mathbf{x}^\star) + \lim_{N \to \infty} \left\{ \frac{p \log N}{2N} - \frac{1}{N} \log \int d\mathbf{y} \; e^{-\frac{1}{2} \sum_{ij} y_i A_{ij} y_j + \mathcal{O}(N^{-\frac{1}{2}}|\mathbf{y}|^3)} \right\}$$

$$= f(\mathbf{x}^\star) = \min_{\mathbf{x} \in \mathbb{R}^p} f(\mathbf{x}) \tag{D.4}$$

Second, for ratios of such integrals one finds if $\lim_{N \to \infty} p/N^{1/3} = 0$ that

$$\lim_{N \to \infty} \frac{\int d\mathbf{x} \; g(\mathbf{x}) e^{-Nf(\mathbf{x})}}{\int d\mathbf{x} \; h(\mathbf{x}) e^{-Nf(\mathbf{x})}} = \lim_{N \to \infty} \frac{I_N[f, g]}{I_N[f, h]}$$

$$= \lim_{N\to\infty} \left\{ \frac{\int \mathbf{dy}\ g(\mathbf{x}^\star + \frac{\mathbf{y}}{\sqrt{N}})\ e^{-\frac{1}{2}\sum_{ij} y_i B_{ij} y_j + \mathcal{O}(N^{-\frac{1}{2}}|\mathbf{y}|^3)}}{\int \mathbf{dy}\ h(\mathbf{x}^\star + \frac{\mathbf{y}}{\sqrt{N}})\ e^{-\frac{1}{2}\sum_{ij} y_i B_{ij} y_j + \mathcal{O}(N^{-\frac{1}{2}}|\mathbf{y}|^3)}} \right\}$$

$$= \lim_{N\to\infty} \frac{g(\mathbf{x}^\star)(2\pi)^{p/2}/\sqrt{\mathrm{Det}\,\mathbf{B}}}{h(\mathbf{x}^\star)(2\pi)^{p/2}/\sqrt{\mathrm{Det}\,\mathbf{B}}} = \frac{g(\mathbf{x}^\star)}{h(\mathbf{x}^\star)}$$

$$= \frac{g(\mathbf{x})}{h(\mathbf{x})}\bigg|_{\arg\min_{\mathbf{x}\in\mathbb{R}^p} f(\mathbf{x})} \tag{D.5}$$

If the function $f(\mathbf{x})$ is complex, the correct procedure to be followed is to deform the integration paths in the complex plane (using Cauchy's theorem) such that along the deformed path the imaginary part of the function $f(\mathbf{x})$ is constant, and preferably (if possible) zero. One then proceeds using Laplace's argument and finds the leading order in N of our integral in the usual manner by extremization of the real part of $f(\mathbf{x})$. In combination, our integrals will thus again be dominated by an extremum of the (complex) function $f(\mathbf{x})$, but since f is complex this extremum need not be a minimum:

$$-\lim_{N\to\infty} \frac{1}{N}\log \int \mathbf{dx}\ e^{-Nf(\mathbf{x})} = \mathrm{extr}_{\mathbf{x}\in\mathbb{R}^p} f(\mathbf{x}) \tag{D.6}$$

$$\lim_{N\to\infty} \frac{\int \mathbf{dx}\ g(\mathbf{x})e^{-Nf(\mathbf{x})}}{\int \mathbf{dx}\ h(\mathbf{x})e^{-Nf(\mathbf{x})}} = \frac{g(\mathbf{x})}{h(\mathbf{x})}\bigg|_{\arg\ \mathrm{extr}_{\mathbf{x}\in\mathbb{R}^p} f(\mathbf{x})} \tag{D.7}$$

Appendix E

Number of sparse graphs with prescribed average degree and degree variance

In this appendix, we calculate the leading orders for $N \to \infty$ of the quantity (5.38), which is the rescaled logarithm of the number of graphs $Z(\langle k \rangle, \langle k^2 \rangle)$ in the sparse scaling regime of the ensemble (5.36) of simple nondirected graphs, in which the first two moments $\langle k \rangle$ and $\langle k^2 \rangle$ of the degree distribution are prescribed as hard constraints.

Before we can evaluate the average over the Erdös–Rényi measure (3.1) in (5.38), with individual link probabilities $\langle k \rangle / (N-1)$, we need to neutralize the square $k_i^2(\mathbf{A})$ in the exponent, which presently prevents us from factorization over links. This is done by inserting $\sum_{k_i} \delta_{k_i, k_i(\mathbf{A})}$ for each node i, and writing the corresponding N new Kronecker deltas in integral form, with integration variables $\boldsymbol{\omega} = (\omega_1, \ldots, \omega_N)$

$$
\frac{1}{N} \log Z(\langle k \rangle, \langle k^2 \rangle) = \frac{1}{2} \langle k \rangle \left[1 + \log \left(\frac{N}{\langle k \rangle} \right) \right] + \mathcal{O}\left(\frac{1}{N} \right) + \frac{1}{N} \log \int_{-\pi}^{\pi} \mathrm{d}\phi \mathrm{d}\phi' \ \mathrm{e}^{\mathrm{i}N(\phi \langle k \rangle + \phi' \langle k^2 \rangle)}
$$

$$
\times \sum_{k_1 \ldots k_N} \int_{-\pi}^{\pi} \frac{\mathrm{d}\boldsymbol{\omega}}{(2\pi)^N} \mathrm{e}^{\mathrm{i}\sum_i \left[(\omega_i - \phi) k_i - \phi' k_i^2 \right]} \left\langle \mathrm{e}^{-\mathrm{i}\sum_i \omega_i \sum_j A_{ij}} \right\rangle_{\mathrm{ER}} \qquad (\text{E.1})
$$

The graph average has now become trivial:

$$
\left\langle \mathrm{e}^{-\mathrm{i}\sum_i \omega_i \sum_j A_{ij}} \right\rangle_{\mathrm{ER}} = \sum_{\mathbf{A} \in G} \mathrm{e}^{-\mathrm{i}\sum_{i<j}(\omega_i + \omega_j) A_{ij}} \prod_{i<j} \left[\frac{\langle k \rangle}{N-1} \delta_{A_{ij}, 1} + \left(1 - \frac{\langle k \rangle}{N-1} \right) \delta_{A_{ij}, 0} \right]
$$

$$
= \prod_{i<j} \left[1 + \frac{\langle k \rangle}{N-1} \left(\mathrm{e}^{-\mathrm{i}(\omega_i + \omega_j)} - 1 \right) \right]
$$

$$
= \mathrm{e}^{\frac{\langle k \rangle}{N-1} \sum_{i<j} [\mathrm{e}^{-\mathrm{i}(\omega_i + \omega_j)} - 1] + \mathcal{O}(N^0)} = \mathrm{e}^{\frac{\langle k \rangle}{2N} \sum_{ij} [\mathrm{e}^{-\mathrm{i}(\omega_i + \omega_j)} - 1] + \mathcal{O}(N^0)}
$$

$$
= \mathrm{e}^{(\langle k \rangle / 2N) \left[\sum_i \mathrm{e}^{-\mathrm{i}\omega_i} \right]^2 - \frac{1}{2} N \langle k \rangle + \mathcal{O}(N^0)} \qquad (\text{E.2})
$$

Hence, we may write $N^{-1} \log Z(\langle k \rangle, \langle k^2 \rangle)$ as

$$
\frac{1}{N} \log Z(\langle k \rangle, \langle k^2 \rangle) = \frac{1}{2} \langle k \rangle \log \left(\frac{N}{\langle k \rangle} \right) + \mathcal{O}\left(\frac{1}{N} \right) + \frac{1}{N} \log \int_{-\pi}^{\pi} \mathrm{d}\phi \mathrm{d}\phi' \ \mathrm{e}^{\mathrm{i}N(\phi \langle k \rangle + \phi' \langle k^2 \rangle)}
$$

$$
\times \sum_{k_1 \ldots k_N} \int_{-\pi}^{\pi} \frac{\mathrm{d}\boldsymbol{\omega}}{(2\pi)^N} \mathrm{e}^{\mathrm{i}\sum_i \left[(\omega_i - \phi) k_i - \phi' k_i^2 \right] + (\langle k \rangle / 2N) \left[\sum_i \mathrm{e}^{-\mathrm{i}\omega_i} \right]^2} \qquad (\text{E.3})
$$

For any (real or complex) number w we may always write

$$e^{\frac{1}{2}w^2} = \int_{-\infty}^{\infty} \frac{dz}{\sqrt{2\pi}} \, e^{zw - \frac{1}{2}z^2} \tag{E.4}$$

Application of this identity to the choice $w = (\langle k \rangle / N)^{\frac{1}{2}} \sum_i e^{-i\omega_i}$ enables us to write

$$
\frac{1}{N} \log Z(\langle k \rangle, \langle k^2 \rangle) = \frac{1}{2}\langle k \rangle \log \left(\frac{N}{\langle k \rangle}\right) + \frac{1}{N} \log \int_{-\infty}^{\infty} \frac{dz}{\sqrt{2\pi}} \, e^{-\frac{1}{2}z^2} \int_{-\pi}^{\pi} d\phi d\phi' \, e^{iN(\phi\langle k\rangle + \phi'\langle k^2\rangle)}
$$
$$
\times \sum_{k_1 \ldots k_N} \int_{-\pi}^{\pi} \frac{d\omega}{(2\pi)^N} e^{i\sum_i \left[(\omega_i - \phi)k_i - \phi' k_i^2\right] + z\sqrt{\langle k\rangle/N} \sum_i e^{-i\omega_i}} + \mathcal{O}\left(\frac{1}{N}\right)
$$
$$
= \frac{1}{2}\langle k \rangle \log \left(\frac{N}{\langle k \rangle}\right) + \frac{1}{N} \log \int_{-\infty}^{\infty} \frac{dz}{\sqrt{2\pi}} \, e^{-\frac{1}{2}(N/\langle k\rangle)z^2} \int_{-\pi}^{\pi} d\phi d\phi' \, e^{iN(\phi\langle k\rangle + \phi'\langle k^2\rangle)}
$$
$$
\times \left[\sum_k e^{-i\phi k - i\phi' k^2} \int_{-\pi}^{\pi} \frac{d\omega}{2\pi} e^{i\omega k + z e^{-i\omega}}\right]^N + \mathcal{O}\left(\frac{1}{N}\right) \tag{E.5}
$$

The complicated part of this expression is now once more of the form that can be handled via steepest descent integration (see Appendix D), and we can therefore write

$$\frac{1}{N} \log Z(\langle k \rangle, \langle k^2 \rangle) = \frac{1}{2}\langle k \rangle \log \left(\frac{N}{\langle k \rangle}\right) + \text{extr}_{z,\phi,\phi'} \, \Psi(z, \phi, \phi') + \epsilon_N \tag{E.6}$$

Here $z \in (-\infty, \infty)$, $\phi, \phi' \in [-\pi, \pi]$, $\lim_{N \to \infty} \epsilon_N = 0$, and

$$\Psi(z, \phi, \phi') = i\phi\langle k\rangle + i\phi'\langle k^2\rangle - \frac{z^2}{2\langle k\rangle} + \log \sum_{k \geq 0} e^{-i\phi k - i\phi' k^2} \int_{-\pi}^{\pi} \frac{d\omega}{2\pi} e^{i\omega k + z e^{-i\omega}} \tag{E.7}$$

We can evaluate the ω-integral in Ψ by expanding $e^{z \exp(-i\omega)}$ as a power series:

$$\int_{-\pi}^{\pi} \frac{d\omega}{2\pi} e^{i\omega k + z e^{-i\omega}} = \sum_{\ell \geq 0} \frac{z^\ell}{\ell!} \int_{-\pi}^{\pi} \frac{d\omega}{2\pi} e^{i\omega(k-\ell)} = \sum_{\ell \geq 0} \frac{z^\ell}{\ell!} \delta_{k\ell} = z^k/k! \tag{E.8}$$

So

$$\Psi(z, \phi, \phi') = i\phi\langle k\rangle + i\phi'\langle k^2\rangle - \frac{z^2}{2\langle k\rangle} + \log \sum_{k \geq 0} \frac{z^k}{k!} e^{-i\phi k - i\phi' k^2} \tag{E.9}$$

We now transform $i\phi = \alpha + \log z$ and $i\phi' = \beta$, and find

$$
\frac{1}{N} \log Z(\langle k \rangle, \langle k^2 \rangle) = \frac{1}{2}\langle k \rangle \log \left(\frac{N}{\langle k \rangle}\right) + \text{extr}_z \left\{ \langle k \rangle \log z - \frac{z^2}{2\langle k\rangle} \right\} + \epsilon_N
$$
$$
+ \text{extr}_{\alpha,\beta} \left\{ \alpha\langle k\rangle + \beta\langle k^2\rangle + \log \sum_{k \geq 0} \frac{1}{k!} e^{-\alpha k - \beta k^2} \right\} \tag{E.10}
$$

We can immediately extremize over z, which gives us $z = \langle k \rangle$, and arrive at

$$
\frac{1}{N} \log Z(\langle k \rangle, \langle k^2 \rangle) = \frac{1}{2}\langle k \rangle \log (N\langle k\rangle) - \frac{1}{2}\langle k \rangle + \epsilon_N
$$
$$
+ \text{extr}_{\alpha,\beta} \left\{ \alpha\langle k\rangle + \beta\langle k^2\rangle + \log \sum_{k \geq 0} \frac{1}{k!} e^{-\alpha k - \beta k^2} \right\} \tag{E.11}
$$

Appendix F
Evolution of graph mobilities for edge-swap dynamics

In this appendix, we evaluate the change $\Delta_{ijk\ell;\alpha}|\Phi_{\mathbf{A}}|$ in the mobility of state $\mathbf{A}$ that is caused by execution of an edge swap $F_{ijk\ell;\alpha}$. Computing these changes should for large graphs be significantly less CPU-costly than calculating the mobility from scratch after each executed move, which according to (6.46) would require $\mathcal{O}(N^4)$ operations after each edge swap. This bookkeeping exercise is not in principle difficult, but can get somewhat messy. Since all degrees are preserved by edge swaps, a move $F_{ijk\ell;\alpha}$ can affect only the three terms in the second line of (6.46), so $\Delta_{ijk\ell;\alpha}|\Phi_{\mathbf{A}}| = \Delta^{(1)}_{ijk\ell;\alpha}|\Phi_{\mathbf{A}}| + \Delta^{(2)}_{ijk\ell;\alpha}|\Phi_{\mathbf{A}}| + \Delta^{(3)}_{ijk\ell;\alpha}|\Phi_{\mathbf{A}}|$. The first term is the simplest to evaluate:

$$\Delta^{(1)}_{ijk\ell;\alpha}|\Phi_{\mathbf{A}}| = -\frac{1}{2}\sum_{rs} k_r(\mathbf{A})k_s(\mathbf{A})\left[(F_{ijk\ell;\alpha}\mathbf{A})_{rs} - A_{rs}\right]$$

$$= \frac{1}{2}\sum_{(r,s)\in\mathcal{S}_{ijk\ell;\alpha}} k_r(\mathbf{A})k_s(\mathbf{A})(2A_{rs}-1) \tag{F.1}$$

Since the sum over (r,s) in the last line involves only the affected pairs in the set $\mathcal{S}_{ijk\ell;\alpha}$, the number of required operations is only $\mathcal{O}(1)$. To keep intermediate calculations for the remaining two terms simple, we introduce an indicator label $I_{(r,s)}$ to tell us whether the adjacency matrix entry (r,s) is affected by the edge swap under consideration, defined as $I_{(r,s)} = 1$ if $(r,s) \in \mathcal{S}_{ijk\ell;\alpha}$, with $I_{(r,s)} = 0$ otherwise. We also introduce its complement $\bar{I}_{(r,s)} = 1 - I_{(r,s)}$. We can now write (6.42, 6.43) as

$$(F_{ijk\ell;\alpha}\mathbf{A})_{rs} = \mathrm{XOR}(A_{rs}, I_{(r,s)}) \tag{F.2}$$

where the XOR operation is defined as usual: $\mathrm{XOR}(0,0) = \mathrm{XOR}(1,1) = 0$, $\mathrm{XOR}(1,0) = \mathrm{XOR}(0,1) = 1$. We note that our edge swaps are defined in such a way that for any trio (r,s,t) of distinct nodes one has[1]

$$I_{(r,s)}I_{(s,t)}I_{(t,r)} = 0 \tag{F.3}$$

In the second term of $\Delta_{ijk\ell;\alpha}|\Phi_{\mathbf{A}}|$ we can use (F.3), symmetrization of sums over indices where possible, as well as the invariance of various terms under node permutations:

---

[1]This goes back to the fact that edge swaps can be seen as inversion of all adjacency matrix entries along a closed path of length four, in which links are alternately present and absent. We can therefore never observe an edge swap inverting the link variables in a closed triangle.

$$\Delta^{(2)}_{ijk\ell;\alpha}|\Phi_{\mathbf{A}}| = \frac{1}{2}\sum_r \left[(F_{ijk\ell;\alpha}\mathbf{A})^3 - \mathbf{A}^3\right]_{rr}$$

$$= \frac{1}{2}\sum_{rst}\left\{\mathrm{XOR}(A_{rs}, I_{(r,s)})\mathrm{XOR}(A_{st}, I_{(s,t)})\mathrm{XOR}(A_{tr}, I_{(t,r)}) - A_{rs}A_{st}A_{tr}\right\}$$

$$= \frac{3}{2}\sum_{rst}\left\{I_{(r,s)}\bar{I}_{(s,t)}\bar{I}_{(t,r)}(1-A_{rs})A_{st}A_{tr} + I_{(r,s)}I_{(s,t)}\bar{I}_{(t,r)}(1-A_{rs})(1-A_{st})A_{tr}\right\}$$

$$+ \frac{1}{2}\sum_{rst}A_{rs}A_{st}A_{tr}\left[I_{(s,t)}I_{(t,r)}+I_{(r,s)}I_{(t,r)}+I_{(r,s)}I_{(s,t)}-I_{(r,s)}-I_{(s,t)}-I_{(t,r)}\right]$$

$$= \frac{3}{2}\sum_{(r,s)\in\mathcal{S}_{ijk\ell;\alpha}}\sum_{t\in\partial_r}\bar{I}_{(r,t)}\left\{(1-A_{rs})\mathrm{XOR}(I_{(s,t)}, A_{st}) - A_{rs}A_{st}\right\} \tag{F.4}$$

Here the neighbourhood ∂_r of a node r is defined as $\partial_r = \{s \leq N|\ A_{rs}=1\}$. The sum over (r, s) in the last line involves only the pairs in the set $\mathcal{S}_{ijk\ell;\alpha}$, so the number of required operations is $\mathcal{O}(1)$ in a sparse graph and $\mathcal{O}(N)$ in a dense one. We finally turn to the third term of $\Delta_{ijk\ell;\alpha}|\Phi_{\mathbf{A}}|$ for edge swaps, again using permutation invariances due to the cyclic nature of the trace operation. In particular, we will benefit from the observation that any factor of the following form will always be zero:

$$I_{(a,b)}I_{(c,d)}I_{(e,f)}A_{ab}A_{cd}A_{ef} \tag{F.5}$$

since if any three index pairs are jointly affected by an edge swap, they cannot all three have a link present. Thus, we find

$$\Delta^{(4)}_{ijk\ell;\alpha}|\Phi_{\mathbf{A}}| = \frac{1}{4}\sum_r \left[(F_{ijk\ell;\alpha}\mathbf{A})^4 - \mathbf{A}^4\right]_{rr}$$

$$= \frac{1}{4}\sum_{rstu}\left\{\mathrm{XOR}(A_{rs}, I_{(r,s)})\mathrm{XOR}(A_{st}, I_{(s,t)})\mathrm{XOR}(A_{tu}, I_{(t,u)})\mathrm{XOR}(A_{ur}, I_{(u,r)})\right.$$

$$\left. - A_{rs}A_{st}A_{tu}A_{ur}\right\}$$

$$= \frac{1}{4}\sum_{rstu}\left\{[I_{rs}(1-A_{rs})+\bar{I}_{rs}A_{rs}][I_{st}(1-A_{st})+\bar{I}_{st}A_{st}][I_{tu}(1-A_{tu})+\bar{I}_{tu}A_{tu}]\right.$$

$$\left.\times[I_{ur}(1-A_{ur})+\bar{I}_{ur}A_{ur}] - A_{rs}A_{st}A_{tu}A_{ur}\right\}$$

$$= \frac{1}{4}\sum_{rstu}\left\{I_{rs}I_{st}I_{tu}I_{ur}[(1-A_{rs})(1-A_{st})(1-A_{tu})(1-A_{ur}) - A_{rs}A_{st}A_{tu}A_{ur}]\right.$$

$$+ 4\bar{I}_{rs}I_{st}I_{tu}I_{ur}A_{rs}[(1-A_{st})(1-A_{tu})(1-A_{ur}) - A_{st}A_{tu}A_{ur}]$$

$$+ 4\bar{I}_{rs}\bar{I}_{st}I_{tu}I_{ur}A_{rs}A_{st}[(1-A_{tu})(1-A_{ur}) - A_{tu}A_{ur}]$$

$$+ 2\bar{I}_{rs}I_{st}\bar{I}_{tu}I_{ur}A_{rs}A_{tu}[(1-A_{st})(1-A_{ur}) - A_{st}A_{ur}]$$

$$\left. + 4\bar{I}_{rs}\bar{I}_{st}\bar{I}_{tu}I_{ur}A_{rs}A_{st}A_{tu}(1-2A_{ur})\right\}$$

$$= \frac{1}{4}\sum_{(u,r)\in\mathcal{S}_{ijk\ell;\alpha}}\sum_{st}\left\{I_{rs}I_{st}I_{tu}(1-4A_{rs}+4A_{rs}A_{st}+2A_{rs}A_{tu})\right.$$

$$+ \sum_{s \in \partial_r} \bar{I}_{rs} \Big[4 I_{st} I_{tu} (1 - A_{st} - A_{tu} - A_{ur} + A_{st} A_{tu} + A_{st} A_{ur} + A_{tu} A_{ur})$$

$$+ 4 \bar{I}_{st} I_{tu} A_{st} (1 - A_{tu} - A_{ur}) + 2 I_{st} \bar{I}_{tu} A_{tu} (1 - A_{st} - A_{ur})$$

$$+ 4 \bar{I}_{st} \bar{I}_{tu} A_{st} A_{tu} (1 - 2 A_{ur}) \Big] \Big\} \tag{F.6}$$

Again this requires $\mathcal{O}(1)$ computations in a sparse graph, and $\mathcal{O}(N)$ in a dense one.

Appendix G
Number of graphs with prescribed degree sequence

Here we calculate the quantity (7.14), which counts the number of nondirected simple graphs with prescribed degree sequence $\mathbf{k} = (k_1, \ldots, k_N)$. We observe that this is just a special case of the more general calculation done in Section 5.1, using the idea of importance sampling. We can therefore simply choose $K = N$, $\mathbf{\Omega}(\mathbf{A}) = \mathbf{k}(\mathbf{A})$ and $\mathbf{\Omega} = \mathbf{k}$ in expression (5.15) and write

$$\frac{1}{N} \log Z(\mathbf{k}) = \frac{1}{2} \langle k \rangle \left[1 + \log \left(\frac{N}{\langle k \rangle} \right) \right] - \log(2\pi)$$

$$+ \frac{1}{N} \log \int_{-\pi}^{\pi} d\boldsymbol{\omega} \, e^{i\boldsymbol{\omega} \cdot \mathbf{k}} \left\langle e^{-i\sum_{i=1}^{N} \omega_i k_i(\mathbf{A})} \right\rangle_{\mathrm{ER}} + \mathcal{O}\left(\frac{1}{N}\right) \qquad (\mathrm{G.1})$$

Here $\langle \ldots \rangle_{\mathrm{ER}}$ denotes averaging over an Erdös–Rènyi ensemble (3.1) with average degree $\langle k \rangle$. This average is calculated in the usual way:

$$\left\langle e^{-i\sum_{i=1}^{N} \omega_i k_i(\mathbf{A})} \right\rangle_{\mathrm{ER}} = \left\langle e^{-i\sum_{i<j}(\omega_i + \omega_j) A_{ij}} \right\rangle_{\mathrm{ER}}$$

$$= \prod_{i<j} \left[1 + \frac{\langle k \rangle}{N-1} (e^{-i(\omega_i + \omega_j)} - 1) \right]$$

$$= e^{\frac{\langle k \rangle}{2N} \sum_{ij} (e^{-i(\omega_i + \omega_j)} - 1) + \mathcal{O}(1)} \qquad (\mathrm{G.2})$$

Hence

$$\frac{1}{N} \log Z(\mathbf{k}) = \frac{1}{2} \langle k \rangle \log \left(\frac{N}{\langle k \rangle} \right) - \log(2\pi)$$

$$+ \frac{1}{N} \log \int_{-\pi}^{\pi} d\boldsymbol{\omega} \, e^{i\boldsymbol{\omega} \cdot \mathbf{k} + \frac{\langle k \rangle}{2N} [\sum_i e^{-i\omega_i}]^2} + \mathcal{O}\left(\frac{1}{N}\right) \qquad (\mathrm{G.3})$$

We can use Gaussian linearization, i.e. the identity $\exp(\frac{1}{2} y^2) = (2\pi)^{-\frac{1}{2}} \int dx \, e^{-\frac{1}{2}x^2 + xy}$, to obtain factorization over sites inside the integral over $\boldsymbol{\omega}$:

$$\int_{-\pi}^{\pi} d\boldsymbol{\omega} \, e^{i\boldsymbol{\omega} \cdot \mathbf{k} + \frac{\langle k \rangle}{2N} [\sum_i e^{-i\omega_i}]^2} = \int \frac{dx}{\sqrt{2\pi}} e^{-\frac{1}{2}x^2} \int_{-\pi}^{\pi} d\boldsymbol{\omega} \, e^{i\boldsymbol{\omega} \cdot \mathbf{k} + x\sqrt{\langle k \rangle / N} \sum_i e^{-i\omega_i}}$$

$$= \int \frac{dx}{\sqrt{2\pi/N}} e^{-\frac{1}{2}Nx^2} \prod_i \int_{-\pi}^{\pi} d\omega \, e^{i\omega k_i + x\sqrt{\langle k \rangle} e^{-i\omega}}$$

.

$$= \int \frac{\mathrm{d}x}{\sqrt{2\pi/N}} e^{-\frac{1}{2}Nx^2} \prod_i \left[\sum_{\ell \geq 0} \frac{(x\sqrt{\langle k \rangle})^\ell}{\ell!} \int_{-\pi}^{\pi} \mathrm{d}\omega \; e^{i\omega(k_i - \ell)} \right]$$

$$= (2\pi)^N \int \frac{\mathrm{d}x}{\sqrt{2\pi/N}} e^{-\frac{1}{2}Nx^2} \prod_i \frac{(x\sqrt{\langle k \rangle})^{k_i}}{k_i!}$$

$$= (2\pi)^N \left(\prod_i \frac{1}{k_i!} \right) \int \frac{\mathrm{d}x}{\sqrt{2\pi/N}} e^{N[\langle k \rangle \log(x\sqrt{\langle k \rangle}) - \frac{1}{2}x^2]} \quad (G.4)$$

Hence

$$\frac{1}{N} \log Z(\mathbf{k}) = \frac{1}{2} \langle k \rangle \log \left(\frac{N}{\langle k \rangle} \right) + \frac{1}{N} \log \left(\prod_i \frac{1}{k_i!} \right)$$

$$+ \frac{1}{N} \log \int \mathrm{d}x \; e^{N[\langle k \rangle \log(x\sqrt{\langle k \rangle}) - \frac{1}{2}x^2]} + \mathcal{O}(N^{-1} \log N) \quad (G.5)$$

The remaining integral over x is evaluated via the steepest descent method (see Appendix D), using $\max_x[\langle k \rangle \log(x\sqrt{\langle k \rangle})) - \frac{1}{2}x^2] = \langle k \rangle \log \langle k \rangle - \frac{1}{2} \langle k \rangle$, and we obtain

$$\frac{1}{N} \log Z(\mathbf{k}) = \frac{1}{2} \langle k \rangle [\log(N \langle k \rangle) - 1] + \frac{1}{N} \log \left(\prod_i \frac{1}{k_i!} \right) + \mathcal{O}(N^{-1} \log N) \quad (G.6)$$

Hence

$$Z(\mathbf{k}) = \left(\prod_i \frac{1}{k_i!} \right) e^{\frac{1}{2}N\langle k \rangle [\log(N \langle k \rangle) - 1] + \mathcal{O}(\log N)} \quad (G.7)$$

Note that in terms of $p(k) = N^{-1} \sum_i \delta_{k,k_i}$ and the Poissonian distribution $\pi(k) = e^{-\langle k \rangle} \langle k \rangle^k / k!$ we can after some simple manipulations also write our final result in a form that allows for clear interpretation:

$$\frac{1}{N} \log Z(\mathbf{k}) = \frac{1}{2} \langle k \rangle [\log(N/\langle k \rangle) + 1] + \sum_k p(k) \log p(k)$$

$$- \sum_k p(k) \log[p(k)/\pi(k)] + \mathcal{O}(N^{-1} \log N) \quad (G.8)$$

The first term is the entropy per node of the Erdös–Rènyi ensemble with average degree $\langle k \rangle$. The second is minus the entropy of the degree distribution, reflecting entropy reduction due to our not imposing the degree distribution but the actual degree sequence. The third term is minus the Kullback–Leibler distance [53] between $p(k)$ and $\pi(k)$, reflecting further entropy reduction if our imposed degree statistics are non-Poissonian.

Appendix H
Degree correlations in ensembles with constrained degrees

Definitions and summation over graphs. We calculate the average degree correlations $\Pi(k, k') = W(k, k')/W(k)W(k')$ of graphs drawn from the following ensemble of simple nondirected graphs, for large N:

$$p(\mathbf{A}|\mathbf{k}, Q) = \frac{1}{Z(\mathbf{k}, Q)} \prod_{i<j} \left[\frac{\langle k \rangle}{N} Q(k_i, k_j) \delta_{A_{ij}, 1} + \left(1 - \frac{\langle k \rangle}{N} Q(k_i, k_j) \right) \delta_{A_{ij}, 0} \right] \delta_{\mathbf{k}, \mathbf{k}(\mathbf{A})}$$

(H.1)

Here $W(k, k') = \sum_{\mathbf{A} \in G} p(\mathbf{A}|\mathbf{k}, Q) W(k, k'|\mathbf{A})$, with $W(k, k'|\mathbf{A})$ as defined in (2.9), and $Q(k, k')$ is assumed to satisfy the conditions (7.31). We use the integral representation $\delta_{nm} = (2\pi)^{-1} \int_{-\pi}^{\pi} d\omega \, e^{i\omega(n-m)}$ to implement the degree constraints. The normalization constant $Z(\mathbf{k}, Q)$ need not be calculated explicitly, as it will follow from the built-in normalization $\sum_{kk'} W(k, k') = 1$. With the shorthands $\boldsymbol{\omega} = (\omega_1, \ldots, \omega_N)$ and $\boldsymbol{\omega} \cdot \mathbf{k} = \sum_i \omega_i k_i$, we may write

$$
\begin{aligned}
W(k, k') &= \frac{1}{N\langle k \rangle} \sum_{\mathbf{A} \in G} p(\mathbf{A}|\mathbf{k}, Q) \sum_{ij} A_{ij} \delta_{k, k_i} \delta_{k', k_j} \\
&= \frac{1}{N\langle k \rangle Z(\mathbf{k}, Q)} \int_{-\pi}^{\pi} \frac{d\boldsymbol{\omega}}{(2\pi)^N} e^{i\boldsymbol{\omega} \cdot \mathbf{k}} \sum_{i<j} (\delta_{k, k_i} \delta_{k', k_j} + \delta_{k', k_i} \delta_{k, k_j}) \sum_{\mathbf{A} \in G} A_{ij} \\
&\quad \times e^{-i \sum_{rs} \omega_r A_{rs}} \prod_{r<s} \left[\frac{\langle k \rangle}{N} Q(k_r, k_s) \delta_{A_{rs}, 1} + \left(1 - \frac{\langle k \rangle}{N} Q(k_r, k_s) \right) \delta_{A_{rs}, 0} \right] \\
&= \frac{1}{N\langle k \rangle Z(\mathbf{k}, Q)} \int_{-\pi}^{\pi} \frac{d\boldsymbol{\omega}}{(2\pi)^N} e^{i\boldsymbol{\omega} \cdot \mathbf{k}} \sum_{i<j} (\delta_{k, k_i} \delta_{k', k_j} + \delta_{k', k_i} \delta_{k, k_j}) \sum_{\mathbf{A} \in G} A_{ij} \\
&\quad \times \prod_{r<s} \left[\frac{\langle k \rangle}{N} Q(k_r, k_s) e^{-i(\omega_r + \omega_s)} \delta_{A_{rs}, 1} + \left(1 - \frac{\langle k \rangle}{N} Q(k_r, k_s) \right) \delta_{A_{rs}, 0} \right] \\
&= \frac{1}{Z(\mathbf{k}, Q)} \int_{-\pi}^{\pi} \frac{d\boldsymbol{\omega}}{(2\pi)^N} e^{i\boldsymbol{\omega} \cdot \mathbf{k} + \mathcal{O}(N^{-1})} \prod_{r<s} \left[1 + \frac{\langle k \rangle}{N} Q(k_r, k_s) [e^{-i(\omega_r + \omega_s)} - 1] \right] \\
&\quad \times \left(\frac{1}{N^2} \sum_{i<j} Q(k_i, k_j) e^{-i(\omega_i + \omega_j)} (\delta_{k, k_i} \delta_{k', k_j} + \delta_{k', k_i} \delta_{k, k_j}) \right)
\end{aligned}
$$

(H.2)

We write the factors inside the square brackets in exponential form, up to non-vanishing orders in N, and use the symmetry of $Q(k, k')$ in the last line. This gives

$$W(k, k') = \frac{1}{Z(\mathbf{k}, Q)} \int_{-\pi}^{\pi} \frac{\mathrm{d}\boldsymbol{\omega}}{(2\pi)^N} e^{\mathrm{i}\boldsymbol{\omega}\cdot\mathbf{k}} \left[\frac{1}{N^2} \sum_{ij} Q(k_i, k_j) e^{-\mathrm{i}(\omega_i+\omega_j)} \delta_{k, k_i} \delta_{k', k_j} \right]$$

$$\times \prod_{r<s} e^{\frac{\langle k \rangle}{N} Q(k_r, k_s)[e^{-\mathrm{i}(\omega_r+\omega_s)} - 1] - \frac{1}{2}\frac{\langle k \rangle^2}{N^2} Q^2(k_r, k_s)[e^{-\mathrm{i}(\omega_r+\omega_s)} - 1]^2} + \mathcal{O}(N^{-1}) \quad \text{(H.3)}$$

Order parameters and saddle point equations. The form of (H.3) suggests that it will be advantageous to define the following shorthand, with $q \in \mathbb{N}$ and $\omega \in [-\pi, \pi]$:

$$P(q, \omega | \mathbf{k}, \boldsymbol{\omega}) = \frac{1}{N} \sum_r \delta_{q, k_r} \delta(\omega - \omega_r) \quad \text{(H.4)}$$

since this enables us to write

$$W(k, k') = \frac{Q(k, k')}{Z(\mathbf{k}, Q)} \int_{-\pi}^{\pi} \frac{\mathrm{d}\boldsymbol{\omega}}{(2\pi)^N} e^{\mathrm{i}\boldsymbol{\omega}\cdot\mathbf{k}} \left[\int_{-\pi}^{\pi} \mathrm{d}\omega \, P(k, \omega|\mathbf{k}, \boldsymbol{\omega}) e^{-\mathrm{i}\omega} \right] \left[\int_{-\pi}^{\pi} \mathrm{d}\omega \, P(k', \omega|\mathbf{k}, \boldsymbol{\omega}) e^{-\mathrm{i}\omega} \right]$$

$$\times e^{\frac{1}{2}\langle k \rangle N \sum_{qq'} \int_{-\pi}^{\pi} \mathrm{d}\omega\mathrm{d}\omega' \, P(q,\omega|\mathbf{k},\boldsymbol{\omega})P(q',\omega'|\mathbf{k},\boldsymbol{\omega})Q(q,q')[e^{-\mathrm{i}(\omega+\omega')} - 1]}$$

$$\times e^{-\frac{1}{2}\langle k \rangle \sum_q \int_{-\pi}^{\pi} \mathrm{d}\omega \, P(q,\omega|\mathbf{k},\boldsymbol{\omega})Q(q,q)[e^{-2\mathrm{i}\omega} - 1]}$$

$$\times e^{-\frac{1}{4}\langle k \rangle^2 \sum_{qq'} \int_{-\pi}^{\pi} \mathrm{d}\omega\mathrm{d}\omega' \, P(q,\omega|\mathbf{k},\boldsymbol{\omega})P(q',\omega'|\mathbf{k},\boldsymbol{\omega})Q^2(q,q')[e^{-\mathrm{i}(\omega+\omega')} - 1]^2} + \mathcal{O}(N^{-1}) \quad \text{(H.5)}$$

We now isolate the distributions $P(q, \omega|\boldsymbol{\omega})$ using by introducing integrals over suitable δ-distributions, similar to the calculation in Appendix C. For each (q, ω) we put

$$1 = \int \mathrm{d}P(q, \omega) \, \delta\left[P(q, \omega) - P(q, \omega|\mathbf{k}, \boldsymbol{\omega}) \right] \quad \text{(H.6)}$$

$$= \int \frac{\mathrm{d}P(q, \omega)\mathrm{d}\hat{P}(q, \omega)}{2\pi/N\Delta} e^{\mathrm{i}N\Delta\hat{P}(q,\omega)P(q,\omega) - \mathrm{i}\Delta\sum_r \delta_{qk_r}\delta(\omega - \omega_r)\hat{P}(q,\omega)} \quad \text{(H.7)}$$

We insert these expressions for each combination (q, ω), where ω is initially discretized with increments Δ. This Δ is sent to zero at the end of the calculation. The result is a path integral representation of $W(k, k')$. With the shorthand $\{\mathrm{d}P\mathrm{d}\hat{P}\} = \lim_{\Delta\to 0} \prod_{q,\omega}[\mathrm{d}P(q, \omega)\mathrm{d}\hat{P}(q, \omega)\Delta/2\pi]$ we obtain the following expression, in which we obtain factorization over sites with regard to the integration over $\boldsymbol{\omega}$

$$W(k, k') = \frac{Q(k, k')}{Z(\mathbf{k}, Q)} \int \{\mathrm{d}P\mathrm{d}\hat{P}\} \, e^{\mathrm{i}N \sum_q \int_{-\pi}^{\pi} \mathrm{d}\omega \, \hat{P}(q,\omega)P(q,\omega)} \int_{-\pi}^{\pi} \left[\prod_r \frac{\mathrm{d}\omega_r}{2\pi} e^{\mathrm{i}[\omega_r k_r - \hat{P}(k_r, \omega_r)]} \right]$$

$$\times \left[\int_{-\pi}^{\pi} \mathrm{d}\omega \, P(k, \omega) e^{-\mathrm{i}\omega} \right] \left[\int_{-\pi}^{\pi} \mathrm{d}\omega \, P(k', \omega) e^{-\mathrm{i}\omega} \right]$$

$$\times e^{\frac{1}{2}\langle k \rangle N \sum_{qq'} \int_{-\pi}^{\pi} \mathrm{d}\omega\mathrm{d}\omega' \, P(q,\omega)P(q',\omega')Q(q,q')[e^{-\mathrm{i}(\omega+\omega')} - 1]}$$

$$\times e^{-\frac{1}{2}\langle k \rangle \sum_q \int_{-\pi}^{\pi} \mathrm{d}\omega \, P(q,\omega)Q(q,q)[e^{-2\mathrm{i}\omega} - 1]}$$

$$\times e^{-\frac{1}{4}\langle k \rangle^2 \sum_{qq'} \int_{-\pi}^{\pi} \mathrm{d}\omega\mathrm{d}\omega' \, P(q,\omega)P(q',\omega')Q^2(q,q')[e^{-\mathrm{i}(\omega+\omega')} - 1]^2} + \mathcal{O}(N^{-1}) \quad \text{(H.8)}$$

Exploiting functional form and normalization. Our intermediate result (H.5) is seen to have the following functional form, which allows us to make use of the normalization of $W(k, k')$ to obtain further simplification:

$$W(k, k') = \frac{Q(k, k')}{Z(\mathbf{k}, Q)} \int \{\mathrm{d}P\mathrm{d}\hat{P}\} \ \mathrm{e}^{N\Psi[P,\hat{P}] + \Phi[P,\hat{P}]} G(k|P)G(k'|P) \qquad (\mathrm{H}.9)$$

in which

$$\Psi[P, \hat{P}] = \mathrm{i} \sum_q \int_{-\pi}^{\pi} \mathrm{d}\omega \ \hat{P}(q, \omega)P(q, \omega) + \frac{1}{N} \sum_r \log \int_{-\pi}^{\pi} \frac{\mathrm{d}\omega}{2\pi} \mathrm{e}^{\mathrm{i}[\omega k_r - \hat{P}(k_r, \omega)]}$$

$$+ \frac{1}{2} \langle k \rangle \sum_{qq'} \int_{-\pi}^{\pi} \mathrm{d}\omega\mathrm{d}\omega' \ P(q, \omega)P(q', \omega')Q(q, q')[\mathrm{e}^{-\mathrm{i}(\omega+\omega')} - 1] \qquad (\mathrm{H}.10)$$

$$\Phi[P, \hat{P}] = -\frac{1}{2} \langle k \rangle \sum_q \int_{-\pi}^{\pi} \mathrm{d}\omega \ P(q, \omega)Q(q, q)[\mathrm{e}^{-2\mathrm{i}\omega} - 1] \qquad (\mathrm{H}.11)$$

$$- \frac{1}{4} \langle k \rangle^2 \sum_{qq'} \int_{-\pi}^{\pi} \mathrm{d}\omega\mathrm{d}\omega' \ P(q, \omega)P(q', \omega')Q^2(q, q')[\mathrm{e}^{-\mathrm{i}(\omega+\omega')} - 1]^2 + \mathcal{O}\left(\frac{1}{N}\right)$$

$$G(k|P) = \int_{-\pi}^{\pi} \mathrm{d}\omega \ P(k, \omega)\mathrm{e}^{-\mathrm{i}\omega} \qquad (\mathrm{H}.12)$$

Working out the consequences of $\sum_{kk'} W(k, k')$ then provides the following simple expression for the normalization factor $Z(\mathbf{k}, Q)$:

$$Z(\mathbf{k}, Q) = \int \{\mathrm{d}P\mathrm{d}\hat{P}\} \ \mathrm{e}^{N\Psi[P,\hat{P}] + \Phi[P,\hat{P}]} \sum_{kk'} Q(k, k')G(k|P)G(k'|P) \qquad (\mathrm{H}.13)$$

With this result our formula for $W(k, k')$ becomes

$$W(k, k') = \frac{\int \{\mathrm{d}P\mathrm{d}\hat{P}\} \ \mathrm{e}^{N\Psi[P,\hat{P}] + \Phi[P,\hat{P}]} \ Q(k, k')G(k|P)G(k'|P)}{\int \{\mathrm{d}P\mathrm{d}\hat{P}\} \ \mathrm{e}^{N\Psi[P,\hat{P}] + \Phi[P,\hat{P}]} \sum_{qq'} Q(q, q')G(q|P)G(q'|P)} \qquad (\mathrm{H}.14)$$

For large N we can use steepest descent integration (see Appendix D) to evaluate the remaining path integrals over P and $\hat{P}$. This reveals that the functions $\Phi[P, \hat{P}]$ simply drop out of formulae, and for $W(k, k')$ we find

$$W(k, k') = \frac{Q(k, k')G(k|P)G(k'|P)}{\sum_{qq'} Q(q, q')G(q|P)G(q'|P)} \qquad (\mathrm{H}.15)$$

in which P is evaluated by extremization of $\Psi[P, \hat{P}]$.

Evaluation of the functional saddle point. Our final task is to calculate the saddle points of (H.10) by functional differentiation with respect to P and $\hat{P}$. The saddle point equations are found to be the following, with $p(q) = N^{-1} \sum_r \delta_{qk_r}$:

$$P(q, \omega) = p(q) \frac{\mathrm{e}^{\mathrm{i}[\omega q - \hat{P}(q,\omega)]}}{\int_{-\pi}^{\pi} \mathrm{d}\omega' \ \mathrm{e}^{\mathrm{i}[\omega' q - \hat{P}(q,\omega')]}} \qquad (\mathrm{H}.16)$$

$$i\hat{P}(q,\omega) = -\langle k \rangle \sum_{q'} Q(q,q') \int_{-\pi}^{\pi} d\omega' \, P(q',\omega')[e^{-i(\omega+\omega')} - 1] \tag{H.17}$$

Upon eliminating $\hat{P}$, we are left with an equation for P only:

$$P(q,\omega) = p(q)\frac{e^{i\omega q + \langle k \rangle e^{-i\omega} F(q)}}{\int_{-\pi}^{\pi} d\omega' \, e^{i\omega' q + \langle k \rangle e^{-i\omega'} F(q)}} \tag{H.18}$$

with $F(q)$ to be solved from

$$F(q) = \sum_{q'} Q(q,q')G(q') \qquad G(q) = p(q)\frac{\int_{-\pi}^{\pi} d\omega \, e^{i\omega(q-1) + \langle k \rangle e^{-i\omega} F(q)}}{\int_{-\pi}^{\pi} d\omega \, e^{i\omega q + \langle k \rangle e^{-i\omega} F(q)}} \tag{H.19}$$

Similar to the calculation in Appendix C, we evaluate the integrals over ϕ via

$$\int_{-\pi}^{\pi} d\omega \, e^{i\omega q + \langle k \rangle e^{-i\omega} F} = \sum_{\ell \geq 0} \frac{\langle k \rangle^{\ell} F^{\ell}}{\ell!} \int_{-\pi}^{\pi} d\omega \, e^{i\omega(q-\ell)}$$

$$= 2\pi \sum_{\ell \geq 0} \frac{\langle k \rangle^{\ell} F^{\ell}}{\ell!} \delta_{q\ell} = \begin{cases} 2\pi \langle k \rangle^q F^q / q! & \text{if } q \geq 0 \\ 0 & \text{otherwise} \end{cases} \tag{H.20}$$

We then find that $G(q) = p(q)q/\langle k \rangle F(q)$, and that $F(q)$ is the solution of

$$F(q) = \sum_{q' \geq 0} p(q')Q(q,q')\frac{q'}{\langle k \rangle F(q')} \tag{H.21}$$

Since the function $G(q)$ in (H.19) is defined as $G(q) = \int_{-\pi}^{\pi} d\omega \, P(q,\omega)e^{-i\omega}$, it is exactly identical to the quantity $G(k|P)$ appearing in (H.15); we therefore obtain our final formula for $W(k,k')$ simply by substituting $G(q) = p(q)q/\langle k \rangle F(q)$ into (H.15):

$$W(k,k') = \frac{Q(k,k')p(k)p(k')kk'/\langle k \rangle^2 F(k)F(k')}{\sum_{qq'} Q(q,q')p(q)p(q')qq'/\langle k \rangle^2 F(q)F(q')}$$

$$= Q(k,k')\frac{p(k)k}{\langle k \rangle F(k)}\frac{p(k')k'}{\langle k \rangle F(k')} \tag{H.22}$$

In the last step, we used the fact that the denominator of the first line equals unity, as a consequence of identity (H.21). The marginals $W(k) = \sum_{k'} W(k,k')$ are

$$W(k) = \frac{p(k)k}{\langle k \rangle F(k)} \sum_{k'} Q(k,k')\frac{p(k')k'}{\langle k \rangle F(k')} = p(k)k/\langle k \rangle \tag{H.23}$$

and the degree correlation kernel $\Pi(k,k') = W(k,k')/W(k)W(k')$ for $N \to \infty$ is

$$\Pi(k,k') = Q(k,k')/F(k)F(k') \tag{H.24}$$

In conclusion, large graphs drawn from the ensemble (H.1) will exhibit the relative degree correlations (H.24), in which the function $F(k)$ obeys (H.21).

Appendix I
Evolution of triangle and square counters due to ordered edge swaps

---

We derive the formulae (7.70, 7.71) for the changes in $\text{Tr}(\mathbf{A}^3)$ and $\text{Tr}(\mathbf{A}^4)$ (which count the number of triangles and squares in $\mathbf{A}$) resulting from execution of the ordered edge swap $F_{ijk\ell}$, as defined in Section 7.5. These are used to keep track of the mobility $|\Phi_{\mathbf{A}}|$, which, in turn, is needed in the move acceptance probabilities (7.73).

Evaluation of $\Delta_{ijk\ell}\text{Tr}(\mathbf{A}^3)$. We first work out the change in $\text{Tr}(\mathbf{A}^3)$ under the edge swap $F_{ijk\ell}$, which is defined as

$$\Delta_{ijk\ell}\text{Tr}(\mathbf{A}^3) = \sum_{mnp}\left[(F_{ijk\ell}\mathbf{A})_{mn}(F_{ijk\ell}\mathbf{A})_{np}(F_{ijk\ell}\mathbf{A})_{pm} - A_{mn}A_{np}A_{pm}\right] \quad (\text{I}.1)$$

All terms in this sum are closed paths of length three, with at least one link corresponding to a pair (r,s) in the set $\mathcal{S}_{ijk\ell} = \{(i,j),(j,k),(k,\ell),(\ell,i),(j,i),(k,j),(\ell,k),(i,\ell)\}$. Closed paths of length three with two or three distinct links in $\mathcal{S}_{ijk\ell}$ must have at least two *consecutive* links in $\mathcal{S}_{ijk\ell}$, leading to terms of the form $A_{pr}A_{rs}$. However, by the nature of edge swaps, any such product would have either $A_{pr}=1-A_{rs}=1$ or $A_{rs}=1-A_{pr}=1$. Hence, closed paths with more than one link in $\mathcal{S}_{ijk\ell}$ cannot contribute to (I.1); for each surviving term in (I.1), two sites must belong to the quadruplet $\{i,j,k,\ell\}$, and the third will not. Thus we obtain

$$\Delta_{ijk\ell}\text{Tr}(\mathbf{A}^3) = 6\,\Delta_{ijk\ell}\sum_{p\notin\{i,j,k,\ell\}}\left(A_{ij}A_{jp}A_{pi} + A_{jk}A_{kp}A_{pj} + A_{k\ell}A_{\ell p}A_{pk} + A_{\ell i}A_{ip}A_{p\ell}\right)$$

$$(\text{I}.2)$$

The factor 6 accounts for: the three options for choosing which index in $\{m,n,p\}$ is not in $\{i,j,k,\ell\}$, and an additional factor two because we chose for the other two indices only index pairs in $\{(i,j),(j,k),(k,\ell),(\ell,i)\}$ (the other four pairs in $\mathcal{S}_{ijk\ell}$ give the same contributions because our graphs are nondirected). Working out the above expression further, given that we know which links are affected by $F_{ijk\ell}$, gives

$$\Delta_{ijk\ell}\text{Tr}(\mathbf{A}^3) = 6\sum_{p\notin\{i,j,k,\ell\}}\Big((1-2A_{ij})A_{jp}A_{pi} + (1-2A_{jk})A_{kp}A_{pj} + (1-2A_{k\ell})A_{\ell p}A_{pk}$$

$$+ (1-2A_{\ell i})A_{ip}A_{p\ell}\Big)$$

$$= 6 \sum_{p \notin \{i,j,k,\ell\}} \left(-A_{jp}A_{pi} + A_{kp}A_{pj} - A_{\ell p}A_{pk} + A_{ip}A_{p\ell} \right)$$

$$= 6 \sum_{p \notin \{i,j,k,\ell\}} (A_{kp} - A_{pi})(A_{pj} - A_{p\ell}) \tag{I.3}$$

Evaluation of $\Delta_{ijk\ell}\mathrm{Tr}(\mathbf{A}^3)$. Here we can reason in the same way, and consider only the contributions from all the squares with at least one link in $\mathcal{S}_{ijk\ell}$, but not two consecutive links in $\mathcal{S}_{ijk\ell}$. This time, we have in principle extra contributions from paths that involve 'backtracking' over links, of which there are two types:

(a) Paths of length two with one link in $\mathcal{S}_{ijk\ell}$, i.e. $A_{ij}A_{jp}A_{pj}A_{ji}$, $A_{jk}A_{kp}A_{pk}A_{kj}$, $A_{k\ell}A_{\ell p}A_{p\ell}A_{\ell k}$ and $A_{\ell i}A_{ip}A_{pi}A_{i\ell}$, all with $p \notin \mathcal{S}_{ijk\ell}$. Each of these paths can be counted in four different ways, depending on the starting point of the path and the orientation.

(b) Paths representing backtracking over a single link in $\mathcal{S}_{ijk\ell}$, i.e. $A_{ij}A_{ji}A_{ij}A_{ji}$, $A_{jk}A_{kj}A_{jk}A_{kj}$, $A_{k\ell}A_{\ell k}A_{k\ell}A_{\ell k}$ and $A_{\ell i}A_{i\ell}A_{\ell i}A_{i\ell}$. Each of these can be counted in two different ways.

In addition to the above backtracking contributions, we have contributions from closed paths along non-repeated links with two non-consecutive links in $\mathcal{S}_{ijk\ell}$:

(c) $A_{ij}A_{j\ell}A_{\ell k}A_{ki}$ and $A_{jk}A_{ki}A_{i\ell}A_{\ell j}$.

We note that the contributions to $\Delta_{ijk\ell}\mathrm{Tr}(\mathbf{A}^3)$ of the above terms all sum to zero:

$$(a) + (b) + (c) = 4\Delta_{ijk\ell}\left(\sum_{p \neq i} A_{ij}^2 A_{jp}^2 + \sum_{p \neq j} A_{jk}^2 A_{kp}^2 + \sum_{p \neq k} A_{k\ell}^2 A_{\ell p}^2 + \sum_{p \neq \ell} A_{\ell i}^2 A_{ip}^2 \right.$$

$$\left. + \sum_{p \neq j} A_{ji}^2 A_{ip}^2 + \sum_{p \neq k} A_{kj}^2 A_{jp}^2 + \sum_{p \neq \ell} A_{\ell k}^2 A_{kp}^2 + \sum_{p \neq i} A_{i\ell}^2 A_{\ell p}^2 \right)$$

$$+ 2\Delta_{ijk\ell}(A_{ij}^4 + A_{jk}^4 + A_{k\ell}^4 + A_{\ell i}^4)$$

$$+ 8\Delta_{ijk\ell}(A_{ij}A_{j\ell}A_{\ell k}A_{ki} + A_{ik}A_{kj}A_{j\ell}A_{\ell i}) = 0 \tag{I.4}$$

Here we used $\Delta_{ijk\ell}A_{j\ell} = \Delta_{ijk\ell}A_{ki} = 0$, $\Delta_{ijk\ell}(A_{ji} + A_{\ell i}) = \Delta_{ijk\ell}(A_{jk} + A_{\ell k}) = 0$, and

$$\Delta_{ijk\ell}(A_{ij}A_{j\ell}A_{\ell k}A_{ki} + A_{ik}A_{kj}A_{j\ell}A_{\ell i}) = A_{ki}A_{j\ell}\Delta_{ijk\ell}(A_{ij}A_{k\ell} + A_{kj}A_{\ell i}) = 0 \tag{I.5}$$

It follows from the above argument that the only contributions to $\Delta_{ijk\ell}\mathrm{Tr}(\mathbf{A}^4)$ come from closed paths of length 4, without any backtracking:

$$\Delta_{ijk\ell}\mathrm{Tr}(\mathbf{A}^4) = 8\Delta_{ijk\ell}\left[\sum_{p \notin \{i,j,k\}} \sum_{r \notin \{i,j,\ell\}} \bar{\delta}_{(pr),(\ell,k)} A_{ij}A_{jp}A_{pr}A_{ri} \right.$$

$$\left. + \sum_{p \notin \{j,k,\ell\}} \sum_{r \notin \{i,j,\ell\}} \bar{\delta}_{(p,r),(i,\ell)} A_{jk}A_{kp}A_{pr}A_{rj} \right.$$

$$+ \sum_{p \notin \{i,k,\ell\}} \sum_{r \notin \{j,k,\ell\}} \bar{\delta}_{(p,r),(j,i)} A_{k\ell} A_{\ell p} A_{pr} A_{rk}$$

$$+ \sum_{p \notin \{i,j,\ell\}} \sum_{r \notin \{i,k,\ell\}} \bar{\delta}_{(p,r),(k,j)} A_{\ell i} A_{ip} A_{pr} A_{r\ell} \Bigg] \qquad (I.6)$$

where the prefactor eight accounts for the eight possible ways of counting a square. Using $\Delta_{ijk\ell} A_{ij} = 1 - 2A_{ij}$, $\Delta_{ijk\ell} A_{jk} = 1 - 2A_{jk}$, $\Delta_{ijk\ell} A_{k\ell} = 1 - 2A_{k\ell}$ and $\Delta_{ijk\ell} A_{\ell i} = 1 - 2A_{\ell i}$ then leads to the final result

$$\Delta_{ijk\ell} \text{Tr}(\mathbf{A}^4) = 8 \Delta_{ijk\ell} \Bigg[- \sum_{p \notin \{i,j,k\}} \sum_{r \notin \{i,j,\ell\}} \bar{\delta}_{(p,r),(\ell,k)} A_{jp} A_{pr} A_{ri}$$

$$+ \sum_{p \notin \{j,k,\ell\}} \sum_{r \notin \{i,j,\ell\}} \bar{\delta}_{(p,r),(i,\ell)} A_{kp} A_{pr} A_{rj} - \sum_{p \notin \{i,k,\ell\}} \sum_{r \notin \{j,k,\ell\}} \bar{\delta}_{(p,r),(j,i)} A_{\ell p} A_{pr} A_{rk}$$

$$+ \sum_{p \notin \{i,j,\ell\}} \sum_{r \notin \{i,k,\ell\}} \bar{\delta}_{(p,r),(k,j)} A_{ip} A_{pr} A_{r\ell} \Bigg] \qquad (I.7)$$

The numerical implementations of $\Delta_{ijk\ell} \text{Tr}(\mathbf{A}^3)$ and $\Delta_{ijk\ell} \text{Tr}(\mathbf{A}^4)$, can be made extremely fast, by a clever use of pointers. In terms of computation costs, evaluating $\Delta_{ijk\ell} \text{Tr}(\mathbf{A}^3)$ requires $k_j + k_k$ computations at each time step, and $\Delta_{ijk\ell} \text{Tr}(\mathbf{A}^4)$ requires $k_k k_j + k_j k_k + k_\ell k_k + k_i k_\ell$ computations at each time step. Hence, the former has a computational cost that scales as $\mathcal{O}(k_{\text{max}})$, whereas for the latter it scales as $\mathcal{O}(k_{\text{max}}^2)$.

Appendix J
Algorithms

Preamble. In this chapter, we provide examples of pseudocode for all the graph generation algorithms presented in the book. We group them into six sections:

- *Markov Chain Monte Carlo (MCMC) algorithms based on link flips.* These are suitable for sampling from models involving only soft constraints, such as exponential random graph models (ERGMs). They are based on repeated applications of link-flip moves, defined in Section 6.3.

- *MCMC algorithms for sampling graphs with hard constraints.* These are algorithms used to sample from network models with hard constraints, in their most general form. In subsequent sections, we specialize these algorithms to deal with hard constraints on the total number of links and on the degree sequence (in the presence or absence of additional soft constraints).

- *MCMC algorithms based on hinge flips.* These are suitable for sampling networks with hard-constrained number of links. They are based on repeated application of hinge-flip moves, as defined in Section 6.3. Choosing the acceptance probability of each move appropriately, one can tailor the algorithm to produce graphs with a number of soft constraints such as the degree distribution (in addition to the hard constraint on the number of links).

- *MCMC algorithms based on edge swaps.* These are suitable for sampling networks with hard-constrained degree sequences. They are based on repeated application of edge-swap moves as defined in Sections 6.4, 7.5 and 7.7. By choosing suitable acceptance probabilities for each move, one can tailor these algorithms to produce graphs with a number of soft constraints, such as degree correlations (in addition to the hard constraint on the degrees).

- *Growth algorithms* These are non-MCMC algorithms, in that they do not target any specific equilibrium measure. They irreversibly create links or join stubs according to recipes which are empirically found to lead to certain network properties (e.g. power-law distributions or high clustering coefficient, etc.)

- *Auxiliary algorithms* These are auxiliary routines that one may wish to call in any of the main algorithms illustrated in the previous sections.

For simplicity, all algorithms are presented with a fixed number of iterations. The reader may wish to implement a break clause based on a predefined internal state of the algorithm, such as the Hamming distance between the initial and the instantaneous state (Algorithm 31), or the value of a specific network observable. If the algorithms are going to be used with a fixed number of iterations, we typically recommend that the number of runs should significantly exceed the total number of edges, in order to ensure that on average enough mutations have acted on each edge to erase features specific to the initial state.

If the system does not show complex behaviour such as bistability, i.e. coexistence of very different phases for the same values of control (i.e. network ensemble) parameters, one can normally span the space of the most relevant configurations within a number of runs of the order of a hundred times the number of edges. The network models presented in Chapter 7 fall within this category, and the algorithms we used to sample from these models and produce the figures in Chapter 7 run typically for 1000 iterations per link.

Conversely, for systems that do exhibit bistability, such as the two-star model and the Strauss model presented in Chapter 4, this number of iterations will only allow the algorithm to explore the configuration space corresponding to one particular phase (either the dense or the sparse one, as decided by the initial conditions). The time required for the system to move from one phase to the other is typically exponentially large in the system size, and the number of iterations required to sample all the most important configurations will have to be accordingly large.

Regardless of the stopping condition used, it is generally good practice to output the Hamming distance between the initial and current state, and to track key observables in order to obtain a sense for whether the algorithm is behaving as expected. In Algorithms 13, 14 and 15 we show this directly.

The term 'draw' (e.g. *draw* $p \in [0, 1]$) is used to mean 'generate uniformly at random an element from the given set'. It is generally obvious from the context if the element is required to be an integer or a real number or another type of object (e.g. a node or a function). It is also taken as a given that the element should be generated uniformly at random from the given distribution, set or interval. The user may wish to investigate the options for random number generators available in their programming environment. The sequence of numbers generated by a random number generator is only an approximation of what is formally understood as *random* because the sequence is typically completely determined by the seed values and will eventually repeat. For very large networks, it may be necessary to incorporate a more sophisticated random generator (e.g. the Mersenne Twister [116]) into the software.

A network **A** will normally be coded as an $N \times N$ matrix. So the instruction *initialize an empty network on N nodes* should be read as setting $A_{ij} = 0$ for $i, j \in \{1, \ldots, N\}$. Conversely, the instruction *set $A_{ij} = 1$* should be understood as indicating that a link has been created from node node j to node i.

When working with a nondirected network, the user may wish to save on memory by only recording the upper triangular part of the connectivity matrix; the lower triangular part will be identical by symmetry. For simplicity, the algorithms are written assuming nondirected networks. It is usually easy to generalize them to directed networks.

An alternative data architecture – designed to leverage the often sparse nature of real-world networks – is to represent the network **A** as a list of links, that is, as a list of pairs of connected nodes $\{(i_1, j_1), (i_2, j_2), \ldots, (i_L, j_L)\}$. Alternatively, one could represent the network $\mathbf{A}$ via a matrix of integers, in which the elements in row i represents the neighbours of node i. In Algorithm 32 we give a routine to build the matrix of neighbours from the adjacency matrix. This saves memory, since it does not require long strings of '0' digits to be saved. Correctly implemented, it can also save on post-initialization run-time. As an example, we show in Algorithm 35 how the matrix of neighbours can be used to efficiently update the traces $\text{Tr}(\mathbf{A}^3)$ under the edge swap $F_{ijk\ell}$ defined in Section 7.5, according to (7.70).

In Algorithm 33 we explicitly describe how to harvest parameters from a model network. This is a general useful strategy to easily determine sensible input parameters $(\mathbf{k}, p(k), Q(k, k')$ etc.) for a graph randomization algorithm.

J.1 Algorithms based on link flips

Algorithm 1: Soft-constrained average number of links: the ER model

---

input : Number of nodes N and target number of links L

output: Network with N nodes and (on average) L links

set $\lambda = \log \frac{L}{N(N-1)/2-L}$

initialise empty $N \times N$ binary matrix $\mathbf{A}$

                 `/* or start with any nondirected graph with `N` nodes */`

for $t = 1 \ldots T$ `/* `$T \gg L$` */`

do

    draw two nodes $i, j \in \{1, \ldots N\}$

    if $A_{ij} = 0$ **then**

        $\mid$ Prob(execute move) $= \min\{1, e^{\lambda}\}$

    else

        $\mid$ Prob(execute move) $= \min\{1, e^{-\lambda}\}$

    end

    draw $r \in [0, 1]$

    if $r <$ Prob(execute move) **then**

        set $A_{ij} \to 1 - A_{ij}$

        set $A_{ji} \to 1 - A_{ji}$ `/* execute move */`

    end

end

Algorithm 2: Soft-constrained degree sequence

---

input : Degree sequence $k_1, k_2, \ldots, k_N$

output: Network $\mathbf{A}$ with average degree sequence $k_1, k_2 \ldots k_N$

solve $\{\lambda_1, \ldots, \lambda_N\}$ from the N equations $k_i = \sum_j \frac{1}{1+e^{-\lambda_i - \lambda_j}}$ by minimizing the function (see discussion in Section 4.1)

$$\sum_{j=1}^{N} [\log(1 + e^{\lambda_i + \lambda_j}) - k_j \lambda_j]$$

with respect to $\lambda_1, \ldots, \lambda_N$ via a suitable minimization algorithm (see, e.g. [150]),

initialise the graph $\mathbf{A}$ to any member $\mathbf{A}_0 \in G$

      `/* `G` is the set of simple nondirected graphs with `N` nodes */`

for $t = 1 \ldots T$ `/* `$T \gg L$`, `L`=number of links of `$\mathbf{A}$` */`
do

   draw two nodes i, j with $1 \le i < j \le N$

   if $A_{ij} = 0$ **then**

      | Prob(execute move) $= \min\{1, e^{\lambda_i + \lambda_j}\}$

   else

      | Prob(execute move) $= \min\{1, e^{-\lambda_i - \lambda_j}\}$

   end

   draw $r \in [0, 1]$

   if $r < $ Prob(execute move) **then**

      set $A_{ij} \to 1 - A_{ij}$

      set $A_{ji} \to 1 - A_{ji}$

   end

end

Algorithm 3: Prescribed degree distribution via degree sequence

---

input : Number of nodes N and target degree distribution $p(k)$

output: Network **A** with average degree distribution $p(k)$

repeat
   | **for** $i = 1 \dots N$ **do**
   | | draw k_i from $p(k)$
   | **end**
until *graphical sequence of degrees* $\{k_i\}$

     /* check graphicality with Erdös--Gallai criterion as given in equation (4.23) */

solve $\{\lambda_1, \dots, \lambda_N\}$ from the N equations $k_i = \sum_j \frac{1}{1+e^{-\lambda_i-\lambda_j}}$ by minimizing the function (see discussion in Section 4.1)

$$\sum_{j=1}^{N} [\log(1 + e^{\lambda_i + \lambda_j}) - k_j \lambda_j]$$

with respect to $\lambda_1, \dots, \lambda_N$ via a suitable minimization algorithm (see e.g. [150]), initialise the graph **A** to any member $\mathbf{A}_0 \in G$

     /* G is the set of simple nondirected graphs with N nodes */

for $t = 1 \dots T$ /* $T \gg L$, L=number of links of **A** */
do
   | draw two nodes i, j with $1 \le i < j \le N$
   | **if** $A_{ij} = 0$ **then**
   | | Prob(execute move) $= \min\{1, e^{\lambda_i + \lambda_j}\}$
   | **else**
   | | Prob(execute move) $= \min\{1, e^{-\lambda_i - \lambda_j}\}$
   | **end**
   | draw $r \in [0, 1]$
   | **if** $r <$ Prob(execute move) **then**
   | | set $A_{ij} \to 1 - A_{ij}$
   | | set $A_{ji} \to 1 - A_{ji}$
   | **end**
end

Algorithm 4: Soft-constraining the first two moments of the degree distribution: the two-star model

---

input : Number of nodes N, target number of links L and two-stars S

output: Network $\mathbf{A}$ with (on average) L links and S two-stars

```
/* but may show degenerate behaviour and fail to converge to
expected value */
```

solve α, β from equations (4.55) by minimizing the function (see discussion in Chapter 4.1)

$$f(\alpha, \beta) - \alpha \frac{2L}{N} - \beta \frac{2(S+L)}{N}$$

with respect to α, β, where $f(\alpha, \beta)$ is as given in (4.54), via a suitable minimization algorithm [150]; alternatively, one can leave α, β as control parameters.

initialise an N node network $\mathbf{A}$

```
/* if parameters correspond to the phase where there is more than
one solution, it is better to choose a starting network with
values of L and S closer to the target values */
```

for $t = 1 \ldots T$ /* $T \gg L$ */
do

     draw two nodes i, j
     if $A_{ij} = 0$ **then**
        | Prob(execute move) $= \min\{1, e^{2\alpha + 2\beta[1 + k_i(\mathbf{A}) + k_j(\mathbf{A})]}\}$
     else
        | Prob(execute move) $= \min\{1, e^{-2\alpha + 2\beta[1 - k_i(\mathbf{A}) - k_j(\mathbf{A})]}\}$
     end
     draw $r \in [0, 1]$
     if $r <$ Prob(execute move) **then**
        | set $A_{ij} \to 1 - A_{ij}$
        | set $A_{ji} \to 1 - A_{ji}$
     end
end

Algorithm 5: Soft-constrained number of links and triangles: the Strauss model

---

input : Number of nodes N, target number of links L and triangles T

output: Network $\mathbf{A}$ with (on average) L links and T triangles

```
        /* but may show degenerate behaviour and fail to converge to
expected value */
```

solve the parameters λ_1, λ_2 from equations (4.63) by minimizing the function (see discussion in Section 4.1)

$$\log Z(\lambda_1, \lambda_2) - \lambda_1 L - \lambda_2 T$$

with respect to λ_1, λ_2, where $Z(\lambda_1, \lambda_2)$ is as given in (4.62), via a suitable minimization algorithm [150]; alternatively, one can leave λ_1, λ_2 as control parameters

initialise an N node network $\mathbf{A}$

```
   /* if parameters correspond to the phase where there is more than
one solution, it is better to choose a starting network with
values of L and T closer to the target values */
```

for $t = 1 \ldots T$ /* $T \gg L$ */
do

$\quad$ draw two indices i, j with $1 \le i < j \le N$

$\quad$ **if** $A_{ij} = 0$ **then**

$\quad\quad \mid$ Prob(execute move) $= \min\{1, e^{\lambda_1 + \lambda_2 \sum_r A_{ir} A_{jr}}\}$

$\quad$ **else**

$\quad\quad \mid$ Prob(execute move) $= \min\{1, e^{-\lambda_1 - \lambda_2 \sum_r A_{ir} A_{jr}}\}$

$\quad$ **end**

$\quad$ draw $r \in [0, 1]$

$\quad$ **if** $r < $ Prob(execute move) **then**

$\quad\quad \mid$ set $A_{ij} \to 1 - A_{ij}$

$\quad\quad \mid$ set $A_{ji} \to 1 - A_{ji}$

$\quad$ **end**

end

Algorithm 6: ERGM to generate a hidden variable ensemble

---

input : Graph on N nodes, with some attribute x_i associated with each node i (if attributes are not discrete, a proper binning procedure must be carried out)

output: Network with same distribution $\rho(x)$ of attributes $\{x_1, \ldots, x_N\}$ and the same statistics of links between each possible flavour of node pairs

define a binning protocol, for converting real-valued densities into discrete probabilities

calculate target $L(x, x') = \sum_{i \neq j} A_{ij} \delta_{xx_i} \delta_{x'x_j} (1 - \frac{1}{2}\delta_{xx'})$ for each combination of x and x' in the input graph, and carry out smoothing

    `/* L records the frequency with which links pairing nodes with`
`flavours x,x' appear in the model graph */`

calculate the frequency $\rho(x) = n(x)/N$ with which each attribute x appears in the model graph, where $n(x) = \sum_i \delta_{x,x_i}$

initialise an empty seed graph on N nodes

for $i = 1 \ldots N$ **do**
  | draw $x_i \in \rho(x)$
end
    `/* allocate each node i a hidden parameter x_i, drawn from ρ(x) */`

for $1 \leq i < j \leq N$ **do**
    if $x_i \neq x_j$ **then**
      | calculate $W(x_i, x_j) = L(x_i, x_j)/\left(N^2 \rho(x_i)\rho(x_j)\right)$
    end
    else
      | calculate $W(x_i, x_i) = L(x_i, x_i)/\left(\frac{1}{2}N\rho(x_i)[N\rho(x_i) - 1]\right)$
    end
    draw $r \in [0, 1]$
    if $r < W(x_i, x_j)$ **then**
    | set $A_{ij} = A_{ji} = 1$
    end
end

Algorithm 7: Temporal networks

---

input : Number of nodes N and number of time steps ℓ
Target number of links L at the initial time $t = 1$
Observables $H(\mathbf{A}(t), \mathbf{A}(t-1))$ of networks at two successive time
steps $t \in [2, \ell]$

output: Temporal graph $\{\mathbf{A}(1), \ldots, \mathbf{A}(\ell)\}$

set $\lambda_1 = \log\left(\frac{L}{N(N-1)/2-L}\right)$
solve $\lambda_2, \ldots, \lambda_\ell$ from equation (10.3)
initialise the ℓ graphs $\{\mathbf{A}(t)\}$ to any member of the set G
/* G is the set of graphs with N nodes */

for $t = 1 \ldots T$ /* $T \gg L$ */
do

    draw uniformly at random a time label $t \in \{1, \ldots, \ell\}$
    draw uniformly at random two indices (i, j) with $i < j$
    calculate $\Delta_{ij,t}(\mathbf{A})$ /* given by equation (10.19) */

    set Prob(execute move) $= \min\left\{1, e^{\Delta_{ij,t}(\mathbf{A})}\right\}$

    draw $r \in [0, 1]$

    if $r <$ Prob(execute move) **then**
        set $A_{ij}(t) \to 1 - A_{ij}(t)$
        set $A_{ji}(t) \to 1 - A_{ji}(t)$
    end
end

Algorithm 8: Multiplex networks

---

input : Number of nodes N and number of layers M

Observables $\Omega_\mu(\vec{\mathbf{A}})$ of multiplex network with $\mu = 1, \ldots, K$

output: Multiplex graph $\vec{\mathbf{A}} = (\mathbf{A}^1, \ldots, \mathbf{A}^M)$

calculate the Lagrange parameters $(\lambda_1, \ldots, \lambda_K)$ from (10.24)

initialise each graph $\mathbf{A}^\alpha$ to any member of the set G, for $\alpha = 1 \ldots M$

        `/* G is the set of multiplex with N nodes and M layers */`

for $t = 1 \ldots T$ `/* T ≫ L, L =number of links */`

do

    draw uniformly random a layer label $\alpha \in \{1, \ldots, M\}$

    draw uniformly at random two indices (i, j) with $i < j$

    calculate $\Delta_{ij,\alpha}(\vec{\mathbf{A}})$ `/* given by equation (10.34) */`

    set Prob(execute move) $= \min\left\{1, e^{\Delta_{ij,\alpha}(\vec{\mathbf{A}})}\right\}$

    draw $r \in [0, 1]$

    if $r < $ Prob(execute move) **then**

        set $A_{ij}^\alpha \rightarrow 1 - A_{ij}^\alpha$

        set $A_{ji}^\alpha \rightarrow 1 - A_{ji}^\alpha$

    end

end

Algorithm 9: Weighted version of an Erdös–Rényi graph

---

    input : Number of nodes N and target average weight of link $\langle w \rangle$

    output: Weighted network **A** with average link weight $\langle w \rangle$

set $p = \langle w \rangle / (1 + \langle w \rangle)$

initialise the graph **A** as empty N node graph

                        `/* expect entries of` A `to be non-negative integers */`

for $1 \leq i < j \leq N$ `/* every pair of vertices */`

do

    set *Indicator* $= 1$

    while *Indicator* $= 1$ **do**

        draw $r \in [0, 1]$

        if $r < p$ **then**

            set $A_{ij} = A_{ij} + 1$

            set $A_{ji} = A_{ji} + 1$

        else

            set *Indicator* $= 0$

        end

    end

end

J.2 MCMC algorithms sampling from graphs with hard constraints

---

Algorithm 10: Switch-and-hold with uniform sampling of moves

---

input : Network $\mathbf{A}$ satisfying certain constraints

output: A randomized version of $\mathbf{A}$ satisfying exactly the same constraints

for $t = 1 \ldots T$ `/* `$T \gg L$`, `L`=number of links of `$\mathbf{A}$` */`
do

> draw move F from the set Φ of all moves
> **if** $\mathbf{A}$ *violates the constraint* **then**
> > | Prob(execute move) $= 1$
>
> **else**
> > **if** $F\mathbf{A}$ *violates the constraint* **then**
> > > | Prob(execute move) $= 0$
> >
> > **end**
> > **else**
> > > | Prob(execute move) $= \min\{1, p(F\mathbf{A})/p(\mathbf{A})\}$
> >
> > **end**
>
> **end**
> draw $r \in [0,1]$
> **if** $r <$ Prob(execute move) **then**
> > | set $\mathbf{A} \rightarrow F\mathbf{A}$ `/* execute move */`
>
> **end**

end

Algorithm 11: MCMC with state-dependent move sets – version 1

---

input : Network **A** satisfying certain constraints

output: A randomized version of **A** satisfying exactly the same constraints

calculate $|\Phi_{\mathbf{A}}|$
```
        /* number of moves which can act on A to transform it into
another network satisfying the constraints */
```

for $t = 1 \ldots T$ `/* `$T \gg L$`, L=number of links of A */`
do

    draw $F \in \Phi_{\mathbf{A}}$ `/* pick an allowed move */`
    calculate $|\Phi_{F\mathbf{A}}|$
    calculate Prob(execute move) $= \min\{1, p(F\mathbf{A})|\Phi_{\mathbf{A}}|/p(\mathbf{A})|\Phi_{F\mathbf{A}}|\}$
    draw $r \in [0,1]$

    if $r < $ Prob(execute move) **then**
       | $\mathbf{A} \rightarrow F\mathbf{A}$ `/* execute move */`
    end
end

Algorithm 12: MCMC with state-dependent move sets – version 2

---

input : Network **A** satisfying a certain constraints

output: A randomized version of **A** satisfying exactly the same constraints

calculate $|\Phi_\mathbf{A}|$
```
            /* number of moves which can act on A to transform it into
another network satisfying the constraints */
```

for $t = 1 \ldots T$ `/* `$T \gg L$`,  L=number of links of A */`
do

> draw $F \in \Phi_\mathbf{A}$ `/* pick an allowed move */`
> calculate $\Delta_F |\Phi_\mathbf{A}| = |\Phi_{F\mathbf{A}}| - |\Phi_\mathbf{A}|$
> calculate $\text{Prob(execute move)} = \min\{1, p(F\mathbf{A})|\Phi_\mathbf{A}|/[p(\mathbf{A})(|\Phi_\mathbf{A}|+\Delta_F|\Phi_\mathbf{A}|)]\}$
> draw $r \in [0,1]$
>
> **if** $r < \text{Prob(execute move)}$ **then**
>> update $\mathbf{A} \to F\mathbf{A}$ `/* execute move */`
>> update $|\Phi_\mathbf{A}| \to |\Phi_\mathbf{A}|+\Delta_F|\Phi_\mathbf{A}|$
>
> **end**

end

J.3   Algorithms based on hinge flips

Random graphs with hard-constrained number of links

Algorithm 13: Homogeneous sampling from applicable moves only (Protocol 3, Chapter 7)

---

input : Number of nodes N and links L

output: Average of observable $\langle\Omega(\mathbf{A})\rangle$ and network $\mathbf{A}$ with N nodes and L links

initialise the graph $\mathbf{A}$ to any member $\mathbf{A}_0 \in G$
   /* G is the set of simple nondirected graphs with N nodes and L
links */

calculate the degrees k_i of $\mathbf{A}$ for $i = 1, \ldots, N$

calculate $\sigma(\mathbf{A}) = \sum_i k_i^2(\mathbf{A})$
   /* the acceptance proabilities will depend on this observable */

set $\langle\Omega(\mathbf{A})\rangle = 0$ /* $\Omega(\mathbf{A})$ is a user-defined observable of interest */

for $t = 1 \ldots T$ /* $T \gg L$ */
do

> compute the observable $\Omega(\mathbf{A})$
> set $\langle\Omega(\mathbf{A})\rangle \rightarrow \langle\Omega(\mathbf{A})\rangle + [\Omega(\mathbf{A}) - \langle\Omega(\mathbf{A})\rangle]/t$ /* running average */
>
> **repeat**
>> draw edge $(i, j) \in \{edges\}$ /* i.e. nodes such that $A_{ij} = 1$ */
>> draw node $k \notin \{i, j\}$ /* three distinct nodes i, j, k */
>
> **until** $A_{ik} \neq 1$
> /* if k is connected to i the hinge flip is not possible   */
>
> calculate $\sigma' = \sigma + 2(1 + k_k - k_j)$
> calculate Prob(execute move) $= ((N-1)L - \sigma)/((N-1)L - \sigma')$
> draw $r \in [0, 1]$
>
> **if** $r < $ Prob(execute move) **then**
>> set $A_{ij} = A_{ji} = 0$ and $A_{ik} = A_{ki} = 1$ /* execute the move */
>> set $\sigma = \sigma'$ /* update observable */
>> set $k_k = k_k + 1$ and $k_j = k_j - 1$ /* update list of node degrees */
>
> **end**

end

Algorithm 14: Inhomogeneous sampling from applicable moves only (Protocol 1, Chapter 7)

---

input : Number of nodes N and links L

output: Average of observable $\langle\Omega(\mathbf{A})\rangle$ and network $\mathbf{A}$ with N nodes and L links

initialise the graph $\mathbf{A}$ to any member $\mathbf{A}_0 \in G$

   `/* G is the set of simple nondirected graphs with N nodes and L links */`

calculate the degrees k_i of $\mathbf{A}$ for $i = 1, \ldots, N$

set $\langle\Omega(\mathbf{A})\rangle = 0$ `/* Ω(A) is a user-defined observable of interest */`

for $t = 1 \ldots T$ `/* T ≫ L */`
do

    compute the observable $\Omega(\mathbf{A})$

    set $\langle\Omega(\mathbf{A})\rangle \to \langle\Omega(\mathbf{A})\rangle + [\Omega(\mathbf{A}) - \langle\Omega(\mathbf{A})\rangle]/t$ `/* running average */`

    repeat

       | draw $j \in \{nodes\}$

    until $k_j \neq 0$

    repeat

       draw node $i \in \partial_j$ `/* pick a neighbour of j */`

       draw node $k \notin \{i, j\}$

    until $A_{ik} \neq 1$

    calculate Prob(execute move) $= k_j/(k_k+1)$

    draw $r \in [0, 1]$

    if $r <$ Prob(execute move) **then**

       set $A_{ij} = A_{ji} = 0$ and $A_{ik} = A_{ki} = 1$ `/* execute the move */`

       set $k_k = k_k + 1$ and $k_j = k_j - 1$ `/* update list of node degrees */`

    end

end

Algorithm 15: Homogeneous sampling from all moves (switch and hold)

---

input : Number of nodes N and links L

output: Average of observable $\langle \Omega(\mathbf{A}) \rangle$ and network $\mathbf{A}$ with N nodes and L links

initialise the graph $\mathbf{A}$ to any member $\mathbf{A}_0 \in G$
   /* G is the set of simple nondirected graphs with N nodes and L links */

set $\langle \Omega(\mathbf{A}) \rangle = 0$ /* $\Omega(\mathbf{A})$ is a user-defined observable of interest   */

for $t = 1 \ldots T$ /* $T \gg L$ */
do
> compute the observable $\Omega(\mathbf{A})$
> $\langle \Omega(\mathbf{A}) \rangle \to \langle \Omega(\mathbf{A}) \rangle + [\Omega(\mathbf{A}) - \langle \Omega(\mathbf{A}) \rangle]/t$ /* running average */
>
> draw edge $(i,j) \in \{edges\}$
> draw node $k \notin \{i,j\}$ /* three distinct nodes i,j,k */
>
> **if** $A_{ik} = 0$ **then**
> > set $A_{ij} = A_{ji} = 0$ and $A_{ik} = A_{ki} = 1$ /* execute the move */
>
> **end**

end

Random graphs with hard-constrained number of links and soft-constrained degree distribution

---

Algorithm 16: Inhomogeneous sampling from applicable moves only (Protocol 1, Chapter 7)

---

input : Number of nodes N, links L and target degree distribution $p(k)$

output: Network **A** with N nodes, L links and degree distribution $p(k)$

initialise the graph **A** to any member $\mathbf{A}_0 \in G$
    /* G is the set of simple nondirected graphs with N nodes and L links */

calculate the degrees k_i of **A** for $i = 1, \ldots, N$

for $t = 1 \ldots T$ /* $T \gg L$ */
do
   | **repeat**
   | | draw uniformly at random a node j
   | **until** $k_j \neq 0$

   | **repeat**
   | | draw node $i \in \partial_j$ /* pick a neighbour of j */
   | | draw node $k \notin \{i, j\}$
   | **until** $A_{ik} \neq 1$

   | calculate Prob(execute move) $= p(k_j - 1)p(k_k + 1)/p(k_j)p(k_k)$
   | draw $r \in [0, 1]$

   | **if** $r <$ Prob(execute move) **then**
   | | set $A_{ij} = A_{ji} = 0$ and $A_{ik} = A_{ki} = 1$ /* execute the move */
   | | set $k_k = k_k + 1$ and $k_j = k_j - 1$ /* update list of node degrees */
   | **end**
end

Algorithm 17: Inhomogeneous sampling from applicable moves only (Protocol 2, Chapter 7)

---

input : Number of nodes N, links L and target degree distribution $p(k)$

output: Network **A** with N nodes, L links and degree distribution $p(k)$

initialize the graph **A** to any member $\mathbf{A}_0 \in G$
   /* G is the set of simple nondirected graphs with N nodes and L links */

calculate the degrees k_i of **A** for $i = 1, \ldots, N$

for $t = 1 \ldots T$ /* $T \gg L$ */
do

    repeat
       | draw nodes i and k from $\{nodes\}$
    until $A_{ik} = 0$ *and* $k_i \neq 0$

    draw node $j \in \partial_i$ /* pick a neighbour of i */

    calculate Prob(execute move) $= p(k_j - 1)p(k_k + 1)(k_k + 1)/\left[p(k_j)p(k_k)k_j\right]$

    draw $r \in [0, 1]$

    if $r <$ Prob(execute move) **then**
       | set $A_{ij} = A_{ji} = 0$ and $A_{ik} = A_{ki} = 1$ /* execute the move */
       | set $k_k = k_k + 1$ and $k_j = k_j - 1$ /* update list of node degrees */
    end
end

Algorithm 18: Homogeneous sampling from applicable moves only (Protocol 3, Chapter 7)

---

input : Number of nodes N, links L and target degree distribution $p(k)$

output: Network $\mathbf{A}$ with N nodes, L links and degree distribution $p(k)$

initialise the graph $\mathbf{A}$ to any member $\mathbf{A}_0 \in G$
      `/* `G` is the set of simple nondirected graphs with `L` links */`

calculate the degrees k_i of $\mathbf{A}$ for $i = 1, \ldots, N$
calculate $\sigma(\mathbf{A}) = \sum_i k_i^2(\mathbf{A})$
      `/* the acceptance proabilities will depend on this observable */`

for $t = 1 \ldots T$ `/* `$T \gg L$` */`
do

   | **repeat**

   | | draw edge (i, j) `/* ordered pair of nodes such that `$A_{ij} = 1$` */`
   | | draw node $k \notin \{i, j\}$ `/* three distinct nodes `i, j, k` */`

   | **until** $A_{ik} \neq 1$
   | `/* if `k` is already connected to `i` the hinge flip is not possible */`

   | calculate $\sigma' = \sigma + 2(1 + k_k - k_j)$

   | calculate Prob(execute move) $=$
   | $((N-1)L - \sigma)p(k_j - 1)p(k_k + 1)(k_k + 1)/((N-1)L - \sigma')p(k_j)p(k_k)k_j$

   | draw $r \in [0, 1]$

   | **if** $r <$ Prob(execute move) **then**

   | | set $A_{ij} = A_{ji} = 0$ and $A_{ik} = A_{ki} = 1$ `/* execute the move */`
   | | set $\sigma = \sigma'$ `/* update observable */`
   | | set $k_k = k_k + 1$ and $k_j = k_j - 1$ `/* update list of node degrees */`

   | **end**

end

J.4   Algorithms based on edge swaps

Random graphs with hard-constrained degrees and soft-constrained degree correlations

---

Algorithm 19: Switch-and-hold MCMC, via uniform sampling of edge-swap moves $F_{ijk\ell;\alpha}$, as defined in Section 6.4

---

input : Nondirected simple network $\mathbf{A}_0$ whose degree sequence and degree correlations we aim to reproduce

output: Network $\mathbf{A}$ with degree sequence matching $\mathbf{A}_0$ and degree correlations averaging to that of $\mathbf{A}_0$; Average of observable $\langle\Omega(\mathbf{A})\rangle$

extract target parameters $\mathbf{k}$, $\langle k\rangle$ and $Q(\cdot,\cdot)$ from $\mathbf{A}_0$ /* Algorithm 33 */

for $1 \le i < j \le N$ do
| set $L_{ij} = \log([\langle k\rangle Q(k_i,k_j)/N]/[1-\langle k\rangle Q(k_i,k_j)/N])$
end

initialize $\mathbf{A}$ to $\mathbf{A}_0$
set $\langle\Omega(\mathbf{A})\rangle = 0$ /* $\Omega(\mathbf{A})$ is a user-defined observable of interest   */

for $t = 1\ldots T$ /* $T \gg L$, L =number of links of $\mathbf{A}$ */
do
| compute the observable $\Omega(\mathbf{A})$
| set $\langle\Omega(\mathbf{A})\rangle \to \langle\Omega(\mathbf{A})\rangle + [\Omega(\mathbf{A}) - \langle\Omega(\mathbf{A})\rangle]/t$ /* running average */
| draw $\alpha \in \{1,2,3\}$
| draw $i<j<k<\ell \in [1,N]$
| /* for sparse networks it is much more efficient to draw links rather than sites, see algorithm 20 */
|
| if $F_{ijk\ell;\alpha}$ *can act on* $\mathbf{A}$ then
| | Prob(execute move) $= \exp[\sum_{r<s,(r,s)\in\mathcal{S}_{ijk\ell;\alpha}} L_{rs}(1-2A_{rs})]$
| | /* Probability of executing the move is the minimum of 1 and Prob(execute move) */
| | draw $r \in [0,1]$
| | if $r <$ Prob(execute move) then
| | | execute $\mathbf{A} \to F_{ijk\ell;\alpha}\mathbf{A}$
| | end
| end
end
end

Algorithm 20: Switch-and-hold MCMC, via uniform sampling of ordered edge-swap moves $F_{ijk\ell}$, as defined in Section 7.5

> **input** : Nondirected simple network $\mathbf{A}_0$ whose degree sequence and degree correlations we aim to reproduce
>
> **output**: Network $\mathbf{A}$ with degree sequence matching $\mathbf{A}_0$ and degree correlations averaging to that of $\mathbf{A}_0$; Average of observable $\langle\Omega(\mathbf{A})\rangle$
>
> extract target parameters $\mathbf{k}$, $\langle k \rangle$ and $Q(\cdot,\cdot)$ from $\mathbf{A}_0$ /* Algorithm 33 */
>
> **for** $1 \le i < j \le N$ **do**
> | set $L_{ij} = \log([\langle k \rangle Q(k_i, k_j)/N]/[1 - \langle k \rangle Q(k_i, k_j)/N])$
> **end**
>
> initialize $\mathbf{A}$ to $\mathbf{A}_0$
> set $\langle\Omega(\mathbf{A})\rangle = 0$ /* $\Omega(\mathbf{A})$ is a user-defined observable of interest   */
>
> **for** $t = 1 \ldots T$ /* $T \gg L$, L =number of links of $\mathbf{A}$ */
> **do**
> > compute the observable $\Omega(\mathbf{A})$
> > set $\langle\Omega(\mathbf{A})\rangle \to \langle\Omega(\mathbf{A})\rangle + [\Omega(\mathbf{A}) - \langle\Omega(\mathbf{A})\rangle]/t$ /* running average */
> >
> > draw two ordered edges (i, j) and (k, ℓ) /* using Algorithm 36 */
> >
> > /* these determine uniquely the candidate ordered edge swap $F_{ijk\ell}$, defined in Section 7.5 */
> >
> > **if** $A_{i\ell} = A_{jk} = 0$ **then**
> > > Prob(execute move) $= \exp[L_{i\ell} + L_{jk} - L_{ij} - L_{k\ell}]$
> > > draw $r \in [0, 1]$
> > >
> > > **if** $r <$ Prob(execute move) **then**
> > > > set $A_{i\ell} = A_{\ell i} = A_{jk} = A_{kj} = 1$ /* execute move */
> > > > set $A_{ij} = A_{ji} = A_{k\ell} = A_{\ell k} = 0$
> > > **end**
> > **end**
> **end**
> **end**

Algorithm 21: Mobility-based MCMC, via uniform sampling of *applicable* edge-swap moves $F_{ijk\ell;\alpha}$, as defined in Section 6.4

---

input : Nondirected simple network $\mathbf{A}_0$ whose degree sequence and degree correlations we aim to reproduce

output: Network $\mathbf{A}$ with degree sequence matching $\mathbf{A}_0$ and degree correlations averaging to that of $\mathbf{A}_0$; Average of observable $\langle\Omega(\mathbf{A})\rangle$

extract target parameters $\mathbf{k}$, $\langle k\rangle$ and $Q(\cdot,\cdot)$ from $\mathbf{A}_0$ `/* Algorithm 33`
`*/`

for $1 \le i < j \le N$ **do**
| set $L_{ij} = \log([\langle k\rangle Q(k_i, k_j)/N]/[1-\langle k\rangle Q(k_i, k_j)/N])$
end
initialize $\mathbf{A}$ to $\mathbf{A}_0$
set $\langle\Omega(\mathbf{A})\rangle = 0$ `/* `$\Omega(\mathbf{A})$` is a user-defined observable of interest`
`*/`
calculate $|\Phi_{\mathbf{A}}|$ `/* how many moves can act on A */`
for $t = 1\ldots T$ `/* `$T \gg L$`, `L`=number of links of A */`
do
| compute the observable $\Omega(\mathbf{A})$
| set $\langle\Omega(\mathbf{A})\rangle \to \langle\Omega(\mathbf{A})\rangle + [\Omega(\mathbf{A}) - \langle\Omega(\mathbf{A})\rangle]/t$ `/* running average */`
| **repeat**
| | draw $\alpha \in \{1,2,3\}$
| | draw $i < j < k < \ell \in [1, N]$
| | `/* for sparse networks it is much more efficient to draw`
| | `links rather than sites, see Algorithm 22`
| | `*/`
| **until** $F_{ijk\ell;\alpha}$ can act on $\mathbf{A}$
| calculate $|\Phi_{F_{ijk\ell;\alpha}\mathbf{A}}|$ `/* via equation (6.46) and a suitable */`
| `/* algorithm such as 35 */`

| set
| $\text{Prob(execute move)} = (|\Phi_{\mathbf{A}}|/|\Phi_{F_{ijk\ell;\alpha}\mathbf{A}}|) \exp[\sum_{r<s,(r,s)\in\mathcal{S}_{ijk\ell;\alpha}} L_{rs}(1-2A_{rs})]$

| draw $r \in [0,1]$
| **if** $r < \text{Prob(execute move)}$ **then**
| | execute $\mathbf{A} \to F_{ijk\ell;\alpha}\mathbf{A}$
| **end**
end

Algorithm 22: Mobility-based MCMC, via uniform sampling of *applicable* ordered edge swaps $F_{ijk\ell}$, as defined in Section 7.5

---

input : Nondirected simple network $\mathbf{A}_0$ whose degree sequence and degree correlations we aim to reproduce

output: Network $\mathbf{A}$ with degree sequence matching $\mathbf{A}_0$ and degree correlations averaging to that of $\mathbf{A}_0$; Average of observable $\langle \Omega(\mathbf{A}) \rangle$

extract target parameters $\mathbf{k}$, $\langle k \rangle$ and $Q(\cdot, \cdot)$ from $\mathbf{A}_0$ /* Algorithm 33 */

for $1 \leq i < j \leq N$ **do**
 | set $L_{ij} = \log([\langle k \rangle Q(k_i, k_j)/N]/[1 - \langle k \rangle Q(k_i, k_j)/N])$
end

initialize $\mathbf{A}$ to $\mathbf{A}_0$
set $\langle \Omega(\mathbf{A}) \rangle = 0$ /* $\Omega(\mathbf{A})$ is a user-defined observable of interest   */

calculate $|\Phi_{\mathbf{A}}|$ /* how many moves can act on $\mathbf{A}$ */

for $t = 1 \ldots T$ /* $T \gg L$, $L =$number of links of $\mathbf{A}$ */
do
 | compute the observable $\Omega(\mathbf{A})$
 | set $\langle \Omega(\mathbf{A}) \rangle \rightarrow \langle \Omega(\mathbf{A}) \rangle + [\Omega(\mathbf{A}) - \langle \Omega(\mathbf{A}) \rangle]/t$ /* running average */
 | draw an applicable move $F_{ijk\ell}$ /* via Algorithm 37 */
 |
 | calculate $\Delta_{ijk\ell}|\Phi_{\mathbf{A}}|$ /* via equations (7.67), (7.70), (7.71)   */
 | /* and a suitable algorithm as 35 */
 |
 | calculate
 | Prob(execute move) $= |\Phi_{\mathbf{A}}|/(|\Phi_{\mathbf{A}}| + \Delta_{ijk\ell}|\Phi_{\mathbf{A}}|) \exp[L_{i\ell} + L_{jk} - L_{ij} - L_{k\ell}]$
 | draw $r \in [0, 1]$
 |
 | **if** $r <$ Prob(execute move) **then**
 | | set $A_{i\ell} = A_{\ell i} = A_{jk} = A_{kj} = 1$
 | | set $A_{ij} = A_{ji} = A_{k\ell} = A_{\ell k} = 0$ /* execute move */
 | | set $|\Phi_{\mathbf{A}}| = |\Phi_{\mathbf{A}}| + \Delta_{ijk\ell}|\Phi_{\mathbf{A}}|$ /* update mobility */
 | **end**
end

Algorithm 23: MCMC based on non-uniform sampling of ordered edge swaps $F_{ijk\ell}$, as defined in Section 7.7

---

input : Nondirected simple network $\mathbf{A}_0$ whose degree sequence and degree correlations we aim to reproduce

output: Network $\mathbf{A}$ with degree sequence matching $\mathbf{A}_0$ and degree correlations averaging to that of $\mathbf{A}_0$; Average of observable $\langle\Omega(\mathbf{A})\rangle$

extract target parameters $\mathbf{k}$, $\langle k \rangle$ and $Q(\cdot,\cdot)$ from $\mathbf{A}_0$ /* Algorithm 33 */

for $1 \le i < j \le N$ **do**
| set $L_{ij} = \log([\langle k \rangle Q(k_i, k_j)/N]/[1-\langle k \rangle Q(k_i, k_j)/N])$
end

initialize $\mathbf{A}$ to $\mathbf{A}_0$
set $\langle\Omega(\mathbf{A})\rangle = 0$ /* $\Omega(\mathbf{A})$ is a user-defined observable of interest   */
calculate $R_i(\mathbf{A}) = N\langle k\rangle - \sum_r A_{ir}k_r - k_i$ for all $i = 1,\ldots,N$

for $t = 1\ldots T$ /* $T \gg L$, L =number of links of $\mathbf{A}$ */
do

    | compute the observable $\Omega(\mathbf{A})$
    | set $\langle\Omega(\mathbf{A})\rangle \to \langle\Omega(\mathbf{A})\rangle + [\Omega(\mathbf{A}) - \langle\Omega(\mathbf{A})\rangle]/t$ /* running average */
    | draw (non-uniformly) a move $F_{ijk\ell}$ /* via Algorithm 38 */

    | calculate $\Delta R_i = k_j - k_\ell$; $\Delta R_j = k_i - k_k$; $\Delta R_k = k_\ell - k_j$; $\Delta R_\ell = k_k - k_i$

$$\mathrm{Prob} = \frac{(R_i+\Delta R_i)^{-1}+(R_j+\Delta R_j)^{-1}+(R_k+\Delta R_k)^{-1}+(R_\ell+\Delta R_\ell)^{-1}}{R_i^{-1}+R_j^{-1}+R_k^{-1}+R_\ell^{-1}}$$
$$\times \exp[L_{i\ell} + L_{jk} - L_{ij} - L_{k\ell}]$$

    | draw $r \in [0,1]$

    | **if** $r < \mathrm{Prob}$ **then**
        | set $A_{i\ell} = A_{\ell i} = A_{jk} = A_{kj} = 1$ /* execute move */
        | set $A_{ij} = A_{ji} = A_{k\ell} = A_{\ell k} = 0$
        | set $R_s = R_s + \Delta R_s$ for $s = i, j, k, \ell$ /* update R_i, R_j, R_k, R_ℓ */
    | **end**
end

J.5 Growth Algorithms

Algorithm 24: Basic preferential attachment

---

input : Size of seed network m_0

output: Network $\mathbf{A}(T)$ with $m_0 + T$ nodes.

initialize a complete seed network $\mathbf{A}(0)$ with m_0 nodes
initialize vector of degrees $(k_1, k_2 \ldots k_{m_0})$ where $k_i = m_0 - 1 \ \forall \ i \in [1, m_0]$

for $t = 1 \ldots T$ **do**

    add new node labelled $m_0 + t$ `/* call new network A(t) */`

    while *indicator* $= 0$ **do**

        draw $i \in [1, (m_0 + t - 1)]$

        draw $r \in [0, 1]$

        if $r \leq \frac{k_i}{\sum_j k_j}$ **then**

            set $A(t)_{i,m_0+t} = A(t)_{m_0+t,i} = 1$

                      `/* create link between node i and new node */`

            update vector of degrees

            set *indicator* $= 1$

        end

    end

end

Algorithm 25: A hidden variable version of preferential attachment

---

input : Distribution $\rho(x)$ for the hidden fitness parameter
Size of seed network m_0

output: Network $\mathbf{A}(T)$ with $m_0 + T$ nodes.

initialize a complete seed network $\mathbf{A}(0)$ with m_0 nodes
/* complete seed network is mathematically simple to analyse, but not essential for the operation of the algorithm */

initialize vector of degrees $(k_1, k_2 \ldots k_{m_0})$ where $k_i = m_0 - 1 \ \forall \ i \in [1, m_0]$

for $i = 1 \ldots m_0$ **do**
| draw $x_i \in \rho(x)$
end

initialize vector of fitnesses $\mathbf{x} = (x_1, x_2 \ldots x_{m_0})$

for $t = 1 \ldots T$ **do**
| add new node labelled $m_0 + t$ $\qquad\qquad$ /* call new network $\mathbf{A}(t)$ */
| draw $x_t \in \rho(x)$ and append to $\mathbf{x}$
| $\qquad\qquad\qquad$ /* hidden fitness parameter for new node */
| **while** $indicator = 0$ **do**
| | draw $i \in [1, (m_0 + t - 1)]$
| | draw $r \in [0, 1]$
| | **if** $r \leq \frac{k_i x_i}{\sum_j k_j x_j}$ **then**
| | | set $A(t)_{i, m_0+t} = A(t)_{m_0+t, i} = 1$
| | | $\qquad\qquad$ /* create link between node i and new node */
| | | update vector of degrees
| | | set $indicator = 1$
| | **end**
| **end**
end

/* to generate a denser graph and create more than one link at each time step the user has to choose whether to include a subroutine to check for double links. A choice of a larger size of seed graph m_0 will reduce the likelihood of multiple links.   */

Algorithm 26: Node-age adjusted preferential attachment

---

input : Size of seed network m_0

output: Network $\mathbf{A}(T)$ with $m_0 + T$ nodes

initialize a complete seed network $\mathbf{A}(0)$ with m_0 nodes
initialize vector of degrees $(k_1, k_2 \ldots k_{m_0})$ where $k_i = m_0 - 1 \ \forall \ i \in [1, m_0]$
initialize vector of ages $\mathbf{x} = (x_1, x_2 \ldots x_{m_0})$ where $x_i = 0 \ \forall \ i \in [1, m_0]$

for $t = 1 \ldots T$ **do**

    set $x_i = x_i + 1 \ \forall \ i \in [1, m_0 + t - 1]$ `/* age each node by 1 */`
    add new node labelled $(m_0 + t)$ `/* call new network A(t) */`
    increase length of age vector by one element
    set $x_{m_0 + t} = 0$ `/* new element is aged zero. */`

    while *indicator* $= 0$ **do**

        draw $i \in [1, (m_0 + t) - 1]$ `/* pick a random existing node */`
        draw $r \in [0, 1]$
        if $r \leq \frac{k_i/x_i}{\sum_j k_j/x_j}$ **then**

            set $A(t)_{i, m_0 + t} = A(t)_{m_0 + t, i} = 1$
                        `/* create link between node i and new node */`
            update vector of degrees
            Set *indicator* $= 1$

        end

    end

end

Algorithm 27: A basic implementation of the stub-joining method: the config-uration model

---

input : Target degree sequence $k_1, \ldots, k_N$

output: Network on N nodes with each node i having degree k_i
/* this network might include double or self links. */

/* define stubs vector */

initialize vector **S** of length $2L = \sum_i k_i$

for $i = 1 \ldots N$ /* populate the stubs vector   */

do

   startingPoint $= \left[\sum_{x=1 \ldots k_i - 1} k_x \right] + 1$

   finishPoint $= \sum_{x=1 \ldots k_i} k_x$

   for $j = startingPoint \ldots finishPoint$ **do**

    | $S_j \leftarrow i$ /* insert value i in position j of vector **S** */

   end

end

/* end up with a vector with the first k_1 entries equal to 1, next k_2 entries equal to 2 and so on */

randomly re-order the elements of **S**

initialize empty graph on N nodes

for $i = 1 \ldots L$ /* move down the stubs vector in steps of two */

do

 | Create link between node S_{2i-1} and S_{2i}

end

/* the procedure of randomizing the stubs vector and then joining them sequentially in pairs is an algorithmic way of describing the instruction to join pairs of stubs at random. */

Algorithm 28: Watts–Strogatz algorithm without deleting edges

---

input : Number of nodes N
               Average degree $\bar{k}$ (even)
               New link probability p

output: Network $\mathbf{A}$ with high clustering and low average path length between
              nodes

```
/* draw a ring graph A with every node connected to its k̄/2
nearest neighbours */
```

initialize $\mathbf{A}$ as an $N \times N$ adjacency matrix

for $1 \leq i < j \leq N$ **do**
    if $0 < |j - i| \leq \bar{k}/2$ **then**
        | $A_{ij} = A_{ji} = 1$
    else
        | $A_{ij} = A_{ji} = 0$
    end
end

```
/* periodic boundary condition. Node N+1 identified with 1 etc.
*/
```

for $t = 1 \dots T$ **do**
```
    /* every iteration of this version will increase k̄ so T should
    not be chosen too large.  */
```
    repeat
        | draw nodes n and $m \in \{nodes\}$
    until $A_{nm} = 0$

    draw $r \in [0, 1]$

    if $r \leq p$ **then**
        | set $A_{nm} = 1$ and $A_{mn} = 1$
    end
end

Algorithm 29: Watts–Strogatz algorithm with rewiring

---

input : Number of nodes N; Average degree $\bar{k}$ (even)
New link probability p

output: Network **A** with high clustering and low average path length between nodes

```
/* draw a ring graph A with every node connected to its k̄/2
nearest neighbours */
```

initialize **A** as an $N \times N$ matrix **A**

for $1 \leq i < j \leq N$ **do**
    if $0 < |j - i| \leq \bar{k}/2$ **then**
        | $A_{ij} = A_{ji} = 1$
    else
        | $A_{ij} = A_{ji} = 0$
    end
end

```
/* periodic boundary condition: node N+1 identified with 1, etc.
*/
```

for $i = 1 \ldots N$
do
    for $s = 1 \ldots \bar{k}/2$ **do**
        draw $r \in [0, 1]$
        if $r \leq p$ **then**
            draw node $m \neq i$ such that $A_{im} = A_{mi} = 0$
            set $A_{i,i+s} = 0$ and $A_{i+s,i} = 0$
            set $A_{i,m} = 1$ and $A_{m,i} = 1$
        end
    end
end

Algorithm 30: Geometric graph algorithm

---

input : Number of nodes N
          Distance threshold for connectivity r

output: Network $\mathbf{A}$

```
/* assume that the space is a d-dimensional cube of side 1 with a
distance metric defined on it */
```

initialize empty $N \times N$ binary matrix $\mathbf{A}$ `/* start with empty graph */`
draw d random numbers $r_d \in [0,1]$
set $p_1 = (r_1 \ldots r_d)$

for $x = 2 \ldots N$ **do**
   draw d random numbers $r_d \in [0,1]$
   set $p_x = (r_1 \ldots r_d)$
   for $y = 1 \ldots (x-1)$ **do**
      if $|p_y - p_x| \leq r$ **then**
         set $A_{xy} = A_{yx} = 1$ `/* draw a link between nodes x and y */`
      end
   end
end

J.6 Auxiliary routines

Algorithm 31: Normalized Hamming distance

---

input : Matrices $\mathbf{A}$ and $\mathbf{A}'$

output: $h \in [0, 1]$ `/* 0 means the matrices are the same */`

set $h = 0$
for $1 \le i < j \le N$ do
  | set $h = h + |A_{ij} - A'_{ij}|$
end
set $h = \frac{2h}{N(N-1)}$ `/* normalize */`

Algorithm 32: Matrix of neighbours and its vector version

---

input : Adjacency matrix $\mathbf{A}$ with $N \times N$ elements

output: Matrix of neighbours $\mathbf{N}$ and its vector version $\mathbf{V}$
Vector $\mathbf{C}$ of cumulative degrees

```
/* N is an Nx(kmax+1) matrix where each row i gives the
neighbours of i (0-th element of row i stores the degree of i) */
```

```
/* V is a L-dimensional vector that stores all the elements of
matrix N (except the 0-th column) on a single row */
```

calculate the degrees k_i of $\mathbf{A}$ for $i = 1, \ldots, N$

calculate the maximum degree $k_{\max}$ and the average degree $\langle k \rangle$ and set
$L = N \langle k \rangle$

define matrix $\mathbf{N}$ with N rows and $k_{\max} + 1$ columns

for $1 \leq i \leq N$ **do**
$\quad$ set $N_{i,0} = 0$;
$\quad$ **for** $1 \leq j \leq N$ **do**
$\quad\quad$ **if** $A_{ij} = 1$ **then**
$\quad\quad\quad$ $N_{i,0} = N_{i,0} + 1$
```
              /* the zero-th column stores the degree sequence      */
```
$\quad\quad\quad$ $N_{i,N_{i,0}} = j$;
$\quad\quad$ **end**
$\quad$ **end**
end

```
                        /* transform matrix N into vector V */
```

define vector $\mathbf{V}$ with L components and vector $\mathbf{C}$ with N components

set $C_i = \sum_{j=0}^{i} k_j$ for all $i = 1, \ldots, N$ $\qquad\qquad$ /* $C_N = L$ */

for $1 \leq i \leq N$ **do**
$\quad$ **for** $1 \leq k \leq k_i$ **do**
$\quad\quad$ $V_{C_{i-1}+k} = N_{ik}$ $\quad$ /* the elements of $\mathbf{V}$ are the rows of matrix $\mathbf{N}$
$\quad\quad$ */
$\quad$ **end**
end

Algorithm 33: Extracting key parameters from model network

---

input : Nondirected simple network **A**

output: Degree sequence **k** and degree correlations kernel $Q(k, k')$

calculate the degrees k_i of **A** for $i = 1, \ldots, N$
calculate $k_{\max}$ and the degree distribution $p(k)$
set $\langle k \rangle = N^{-1} \sum_i k_i$

initialize all the elements of $k_{\max} \times k_{\max}$ matrix $W^*(\cdot, \cdot)$ to zero
    `/* matrix to record un-normalized frequency of connected pairs of degrees appearing in model network */`

for $i, j = 1 \ldots N$ **do**
    if $A_{ij} = 1$ **then**
        set $W^*(k_i, k_j) = W^*(k_i, k_j) + 1$
        `/* for every link record degree of starting and end point by incrementing relevant element in matrix W*(·,·) */`
    end
end

initialize $k_{\max} \times k_{\max}$ matrix $Q(\cdot, \cdot)$
    `/* transform degree-degree correlation matrix into the canonical kernel Q(·,·) */`

for $k, k' = 1 \ldots k^{\max}$ **do**
    set $W(k, k') = W^*(k, k') / \left[N \langle k \rangle \right]$
    set $Q(k, k') = W(k, k') / \left[p(k) p(k') \right]$
end
    `/* alternative is to directly input N, k and W(k,k') checking that the proposed degree sequence is graphical and the proposed degree-degree correlation is correctly normalized */`

Algorithm 34: Computing the relative degree correlations $\Pi(k, k')$ from the kernel $Q(k, k')$

---

input : Kernel $Q(k, k')$, degree distribution $p(k)$ and desired precision
$\quad\quad\quad$ /* precision is a small number, e.g. 10^{-5} or smaller */

output: Matrix $\Pi(k, k')$ and vector $\mathbf{F} = (F_1, \ldots, k_{\max})$ $\quad$ /* $F_k = F(k)$ */

$\quad\quad\quad$ /* $F(k)$ is needed to compute $\Pi(k, k')$ via equation (7.33) */
$\quad\quad\quad$ /* F_k computed from equation (7.34), of the form $F_k = R_k(\mathbf{F})$ */
$\quad\quad\quad$ /* initialize $\{F_k\}$ to some value and update them until
$F_k - R_k(\mathbf{F}) = 0$ for all k */

set $F_i = F_i' = 1$ for all $i = 1, \ldots, k_{\max}$ $\quad\quad$ /* F_i' is updated F_i */
set $E_{\min} = \infty$ $\quad\quad\quad\quad\quad$ /* minimum distance between $\mathbf{F}$ and $\mathbf{F}'$ */
set $\epsilon = 1/\pi$ $\quad\quad$ /* step of update ϵ is a number smaller than unit */

while $E_{\min} >$ precision **do**
$\quad$ $P = 0$ $\quad\quad\quad\quad\quad$ /* penalty for violating any constraint */
$\quad$ **for** $1 \leq i \leq N$ **do**
$\quad\quad$ calculate $R_i(\mathbf{F})$ $\quad\quad\quad\quad$ /* RHS of equation (7.34) */
$\quad\quad$ $F_i' = F_i - \epsilon(F_i - R_i(\mathbf{F}))$ $\quad\quad\quad\quad$ /* update F_i */
$\quad\quad$ /* want $R_i = F_i$, so increase (decrease) F_i if $F_i <(>)R_i$ $\quad$ */
$\quad\quad$ /* step proportional to discrepancy $\quad\quad\quad\quad\quad\quad$ */

$\quad\quad$ $E_i = F_i' - F_i$
$\quad\quad$ **if** $F_i' < 0$ **then**
$\quad\quad\quad\quad\quad\quad\quad\quad\quad\quad\quad\quad$ /* constraint violated */
$\quad\quad\quad$ set $P = P + \text{LargeNumber}$ $\quad\quad\quad$ /* large penalty */
$\quad\quad\quad$ set $\epsilon = \epsilon/\pi$ $\quad\quad\quad\quad\quad\quad$ /* make step smaller */
$\quad\quad\quad$ set $F_i' = F_i$ $\quad\quad\quad\quad\quad\quad$ /* erase last update */
$\quad\quad$ **end**
$\quad$ **end**
$\quad$ $E = \sqrt{\frac{\sum_i E_i^2}{k_{\max}}} + P$ $\quad$ /* distance between $\mathbf{F}$ and $\mathbf{F}'$ plus penalties */

$\quad$ **if** $E < E_{\min}$ **then**
$\quad\quad\quad\quad\quad\quad\quad\quad\quad$ /* if update decreases the distance */
$\quad\quad$ $F_i = F_i'$ for all $i = 1, \ldots, k_{\max}$ $\quad\quad\quad$ /* carry out update */
$\quad\quad$ $E_{\min} = E$ $\quad\quad\quad\quad\quad\quad$ /* update minimum distance */
$\quad$ **end**
end

calculate $\Pi(k, k') = Q(k, k')/[F(k)F(k')]$ for all $k, k' \in [1, k_{\max}]$

Algorithm 35: Computing trace variation $\Delta_{ijk\ell}\mathrm{Tr}(\mathbf{A}^3)$ from equation (7.70)

---

input : Matrix $\mathbf{A}$, Move $F_{ijk\ell}$

output: $\Delta_{ijk\ell}\mathrm{Tr}(\mathbf{A}^3)$ `/* trace variation under move` $F_{ijk\ell}$ `*/`

sum$=0$

calculate the matrix $\mathbf{N}$ of neighbours using Algorithm 32

for $n_p = 1 \cdots k_j$ **do**
   $p = N_{jn_p}$;
   if $p \neq i,j,k,\ell$ **then**
      if $A_{pk} = 1$ sum++
      if $A_{ip} = 1$ sum$--$
   end
end
for $n_p = 1 \cdots k_\ell$ **do**
   $p = N_{\ell n_p}$
   if $p \neq i,j,k,\ell$ **then**
      if $A_{pk} = 1$ sum$--$
      if $A_{ip} = 1$ sum++;
   end
end

Algorithm 36: Sampling uniformly a pair of ordered edges

---

input : Adjacency matrix $\mathbf{A}$

output: Candidate move $F_{ijk\ell}$ (as defined in Section 7.5)

compute matrix $\mathbf{N}$ of neighbours, vector $\mathbf{V}$ of ordered links and vector $\mathbf{C}$ of cumulative degrees   /* see Algorithm 32 */

set $L = C_N$ /* total number of (ordered) edges */

                                                 /* draw an ordered edge $i - j$ */

draw $J \in \{1, \ldots, L\}$ /* J denotes a position along vector $\mathbf{V}$, i.e. an ordered edge */

                          /* next search for nodes i, j attached to the drawn link */

$j = V_J$ /* j is simply the J-th element of $\mathbf{V}$ */

          /* start from segment of $\mathbf{V}$ referring to node 1 and search i */

$i = 1$
while $C_i < J$ **do**
 | $i + +$
end
set $\tilde{J} = J - C_{i-1}$ /* position of j among neighbours of i, useful to update $\mathbf{N}$ after move is applied, e.g. via $N_{i\tilde{j}} = \ell$ */

                                        /* draw an ordered edge $k - \ell$ */

draw $K \in \{1, \ldots, L\}$ /* K denotes a position along vector $\mathbf{V}$ */
$k = V_K$
$\ell = 1$
while $C_\ell < K$ **do**
 | $\ell + +$
end
set $\tilde{K} = K - C_{\ell-1}$ /* position of k among neighbours of ℓ, useful to update $\mathbf{N}$ after move is applied, e.g. via $N_{\ell\tilde{K}} = i$ */

Algorithm 37: Sampling uniformly a pair of *switchable* ordered edges

---

input : Adjacency matrix $\mathbf{A}$

output: Two ordered edges $(i, j), (k, \ell)$ such that $k \neq i, j$; $\ell \neq i, j$;
$\qquad A_{i\ell} = A_{jk} = 0$

compute matrix $\mathbf{N}$ of neighbours, compute vector $\mathbf{V}$ of ordered links and $\mathbf{C}$ of
cumulative degrees $\qquad\qquad\qquad\qquad$ `/* see Algorithm 32 */`

set $L = C_N$ $\qquad\qquad\qquad\qquad$ `/* total number of (ordered edges) */`

set switchable $= 0$
while switchable $= 0$ **do**
$\quad$ `/* draw an ordered edge` $i - j$ `(see also comments in Algorithm`
$\quad$ `36) */`

$\quad$ draw $J \in \{1, \ldots, L\}$
$\quad j = V_J$
$\quad i = 1$
$\quad$ **while** $C_i < J$ **do**
$\quad | \quad i{+}{+}$
$\quad$ **end**
$\quad \tilde{J} = J - C_{i-1}$ $\quad$ `/* position of` j `along the neighbours of` i`, useful`
$\quad$ `to update` $\mathbf{N}$ `after move is applied, see Algorithm 36 */`

$\qquad\qquad\qquad\qquad$ `/* draw an ordered edge` $k - \ell$ `*/`

$\quad$ draw $K \in \{1, \ldots, L\}$ $\quad$ `/*` K `denotes a position along vector` $\mathbf{V}$ `*/`
$\quad k = V_K$
$\quad \ell = 1$
$\quad$ **while** $C_\ell < K$ **do**
$\quad | \quad \ell{+}{+}$
$\quad$ **end**
$\quad \tilde{K} = K - C_{\ell-1}$ `/* position of` k `among the neighbours of` ℓ`, useful`
$\quad$ `to update` $\mathbf{N}$ `after move is applied; see Algorithm 36 */`

$\quad$ **if** $k \neq i, j$; $\ell \neq i, j$; $A_{jk} = A_{i\ell} = 0$ **then**
$\quad | \quad$ set switchable$= 1$
$\quad$ **end**
end

Algorithm 38: Non-uniform sampling of a pair of ordered edges

---

input : Adjacency matrix $\mathbf{A}$

output: Two ordered edges $(i,j), (k,\ell)$ such that $\ell \neq i,j$ and $A_{i\ell} = 0$

compute matrix $\mathbf{N}$ of neighbours, compute vectors $\mathbf{V}$ and $\mathbf{C}$ of ordered links
and cumulative degrees   /* see Algorithm 32 */

    /* draw an ordered edge $i - j$ (see also comments in Algorithm 36)
*/

set $L = C_N$; draw $J \in \{1, \ldots, L\}$
$j = V_J$
$i = 1$
while $C_i < J$ **do**
 | $\quad i++$
end
$\tilde{J} = J - C_{i-1}$

    /* draw an ordered edge $k - \ell$, such that $\ell \neq i,j$ and $A_{i\ell} = 0$ */

for $1 \leq a \leq k_i$ **do**
 | set $L = L - k_{N_{ia}}$ /* count number of links that are not connected
 | to neighbours of i.... */
end
set $L = L - k_i$ /* .. or i itself */

draw $A \in \{1, \ldots, L\}$ /* draw (k,ℓ) with $\ell \neq i,j$ and $A_{i\ell} = 0$ */

set $K = 0$ /* K is position of drawn link on $\mathbf{V}$ i.e. among all
links, while A is position among allowed links */

$\ell = 1$ /* start from segment of $\mathbf{V}$ relative to node 1 */
repeat
 | **if** $\ell = i$ *or* $A_{i\ell} = 1$ **then**
 | | /* if in a segment not allowed, i.e. referring to i or
 | | neighbours of i */
 | | $K = K + k_\ell$ /* go to next segment */
 | | $\ell++$
 | **end**
 | **else**
 | | /* when on allowed segments   */
 | | $K++$ /* move along entries of $\mathbf{V}$ */
 | | $A--$ /* decrease number of allowed entries to visit */
 | **end**
until $A >= 1$

                                  /* found ℓ */
set $k = V_K$ /* found k */
$\tilde{K} = A - C_{\ell-1}$ /* position of ℓ along the neighbours of k */

References

[1] Agliari, Elena, Annibale, Alessia, Barra, Adriano, Coolen, Anthonius, and Tantari, Daniele (2013). Immune networks: Multitasking capabilities near saturation. *Journal of Physics A: Mathematical and Theoretical*, **46**(41), 415003.

[2] Airoldi, Edoardo M, Blei, David M, Fienberg, Stephen E, and Xing, Eric P (2008). Mixed membership stochastic blockmodels. *Journal of Machine Learning Research*, **9**(Sep), 1981–2014.

[3] Albert, Réka and Barabási, Albert-László (1999). Emergence of scaling in random networks. *Science*, **286**(5439), 509–512.

[4] Albert, Réka and Barabási, Albert-László (2000). Topology of evolving networks: Local events and universality. *Physical Review Letters*, **85**(24), 5234.

[5] Albert, Réka and Barabási, Albert-László (2002). Statistical mechanics of complex networks. *Reviews of Modern Physics*, **74**(1), 47.

[6] Altieri, Dario C (2008). Survivin, cancer networks and pathway-directed drug discovery. *Nature Reviews Cancer*, **8**(1), 61–70.

[7] Anderson, Carolyn J, Wasserman, Stanley, and Croach, Bradley (2002). A $p^\star$ primer: Logit models for social networks. *Social Networks*, **21**, 37–66.

[8] Anderson, James (2006). *Hyperbolic Geometry*. Springer Science & Business Media, London.

[9] Annibale, Alessia, Coolen, Anthonius, Fernandes, Luis, Fraternali, Franca, and Kleinjung, Jens (2009). Tailored graph ensembles as proxies or null models for real networks I: Tools for quantifying structure. *Journal of Physics A: Mathematical and Theoretical*, **42**(48), 485001.

[10] Annibale, Alessia and Courtney, Owen T (2015). The two-star model: Exact solution in the sparse regime and condensation transition. *Journal of Physics A: Mathematical and Theoretical*, **48**(36), 365001.

[11] Austad, Haakon and Friel, Nial (2010). Deterministic Bayesian inference for the $p^\star$ model. *AISTATS*, **9**, 41–48.

[12] Baek, Yongjoo, Kim, Daniel, Ha, Meesoon, and Jeong, Hawoong (2012). Fundamental structural constraint of random scale-free networks. *Physical Review Letters*, **109**(11), 118701.

[13] Banks, David and Carley, Kathleen M (1996). Models of social network evolution. *Journal of Mathematical Sociology*, **21**(1-2), 173–196.

[14] Barrat, Alain, Barthélemy, Marc, Pastor-Satorras, Romualdo, and Vespignani, Alessandro (2004). The architecture of complex weighted networks. *Proceedings of the National Academy of Sciences*, **101**(11), 3747–3752.

[15] Barrat, Alain, Barthélemy, Marc, and Vespignani, Alessandro (2004). Weighted evolving networks: Coupling topology and weight dynamics. *Physical Review Letters*, **92**(22), 228701.

[16] Bashor, Caleb J, Horwitz, Andrew A, Peisajovich, Sergio G, and Lim, Wendell A (2010). Rewiring cells: Synthetic biology as a tool to interrogate the organizational principles of living systems. *Annual Review of Biophysics*, **39**(1), 515–537.

[17] Bender, Edward and Canfield, E Rodney (1978). The asymptotic number of labelled graphs with given degree sequences. *Journal of Combinatorial Theory Series A*, **24**, 296.

[18] Besag, Julian (1975). Statistical analysis of non-lattice data. *The Statistician*, **24**, 179–195.

[19] Bettstetter, Christian (2002). On the minimum node degree and connectivity of a wireless multihop network. In *Proceedings of the 3rd ACM International Symposium on Mobile Ad Hoc Networking & Computing*, pp. 80–91. ACM.

[20] Bianconi, Ginestra (2013). Statistical mechanics of multiplex networks: Entropy and overlap. *Physical Review E*, **87**(6), 062806.

[21] Bianconi, Ginestra and Barabási, Albert-László (2001). Competition and multiscaling in evolving networks. *Europhysics Letters*, **54**, 436.

[22] Boccaletti, Stefano, Bianconi, Ginestra, Criado, Regino, Del Genio, Charo, Gómez-Gardeñes, Jesus, Romance, Miguel, Sendina-Nadal, Irene, Wang, Zhen, and Zanin, Massimiliano (2014). The structure and dynamics of multilayer networks. *Physics Reports*, **544**(1), 1–122.

[23] Boccaletti, S., Latora, V., Moreno, Y., Chavez, M., and Hwang, D.-U. (2006). Complex networks: Structure and dynamics. *Physics Reports*, **424**(45), 175 – 308.

[24] Bodirsky, Manuel, Gröpl, Clemens, and Kang, Mihyun (2003). Generating labeled planar graphs uniformly at random. In *Automata, Languages and Programming*, pp. 1095–1107. Springer, Berlin, Heidelberg.

[25] Bogacz, Leszek, Burda, Zdzisław, Janke, Wolfhard, and Waclaw, Bartlomiej (2005). A program generating homogeneous random graphs with given weights. *Computer Physics Communications*, **173**(3), 162–174.

[26] Bollobás, Béla (2001). *Random Graphs* (2 edn). Cambridge University Press, Cambridge.

[27] Bollobás, Béla, Borgs, Christian, Chayes, Jennifer, and Riordan, Oliver (2003). Directed scale-free graphs. In *Proceedings of the Fourteenth Annual ACM-SIAM Symposium on Discrete Algorithms*, pp. 132–139. SIAM.

[28] Bollobás, Béla, Riordan, Oliver, Spencer, Joel, Tusnády, Gábor et al. (2001). The degree sequence of a scale-free random graph process. *Random Structures and Algorithms*, **18**(3), 279–290.

[29] Borgs, Christian, Chayes, Jennifer, Daskalakis, Constantinos, and Roch, Sebastien (2007). First to market is not everything: An analysis of preferential attachment with fitness. In *Proceedings of the Thirty-Ninth Annual ACM Symposium on Theory of Computing*, pp. 135–144. ACM.

[30] Bornholdt, Stefan and Schuster, Heinz Georg (2002). *Handbook of Graphs & Networks*. Wiley Online Library.

[31] Boyer, John M and Myrvold, Wendy J (2004). On the cutting edge: Simplified $\mathcal{O}(n)$ planarity by edge addition. *Journal of Graph Algorithms and Applications*, **8**(2), 241–273.

[32] Britton, Tom, Deijfen, Maria, and Martin-Löf, Anders (2006). Generating simple random graphs with prescribed degree distribution. *Journal of Statistical Physics*, **124**(6), 1377–1397.

[33] Brot, Hilla, Honig, Michal, Muchnik, Lev, Goldenberg, Jacob, and Louzoun, Yoram (2013). Edge removal balances preferential attachment and triad closing. *Physical Review E*, **88**, 042815.

[34] Burda, Zdzisław, Correia, Joao, and Krzywicki, Andrzej (2001). Statistical ensemble of scale-free random graphs. *Physical Review E*, **64**, 046118.

[35] Burda, Zdzisław, Jurkiewicz, Jerzy, and Krzywicki, Andrzej (2004). Perturbing general uncorrelated networks. *Physical Review E*, **70**, 026106.

[36] Caimo, Alberto and Friel, Nial (2011). Bayesian inference for exponential random graph models. *Social Networks*, **33**, 41–55.

[37] Caimo, Alberto and Friel, Nial (2013). Bayesian model selection for exponential random graph models. *Social Networks*, **35**, 11–24.

[38] Caldarelli, Guido, Capocci, Andrea, De Los Rios, Paolo, and Muñoz, Miguel A (2002). Scale-free networks from varying vertex intrinsic fitness. *Physical Review Letters*, **89**(25), 258702.

[39] Caldarelli, Guido, De Los Rios, Paolo, and Pietronero, Luciano (2003). Generalized network growth: From microscopic strategies to the real internet properties. *Preprint arXiv:0307610*.

[40] Caldarelli, Guido and Vespignani, Alessandro (2007). *Large Scale Structure and Dynamics of Complex Networks*. World Scientific Singapore, Singapore.

[41] Callaway, Duncan S, Hopcroft, John E, Kleinberg, Jon M, Newman, Mark EJ, and Strogatz, Steven H (2001). Are randomly grown graphs really random? *Physical Review E*, **64**(4), 041902.

[42] Camacho, Juan, Guimerà, Roger, and Nunes Amaral, Luís (2002). Robust patterns in food web structure. *Physical Review Letters*, **88**(22), 228102.

[43] Camerano, Lorenzo (1880). Desequilibrio dei viventi merc la reciproca distruzione. *Atti della Reale Accademia delle Scienze di Torino*, **15**, 393414.

[44] Catanzaro, Michele, Boguñá, Marián, and Pastor-Satorras, Romualdo (2005). Generation of uncorrelated random scale-free networks. *Physical Review E*, **71**(2), 027103.

[45] Chen, Ningyuan, Olvera-Cravioto, Mariana et al. (2013). Directed random graphs with given degree distributions. *Stochastic Systems*, **3**(1), 147–186.

[46] Chung, Fan, Handjani, Shirin, and Jungreis, Doug (2003). Generalizations of Polya's urn problem. *Annals of Combinatorics*, **7**(2), 141–153.

[47] Chung, Fan and Lu, Linyuan (2002). Connected components in random graphs with given expected degree sequence. *Annals of Combinatorics*, **6**(2), 125.

[48] Chung, Fan, Lu, Linyuan, Dewey, T Gregory, and Galas, David J (2003). Duplication models for biological networks. *Journal of Computational Biology*, **10**(5), 677–687.

[49] Clark, Brent N, Colbourn, Charles J, and Johnson, David S (1990). Unit disk graphs. *Discrete Mathematics*, **86**(1–3), 165–177.

[50] Cohen, Joel E (1995). Random graphs in ecology. *Topics in Contemporary Probability and Its Applications*, 233–260.

[51] Collevecchio, Andrea, Cotar, Codina, LiCalzi, Marco et al. (2013). On a preferential attachment and generalized Polya's urn model. *The Annals of Applied Probability*, **23**(3), 1219–1253.

[52] Coolen, Anthonius, De Martino, Andrea, and Annibale, Alessia (2009). Constrained Markovian dynamics of random graphs. *Journal of Statistical Physics*, **136**(6), 1035–1067.

[53] Cover, Thomas M and Thomas, Joy A (1991). Information theory and statistics. *Elements of Information Theory*, 279–335.

[54] Criado, Regino, Flores, Julio, García del Amo, Alejandro, Gómez-Gardeñes, Jesús, and Romance, Miguel (2012). A mathematical model for networks with structures in the mesoscale. *International Journal of Computer Mathematics*, **89**(3), 291–309.

[55] D'Agostino, Gregorio and Scala, Antonio (2014). *Networks of Networks: The Last Frontier of Complexity*. Springer International Publishing, Switzerland.

[56] Decelle, Aurelien, Krzakala, Florent, Moore, Cristopher, and Zdeborová, Lenka (2011). Asymptotic analysis of the stochastic block model for modular networks and its algorithmic applications. *Physical Review E*, **84**(6), 066106.

[57] Deijfen, Maria and Lindholm, Mathias (2009). Growing networks with preferential deletion and addition of edges. *Physica A: Statistical Mechanics and its Applications*, **388**(19), 4297–4303.

[58] Del Genio, Charo I, Gross, Thilo, and Bassler, Kevin E (2011). All scale-free networks are sparse. *Physical Review Letters*, **107**(17), 178701.

[59] Denise, Alain, Vasconcellos, Marcio, and Welsh, Dominic JA (1996). The random planar graph. *Congressus Numerantium*, **113**, 61–80.

[60] Diesner, Jana, Frantz, Terrill L, and Carley, Kathleen M (2005). Communication networks from the Enron email corpus. *Computational & Mathematical Organization Theory*, **11**(3), 201–228.

[61] Dorogovtsev, Sergey N and Mendes, José FF (2000). Evolution of networks with aging of sites. *Physical Review E*, **62**, 1842–1845.

[62] Dorogovtsev, Sergey N and Mendes, José FF (2002). Evolution of networks. *Advances in Physics*, **51**(4), 1079–1187.

[63] Dorogovtsev, Sergey N and Mendes, José FF (2013). *Evolution of networks: From Biological Nets to the Internet and WWW*. Oxford University Press, Oxford.

[64] Dorogovtsev, Sergey N, Mendes, José FF, and Samukhin, Alexander N (2003). Principles of statistical mechanics of uncorrelated random networks. *Nuclear Physics B*, **666**(3), 396–416.

[65] Durrett, Richard (2007). *Random Graph Dynamics*. Volume 200. Cambridge University Press, Cambridge.

[66] Eggleton, Roger and Holton, Derek Allan (1981). *Simple and Multigraphic Realizations of Degree Sequences*. Springer, Berlin, Heidelberg.

[67] Erdös, Paul and Gallai, Tibor (1960). Graphen mit punkten vorgeschriebenen grades. *Matematikai Lapok*, **11**, 264–274.

[68] Evans, Martin, Majumdar, Satya N, and Zia, Royce (2006). Canonical analysis of condensation in factorised steady states. *Journal of Statistical Physics*, **123**(2), 357–390.

[69] Fagan, William F (1997). Omnivory as a stabilizing feature of natural communities. *The American Naturalist*, **150**(5), 554–567.

[70] Faloutsos, Michalis, Faloutsos, Petros, and Faloutsos, Christos (1999). On power-law relationships of the internet topology. *SIGCOMM Computer Communication Review*, **29**(4), 251–262.

[71] Fernandes, Luis, Annibale, Alessia, Kleinjung, Jens, Coolen, Anthonius, and Fraternali, Franca (2010). Protein networks reveal detection bias and species consistency when analysed by information-theoretic methods. *PLoS One*, **5**(8), e12083.

[72] Feynman, Richard P (1998). *Statistical Mechanics: A Set of Lectures* (2 edn). Perseus Books Group, New York.

[73] Flaxman, Abraham D, Frieze, Alan M, and Vera, Juan (2006). A geometric preferential attachment model of networks. *Internet Mathematics*, **3**(2), 187–205.

[74] Force, Allan, Lynch, Michael, Pickett, F Bryan, Amores, Angel, Yan, Yi-lin, and Postlethwait, John (1999). Preservation of duplicate genes by complementary, degenerative mutations. *Genetics*, **151**(4), 1531–1545.

[75] François, Paul and Hakim, Vincent (2004). Design of genetic networks with specified functions by evolution in silico. *Proceedings of the National Academy of Sciences*, **101**(2), 580–585.

[76] Frank, Ove and Strauss, David (1986). Markov graphs. *Journal of the American Statistical Association*, **81**(395), 832–842.

[77] Fronczak, Agata and Fronczak, Piotr (2005). Networks with given two-point correlations: Hidden correlations from degree correlations. *Physical Review E*, **74**(2), 026121.

[78] Funk, Sebastian and Jansen, Vincent AA (2010). Interacting epidemics on overlay networks. *Physical Review E*, **81**(3), 036118.

[79] Garcia-Domingo, Josep L, Juher, David, and Saldaña, Joan (2008). Degree correlations in growing networks with deletion of nodes. *Physica D: Nonlinear Phenomena*, **237**(5), 640–651.

[80] Garlaschelli, Diego (2009). The weighted random graph model. *New Journal of Physics*, **11**(7), 073005.

[81] Gotelli, Nicholas J and Ulrich, Werner (2012). Statistical challenges in null model analysis. *Oikos*, **121**(2), 171–180.

[82] Gradshteyn, Israil' Solomonovich and Ryzhik, Iosif Moiseevich (2007). *Table of Integrals, Series, and Products*. Academic Press, San Diego MA, London UK, Oxford UK.

[83] Grimmett, Geoffrey and Stirzaker, David (2001). *Probability and Random Processes*. Oxford University Press, Oxford.

[84] Hajra, Kamalika Basu and Sen, Parongama (2005). Aging in citation networks. *Physica A: Statistical Mechanics and its Applications*, **346**(1), 44–48.

[85] Handcock, Mark S., Robins, Garry, Snijders, Tom, Moody, Jim, and Besag, Julian (2003). Assessing degeneracy in statistical models of social networks. *Journal of the American Statistical Association*, **76**, 33–50.

[86] Hanneke, Steve, Fu, Wenjie, Xing, Eric P et al. (2010). Discrete temporal models of social networks. *Electronic Journal of Statistics*, **4**, 585–605.

[87] Holland, Paul W, Laskey, Kathryn Blackmond, and Leinhardt, Samuel (1983). Stochastic blockmodels: First steps. *Social Networks*, **5**(2), 109–137.

[88] Holland, Paul W and Leinhardt, Samuel (1981). An exponential family of probability distributions for directed graphs. *Journal of the American Statistical Association*, **76**(373), 54–57.

[89] Holme, Petter (2005). Network reachability of real-world contact sequences. *Physical Review E*, **71**(4), 046119.

[90] Holme, Petter and Kim, Beom Jun (2002). Growing scale-free networks with tunable clustering. *Physical Review E*, **65**(2), 026107.

[91] Holme, Petter and Saramki, Jari (2012). Temporal networks. *Physics Reports*, **519**(3), 97–125.

[92] Holyoak, Marcel and Sachdev, Sambhav (1998). Omnivory and the stability of simple food webs. *Oecologia*, **117**(3), 413–419.

[93] Itzkovitz, Shalev, Milo, Ron, Kashtan, Nadav, Ziv, Guy, and Alon, Uri (2003). Subgraphs in random networks. *Physical Review E*, **68**(2), 026127.

[94] Jonasson, Johan (1999). The random triangle model. *Journal of Applied Probability*, **36**(3), 852–867.

[95] Karl, Holger and Willig, Andreas (2007). *Protocols and Architectures for Wireless Sensor Networks*. John Wiley & Sons, Chichister, UK.

[96] Kellis, Manolis, Birren, Bruce W, and Lander, Eric S (2004). Proof and evolutionary analysis of ancient genome duplication in the yeast Saccharomyces cerevisiae. *Nature*, **428**(6983), 617–624.

[97] Kim, Ethan DH, Sabharwal, Ashish, Vetta, Adrian R, and Blanchette, Mathieu (2010). Predicting direct protein interactions from affinity purification mass spectrometry data. *Algorithms for Molecular Biology*, **5**(1), 1–17.

[98] Kim, Jung Yeol and Goh, Kwang-Il (2013). Coevolution and correlated multiplexity in multiplex networks. *Physical Review Letters*, **111**(5), 058702.

[99] Kinney, Ryan, Crucitti, Paolo, Albert, Réka, and Latora, Vito (2005). Modeling cascading failures in the North American power grid. *The European Physical Journal B*, **46**(1), 101–107.

[100] Kiss, Istvan Z and Green, Darren M (2008). Comment on properties of highly clustered networks. *Physical Review E*, **78**(4), 048101.

[101] Klein-Hennig, Hendrike and Hartmann, Alexander K (2012). Bias in generation of random graphs. *Physical Review E*, **85**(2 Pt 2), 026101–026101.

[102] Kleinberg, Jon (1999). Authoritative sources in a hyperlinked environment. *Journal of the ACM*, **46**(5), 604–632.

[103] Kleinberg, Jon (2002). Small-world phenomena and the dynamics of information. *Advances in Neural Information Processing Systems*, **1**, 431–438.

[104] Krapivsky, Paul and Krioukov, Dmitri (2008). Scale-free networks as preasymptotic regimes of superlinear preferential attachment. *Physical Review E*, **78**(2), 026114.

[105] Krapivsky, Paul and Redner, Sidney (2001). Organization of growing random networks. *Physical Review E*, **63**(6), 066123.

[106] Krapivsky, Paul and Redner, Sidney (2005, Mar). Network growth by copying. *Physical Review E*, **71**, 036118.

[107] Krapivsky, Paul, Redner, Sidney, and Leyvraz, Francois (2000). Connectivity of growing random networks. *Physical Review Letters*, **85**(21), 4629.

[108] Krioukov, Dmitri, Papadopoulos, Fragkiskos, Kitsak, Maksim, Vahdat, Amin, and Boguñá, Marián (2010). Hyperbolic geometry of complex networks. *Physical Review E*, **82**, 036106.

[109] Krivelevich, Michael and Vilenchik, Dan (2006). Semirandom models as benchmarks for coloring algorithms. In *ANALCO*, pp. 211–221. SIAM.

[110] Kumar, Ravi, Raghavan, Prabhakar, Rajagopalan, Sridhar, Sivakumar, Dandapani, Tompkins, Andrew, and Upfal, Eli (2000). The Web as a graph. In *Proceedings of the Nineteenth ACM SIGMOD-SIGACT-SIGART Symposium on Principles of Database Systems*, pp. 1–10. ACM.

[111] Lee, Kyu-Min, Kim, Jung Yeol, Cho, Won-Kuk, and Goh, Kwang-Il (2012). Correlated multiplexity and connectivity of multiplex random networks. *New Journal of Physics*, **14**(3), 3027.

[112] Li, Ji, Andrew, Lachlan, Foh, Chuan, Zukerman, Moshe, and Chen, Hsiao-Hwa (2009). Connectivity, coverage and placement in wireless sensor networks. *Sensors*, **9**(10), 7664–7693.

[113] Li, Lun, Alderson, David, Doyle, John C, and Willinger, Walter (2005). Towards a theory of scale-free graphs: Definition, properties, and implications. *Internet Mathematics*, **2**(4), 431–523.

[114] Marceau, Vincent, Noël, Pierre-André, Hébert-Dufresne, Laurent, Allard, Antoine, and Dubé, Louis J (2011). Modeling the dynamical interaction between epidemics on overlay networks. *Physical Review E*, **84**(2), 026105.

[115] Maslov, Sergei and Sneppen, Kim (2002). Specificity and stability in topology of protein networks. *Science*, **296**(5569), 910–913.

[116] Matsumoto, Makoto and Nishimura, Takuji (1998). Mersenne twister: A 623-dimensionally equidistributed uniform pseudo-random number generator. *ACM Transactions on Modeling and Computer Simulation*, **8**(1), 3–30.

[117] McCann, Kevin and Hastings, Alan (1997). Re–evaluating the omnivory–stability relationship in food webs. *Proceedings of the Royal Society of London B: Biological Sciences*, **264**(1385), 1249–1254.

[118] McCann, Kevin, Hastings, Alan, and Huxel, Gary R (1998). Weak trophic interactions and the balance of nature. *Nature*, **395**(6704), 794–798.

[119] McKay, Brendan D (1981). The expected eigenvalue distribution of a large regular graph. *Linear Algebra and Its Applications*, **40**, 203–216.

[120] Meinert, Sascha and Wagner, Dorothea (2011). *An Experimental Study on Generating Planar Graphs*. Springer, Berlin, Heidelberg.

[121] Melián, Carlos J and Bascompte, Jordi (2002). Food web structure and habitat loss. *Ecology Letters*, **5**(1), 37–46.

[122] Melián, Carlos J and Bascompte, Jordi (2004). Food web cohesion. *Ecology*, **85**(2), 352–358.

[123] Melnik, Sergey, Porter, Mason A, Mucha, Peter J, and Gleeson, James P (2014). Dynamics on modular networks with heterogeneous correlations. *Chaos: An Interdisciplinary Journal of Nonlinear Science*, **24**(2), 023106.

[124] Menzel, Donald H (1960). *Fundamental Formulas of Physics*. Volume 1. Courier Corporation, New York.

[125] Milgram, Stanley (1967). The small world problem. *Psychology Today*, **2**(1), 60–67.

[126] Miller, Joel C (2009). Percolation and epidemics in random clustered networks. *Physical Review E*, **80**(2), 020901.

[127] Milo, Ron, Kashtan, Nadav, Itzkovitz, Shalev, Newman, Mark EJ, and Alon, Uri (2004). On the uniform generation of random graphs with prescribed degree sequences. *Preprint arXiv:0312028*.

[128] Molloy, Michael and Reed, Bruce (1995). A critical point for random graphs with a given degree sequence. *Random Structures and Algorithms*, **6**(2–3), 161–180.

[129] Newman, Mark EJ (2002). Assortative mixing in networks. *Physical Review Letters*, **89**, 208701.

[130] Newman, Mark EJ (2003). Properties of highly clustered networks. *Physical Review E*, **68**(2), 026121.

[131] Newman, Mark EJ (2010). *Networks: An Introduction*. Oxford University Press, Oxford.

[132] Newman, Mark EJ, Watts, Duncan J, and Strogatz, Steven H (2002). Random graph models of social networks. *Proceedings of the National Academy of Sciences*, **99**(suppl 1), 2566–2572.

[133] Nicosia, Vincenzo, Bianconi, Ginestra, Latora, Vito, and Barthélemy, Marc (2013). Growing multiplex networks. *Physical Review Letters*, **111**(5), 058701.

[134] Nicosia, Vincenzo, Bianconi, Ginestra, Latora, Vito, and Barthélemy, Marc (2014). Nonlinear growth and condensation in multiplex networks. *Physical Review E*, **90**(4–1), 042807.

[135] Nishimori, Hidetoshi (2001). *Statistical Physics of Spin Glasses and Information Processing*. Oxford University Press, Oxford.

[136] Ohnishi, Takaaki, Takayasu, Hideki, and Takayasu, Misako (2010). Network motifs in an inter-firm network. *Journal of Economic Interaction and Coordination*, **5**(2), 171–180.

[137] Ohno, Susumu (1970). *Evolution by Gene Duplication*. London: George Allen & Unwin Ltd. Berlin, Heidelberg and New York: Springer-Verlag.

[138] Onat, Furuzan Atay, Stojmenovic, Ivan, and Yanikomeroglu, Halim (2008). Generating random graphs for the simulation of wireless ad hoc, actuator, sensor, and internet networks. *Pervasive and Mobile Computing*, **4**(5), 597–615.

[139] Pagani, Giuliano Andrea and Aiello, Marco (2013). The power grid as a complex network: A survey. *Physica A: Statistical Mechanics and Its Applications*, **392**, 2688–2700.

[140] Park, Juyong and Newman, Mark EJ (2004). Statistical mechanics of networks. *Physical Review E*, **70**, 066117.

[141] Park, Juyong and Newman, Mark EJ (2005). Solution for the properties of a clustered network. *Physical Review E*, **72**, 026136.

[142] Pastor-Satorras, Romualdo, Smith, Eric, and Solé, Ricard V (2003). Evolving protein interaction networks through gene duplication. *Journal of Theoretical Biology*, **222**(2), 199–210.

[143] Pastor-Satorras, Romualdo and Vespignani, Alessandro (2007). *Evolution and Structure of the Internet: A Statistical Physics Approach*. Cambridge University Press, Cambridge.

[144] Pattison, Philippa E and Robins, Garry (2002). Neighbourhood-based models for social network. *Sociological Methodology*, **32**, 301–337.

[145] Penrose, Mathew D (1999). On k-connectivity for a geometric random graph. *Random Structures and Algorithms*, **15**(2), 145–164.

[146] Pérez-Vicente, Conrad and Coolen, Anthonius (2008). Spin models on random graphs with controlled topologies beyond degree constraints. *Journal of Physics A: Mathematical and Theoretical*, **41**(25), 255003.

[147] Pimm, Stuart and Lawton, John H (1978). On feeding on more than one trophic level. *Nature*, **275**(5680), 542–544.

[148] Polis, Gary A and Strong, Donald R (1996). Food web complexity and community dynamics. *American Naturalist*, 813–846.

[149] Prasad, TS Keshava, Goel, Renu, Kandasamy, Kumaran, Keerthikumar, Shivakumar, Kumar, Sameer, Mathivanan, Suresh, Telikicherla, Deepthi, Raju, Rajesh, Shafreen, Beema, Venugopal, Abhilash et al. (2009). Human protein reference database2009 update. *Nucleic Acids Research*, **37**(suppl 1), D767–D772.

[150] Press, William H (2007). *Numerical Recipes 3rd edition: The Art of Scientific Computing*. Cambridge University Press, Cambridge.

[151] Rao, A. Ramachandra, Jana, Rabindranath, and Bandyopadhyay, Suraj (1996). A Markov Chain Monte Carlo method for generating random (0, 1)-matrices with given marginals. *Sankhya: The Indian Journal of Statistics, Series A*, **58**(2), 225–242.

[152] Rényi, Alfréd and Erdös, Paul (1959). On random graphs. *Publicationes Mathematicae*, **6**(290–297), 5.

[153] Rikvold, Per A, Hamad, Ibrahim A, Israels, Brett, and Poroseva, Svetlana V (2012). Modeling power grids. *Physics Procedia*, **34**, 119–123.

[154] Roberts, Ekaterina, Annibale, Alessia, and Coolen, Anthonius (2014). Controlled Markovian dynamics of graphs: Unbiased generation of random graphs with prescribed topological properties. In *Nonlinear Maps and Their Applications*, pp. 25–34. Springer, New York.

[155] Roberts, Ekaterina and Coolen, Anthonius (2012). Unbiased degree-preserving randomization of directed binary networks. *Physical Review E*, **85**(4), 046103.

[156] Robins, Garry, Elliott, Peter, and Pattison, Philippa E (2001). Network models for social selection processes. *Social Networks*, **23**(1), 1–30.

[157] Robins, Garry, Pattison, Philippa E, Kalish, Yuval, and Lusher, Dean (2007). An introduction to exponential random graph (p^*) models for social networks. *Social Networks*, **29**(2), 173–191.

[158] Rogers, Tim, Pérez-Vicente, Conrad J, Takeda, Koujin, and Pérez Castillo, Isaac (2010). Spectral density of random graphs with topological constraints. *Journal of Physics A: Mathematical and Theoretical*, **43**(19), 195002.

[159] Sanil, Ashish, Banks, David, and Carley, Kathleen (1995). Models for evolving fixed node networks: Model fitting and model testing. *Social Networks*, **17**(1), 65–81.

[160] Santi, Paolo (2005). Topology control in wireless ad hoc and sensor networks. *ACM Computing Surveys (CSUR)*, **37**(2), 164–194.

[161] Saramäki, Jari and Kaski, Kimmo (2004). Scale-free networks generated by random walkers. *Physica A Statistical Mechanics and Its Applications*, **341**, 80–86.

[162] Saumell-Mendiola, Anna, Serrano, M Ángeles, and Boguñá, Marián (2012). Epidemic spreading on interconnected networks. *Physical Review E*, **86**(2), 026106.

[163] Schaeffer, Satu Elisa (2007). Graph clustering. *Computer Science Review*, **1**(1), 27–64.

[164] Shen-Orr, Shai S, Milo, Ron, Mangan, Shmoolik, and Alon, Uri (2002). Network motifs in the transcriptional regulation network of *Escherichia coli*. *Nature Genetics*, **31**(1), 64–68.

[165] Snijders, Tom AB (2001). The statistical evaluation of social network dynamics. *Sociological Methodology*, **31**(1), 361–395.

[166] Snijders, Tom AB (2002). Markov Chain Monte Carlo estimation of exponential random graph model. *Journal of Social Structure*, **3**(2), 1–40.

[167] Snijders, Tom AB, Pattison, Philipa E, Robins, Garry, and Handcock, Mark S (2006). New specifications for exponential random graph models. *Sociological Methodology*, **36**(1), 99–153.

[168] Söderberg, Bo (2003). Properties of random graphs with hidden color. *Physical Review E*, **68**, 026107.

[169] Strauss, David (1986). On a general class of models for interactions. *SIAM Review*, **28**(4), 513–527.

[170] Strauss, David and Ikeda, Michael (1990). Pseudolikelihood estimation for social networks. *Journal of the American Statistical Association*, **85**(409), 204–212.

[171] Stumpf, Michael PH, Balding, David J, and Girolami, Mark (2011). *Handbook of Statistical Systems Biology*. Wiley Online Library.

[172] Taylor, John S and Raes, Jeroen (2004). Duplication and divergence: The evolution of new genes and old ideas. *Annual Review of Genetics*, **38**, 615–643.

[173] Taylor, Richard (1981). *Constrained Switchings in Graphs*. Springer.

[174] Touchette, Hugo (2009). The large deviation approach to statistical mechanics. *Physics Reports*, **478**(1), 1–69.

[175] Vázquez, Alexei (2000). Knowing a network by walking on it: Emergence of scaling. *Preprint arXiv:0006132*.

[176] Vázquez, Alexei, Flammini, Alessandro, Maritan, Amos, and Vespignani, Alessandro (2002). Modeling of protein interaction networks. *Complexus*, **1**(1), 38–44.

[177] Vázquez, Alexei, Racz, Balazs, Lukacs, Andras, and Barabási, Albert-László (2007). Impact of non-Poissonian activity patterns on spreading processes. *Physical Review Letters*, **98**(15), 158702.

[178] von Looz, Moritz, Meyerhenke, Henning, and Prutkin, Roman (2015). Generating random hyperbolic graphs in subquadratic time. In *Algorithms and Computation*, pp. 467–478. Springer, Berlin, Heidelberg.

[179] Wang, Yuchung J and Wong, George Y (1987). Stochastic blockmodels for directed graphs. *Journal of the American Statistical Association*, **82**(397), 8–19.

[180] Wang, Zhifang, Scaglione, Anna, and Thomas, Robert J (2010). Generating statistically correct random topologies for testing smart grid communication and control networks. *IEEE Transactions on Smart Grid*, **1**(1), 28–39.

[181] Wasserman, Stanley and Pattison, Philipa E (1996). Logit models and logistic regression for social networks: An introduction to Markov graphs and $p^\star$. *Psychometrika*, **61**(3), 401–425.

[182] Wasserman, Stanley and Robins, Garry (2005). An introduction to random graphs, dependence graphs, and $p^\star$. *Models and Methods in Social Network Analysis*, **27**, 148–161.

[183] Watts, Duncan J and Strogatz, Steven H (1998). Collective dynamics of 'small-world' networks. *Nature*, **393**(6684), 440–442.

[184] Weeks, Jeff. Kaleidotile. http://www.geometrygames.org/.

[185] Wilson, J Bastow (1996). The myth of constant predator:prey ratios. *Oecologia*, **106**(2), 272–276.

[186] Xu, Yan, Gurfinkel, Aleks Jacob, and Rikvold, Per A (2014). Architecture of the Florida power grid as a complex network. *Physica A: Statistical Mechanics and its Applications*, **401**, 130–140.

[187] Yook, Soon-Hyung, Jeong, Hawoong, Barabási, Albert-László, and Tu, Yuhai (2001). Weighted evolving networks. *Physical Review Letters*, **86**(25), 5835.

Index